Lecture Notes in Mathematics

Charles Favre
Mattias Jonsson

The Valuative Tree

Authors

Charles Favre

CNRS
Institut de Mathématique de Jussieu
Université Denis Diderot
Case 7012
2, place Jussieu
75251 Paris Cedex 05
France
e-mail: favre@math.jussieu.fr

Mattias Jonsson

Department of Mathematics
University of Michigan
Ann Arbor, MI 48109-1109
U.S.A.
e-mail: mattiasj@umich.edu

Cover picture: Arriving Ambrosden by Heather Power

Library of Congress Control Number: 2004110772

Mathematics Subject Classification (2000): 14H20, 13A18, 54F50

ISSN 0075-8434
ISBN 3-540-22984-1 Springer Berlin Heidelberg New York
DOI: 10.1007/b100262

Springer is a part of Springer Science + Business Media http://www.springeronline.com

Printed in Germany

Typesetting: Camera-ready TEX output by the authors

41/3142/du - 543210 - Printed on acid-free paper

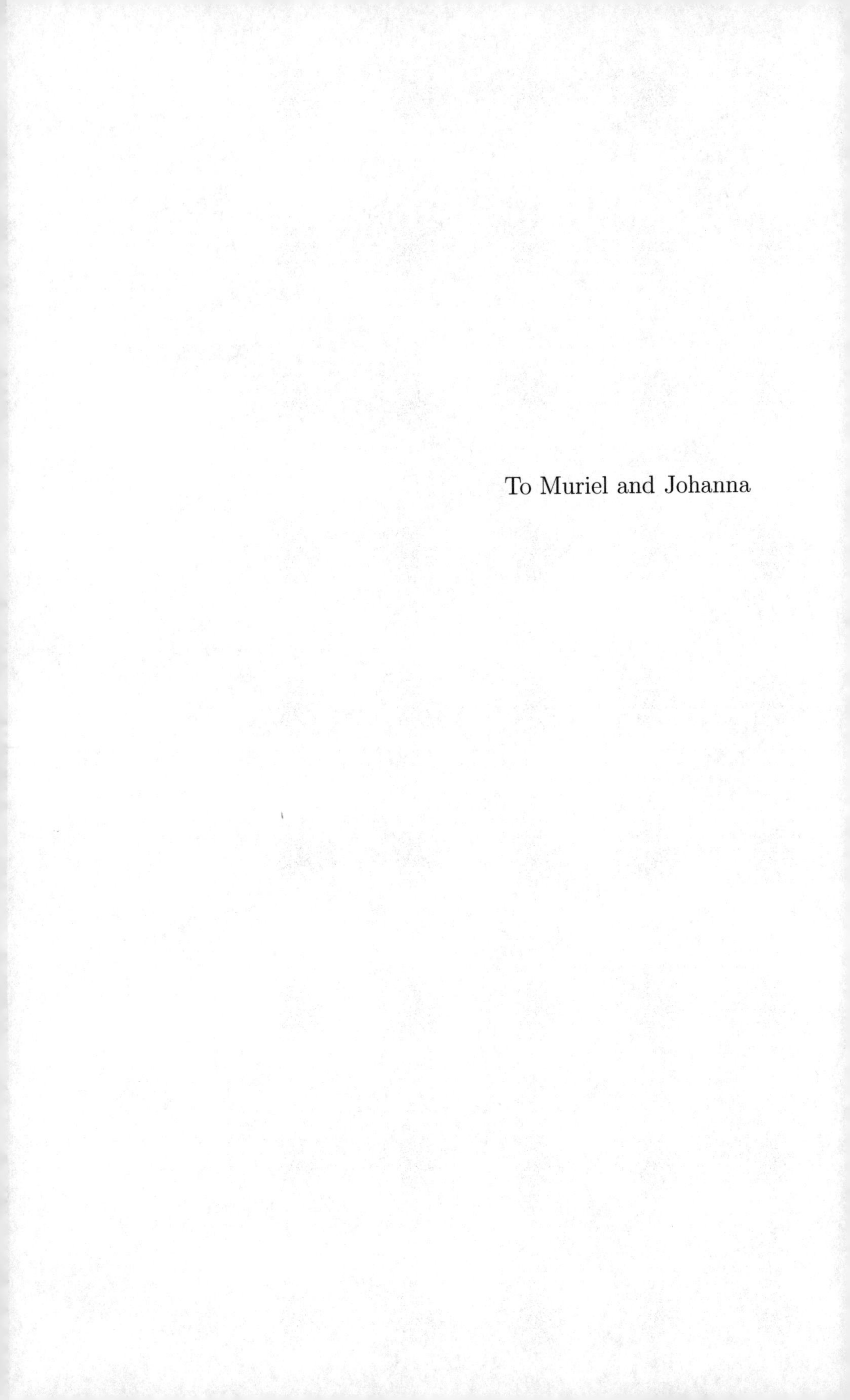

To Muriel and Johanna

Preface

This book grew out of a common passion for a beautiful natural object that we decided to call the "valuative tree". Motivated by questions stemming from complex dynamics and complex analysis, we realized that we needed to understand the link between valuations, which are purely algebraic objects, and more geometric or analytic constructions such as blowups or Lelong numbers. More precisely, we looked at the structure of a special set of valuations, and we found that this space had a very rich and delicate topological structure. We hope that the reader will share our enthusiasm while progressively exploring this space into its finer details all along this book.

This monograph has benefited from the help of many people. The first author wishes to warmly thank Bernard Teissier for his constant support and help, and Patrick Popescu-Pampu, Mark Spivakovsky and Michel Vaquié for fruitful discussions. The second author expresses his gratitude to Jean-François Lafont, Robert Lazarsfeld and Karen Smith. We both thank the referees for a number of useful suggestions.

Our work was done at several places, including the Institut de Mathématiques in Paris, the Department of Mathematics of the University of Michigan in Ann Arbor, RIMS in Kyoto and IMPA in Rio de Janeiro. We are grateful to these institutions for having provided us an excellent and motivating atmosphere for working on our project. During the writing period, the first author was supported by the CNRS, whereas the second author was supported by STINT and by NSF Grant No DMS-0200614.

June 2004

Structure of the Book

Before embarking to a journey into the valuative tree, we describe below the structure of the volume. A plain arrow linking chapter A to chapter B indicates that the understanding of B relies heavily on a previous lecture of A. A dashed arrow indicates a looser link between both chapters.

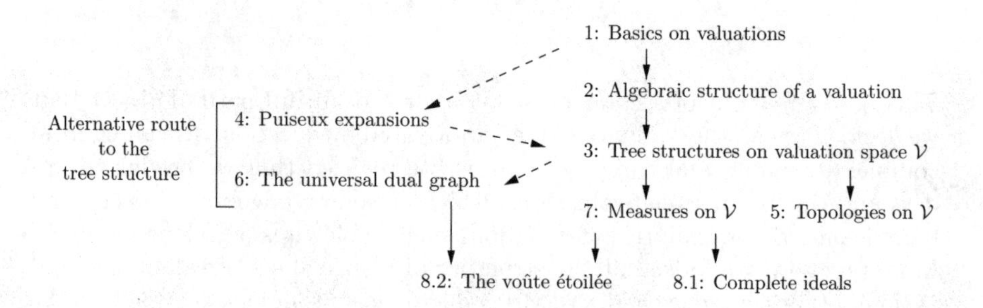

The Valuative Tree

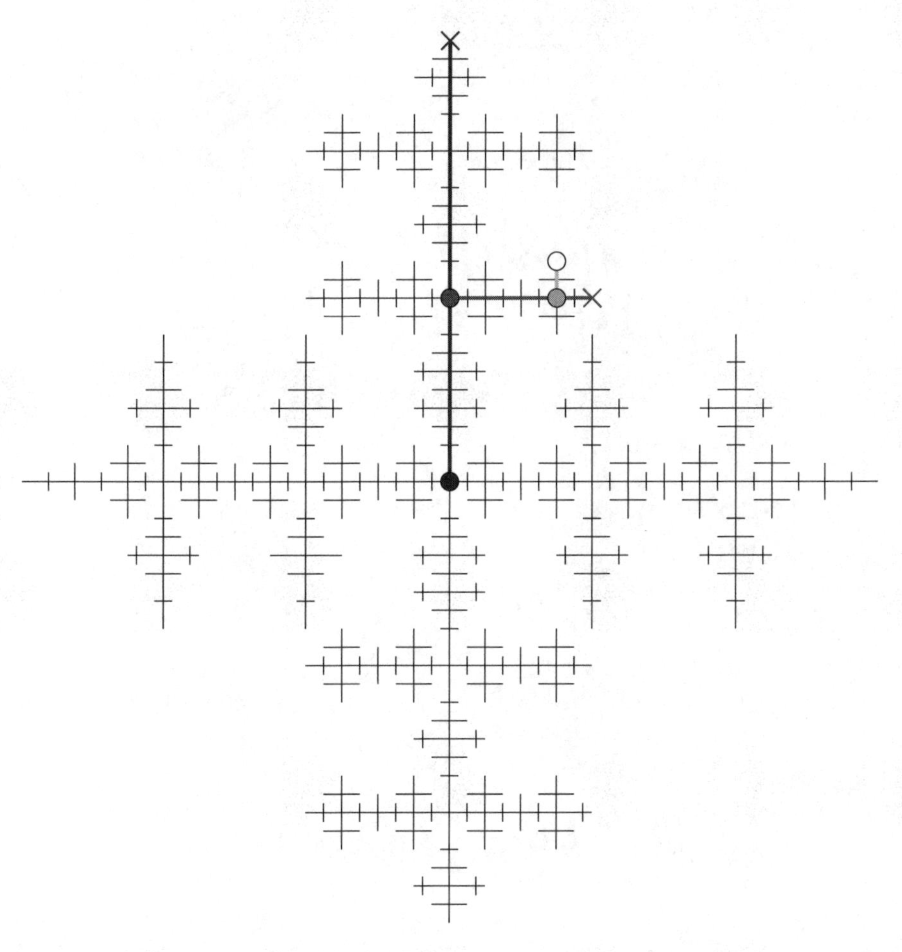

● ν_{m} ✕ ν_y — $m = 1$

● $\nu_{y,3/2} = \nu_{\phi,3/2}$ ✕ $\nu_{y^2-x^3}$ — $m = 2$

● $\nu_{y^2-x^3,10/3} = \nu_{\phi,10/3}$ — $m = 6$

○ ν_ϕ

$$\phi = (y^2 - x^3)^3 - x^{10}$$

Contents

Introduction

The purpose of this monograph is to give a new approach to singularities in a local, two-dimensional setting. Our method enables us to study curves, analytic ideals in $R = \mathbf{C}[[x, y]]$, and plurisubharmonic functions in a unified way. It is also general and powerful enough so that it can be applied to other situations, such as the dynamics of fixed point germs $f : (\mathbf{C}^2, 0) \circlearrowleft$.

Singularities of curves, ideals and plurisubharmonic functions can be analyzed by performing a composition of point blowups and considering the order of vanishing of the pullbacks along irreducible components of the exceptional divisor.[1] These orders of vanishing define real-valued functions on the ring R called *divisorial valuations*. It is a classical fact that the singularity of a curve or ideal is completely determined by the values of *all* divisorial valuations.

This naturally leads us to look at the set of all divisorial valuations. Indeed our aim is to describe in detail the structure of a slightly larger set $\mathcal{V}$ that we call *valuation space*. The elements of $\mathcal{V}$ are functions $\nu : R \to \overline{\mathbf{R}}_+ = [0, \infty]$, with $\nu(c) = 0$ for all $c \in \mathbf{C}^*$, satisfying the standard axioms of valuations: $\nu(\phi\psi) = \nu(\phi) + \nu(\psi)$; $\nu(\phi + \psi) \geq \min\{\nu(\phi), \nu(\psi)\}$ for all $\phi, \psi \in R$. We normalize them by $\nu(\mathfrak{m}) = \min\{\nu(x), \nu(y)\} = 1$. The central theme of our work is that valuation space has a natural structure of a *tree* modeled on the real line, and that this structure can be used to efficiently encode singularities of various kinds.

We distinguish between three different types of tree structures. A (rooted) *nonmetric tree* is a poset having a unique minimal element (its root) in which all sets of the form $\{\sigma\ ;\ \sigma \leq \tau\}$ are isomorphic to real intervals, i.e. there exists an order-preserving bijection of each such set onto a real interval. A *parameterized tree* is a nonmetric tree in which all these bijections are chosen in a compatible way. Finally, almost equivalent to parameterized trees are

[1]In the case of a plurisubharmonic function one has to replace "order of vanishing" by "generic Lelong number".

metric trees; these are metric spaces in which any two points are joined by a unique path isometric to a real interval.[2]

The nonmetric tree structure on $\mathcal{V}$ arises as follows. For $\nu, \mu \in \mathcal{V}$, we declare $\nu \leq \mu$ when $\nu(\phi) \leq \mu(\phi)$ for all $\phi \in R$. Our normalization $\nu(\mathfrak{m}) = 1$ implies that the *multiplicity valuation* $\nu_{\mathfrak{m}}$ sending ϕ to its multiplicity $m(\phi)$ at the origin is dominated by any other valuation. This natural order defines a nonmetric tree structure on $\mathcal{V}$, rooted at $\nu_{\mathfrak{m}}$ (Theorem 3.14).

As for the other two tree structures, any irreducible (formal local) curve C defines a *curve valuation* $\nu_C \in \mathcal{V}$: $\nu_C(\phi)$ is the normalized intersection number between the curves C and $\{\phi = 0\}$. A curve valuation is a maximal element under $\leq$ and the segment $[\nu_{\mathfrak{m}}, \nu_C]$ is isomorphic, as a totally ordered set, to the interval $[1, \infty]$. We construct an increasing function $\alpha : \mathcal{V} \to [1, \infty]$ that restricts to a bijection of $[\nu_{\mathfrak{m}}, \nu_C]$ onto $[1, \infty]$ for any C; as a consequence, α defines a parameterization of $\mathcal{V}$. The number $\alpha(\nu)$ is called the *skewness* of ν.[3] It is defined by the formula $\alpha(\nu) = \sup_\phi \nu(\phi)/m(\phi)$.

In addition to the partial ordering and skewness parameterization just described, the valuative tree also carries an important *multiplicity* function. The multiplicity of a valuation ν is equal to the infimum of the multiplicity of all curves whose associated curve valuations dominate ν in the partial ordering. Thus the multiplicity function is an increasing function on $\mathcal{V}$ with values in $\overline{\mathbf{N}} = \mathbf{N} \cup \{\infty\}$. A second important parameterization of $\mathcal{V}$, *thinness*, can be defined in terms of skewness and multiplicity.[4] We shall refer loosely to the combination of the partial ordering, the parameterizations by skewness and thinness, and the multiplicity function as the *tree structure on* $\mathcal{V}$.

There are four types of inhabitants of the valuative tree $\mathcal{V}$. The interior points, i.e. the points that are not maximal in the partial ordering, are valuations that become monomial (i.e. determined by their values on a pair of local coordinates) after a finite sequence of blowups. We call them *quasimonomial.* They include all divisorial valuations but also all *irrational* valuations such as the monomial valuation defined by $\nu(x) = 1$, $\nu(y) = \sqrt{2}$. The other points in $\mathcal{V}$, i.e. the ends of the valuative tree, are curve valuations and *infinitely singular* valuations, which can be characterized as the valuations with infinite multiplicity.

There is in fact a fifth type of valuations. These valuations cannot be defined as real-valued functions, but define functions on R with values in $\mathbf{R}_+ \times \mathbf{R}_+$ (endowed with the lexicographic order). In fact they do have a natural place in the valuative tree, as tree tangent vectors at points corresponding to divisorial valuations (see Theorem B.1). Geometrically they are curve val-

[2]Metric trees are often called $\mathbf{R}$-trees in the literature.

[3]The skewness is the inverse of the volume of a valuation as defined in [ELS] (see Remark 3.33).

[4]The thinness is also very precisely related to the Jacobian ideal of the valuation (see Remark 3.50).

uations where the curve is defined by an exceptional divisor. We hence call them *exceptional curve valuations.*

The valuative tree is a beautiful object which may be viewed in a number of different ways. Each corresponds to a particular interpretation of a valuation, and each gives a new insight into it. Some of them will hopefully lead to generalizations in a broader context. Let us describe four such points of views. See also the diagrams on page 7.

The first way consists of identifying valuations with balls of curves. For any two irreducible curves C_1, C_2 set $d(C_1, C_2) = m(C_1)m(C_2)/C_1 \cdot C_2$ where $m(C_i)$ is the multiplicity of C_i and $C_1 \cdot C_2$ is the intersection multiplicity of C_1 and C_2. It is a nontrivial fact that d defines an ultrametric on the set $\mathcal{C}$ of all irreducible formal curves (c.f. [Ga]). This fact allows us to associate to $(\mathcal{C}, d)$ a tree $\mathcal{T}_{\mathcal{C}}$ by declaring a point in $\mathcal{T}_{\mathcal{C}}$ to be a closed ball in $\mathcal{C}$. The tree structure on $\mathcal{T}$ is given by reverse inclusion of balls (partial ordering), inverse radii of balls (parameterization) and minimum multiplicity of curves in a ball (multiplicity). Theorem 3.57 states that the tree $\mathcal{T}_{\mathcal{C}}$ is isomorphic to the valuative tree $\mathcal{V}$ with its ends removed (i.e. to the set $\mathcal{V}_{\mathrm{qm}}$ of quasimonomial valuations).

A second way is through Puiseux series. Just as irreducible curves can be represented by Puiseux series, the elements in $\mathcal{V}$ are represented by valuations on the power series ring in one variable with Puiseux series coefficients. The set $\widehat{\mathcal{V}}_x$ of all such (normalized) valuations has a natural tree structure and a suitably defined restriction map from $\widehat{\mathcal{V}}_x$ to $\mathcal{V}$ recovers the tree structure on $\mathcal{V}$. In fact, $\mathcal{V}$ is naturally the orbit space of $\widehat{\mathcal{V}}_x$ under the action by the relevant Galois group (Theorem 4.17). This approach can also be viewed in the context of Berkovich spaces and Bruhat-Tits buildings. As a nonmetric tree, $\mathcal{V}$ embeds as the closure of a disk in the Berkovich projective line over the field of Laurent series in one variable. The metric on $\mathcal{V}_{\mathrm{qm}}$ induced by thinness then also arises from an identification of a subset of the Berkovich projective line with the Bruhat-Tits building of PGL_2 (see Section 4.6).

The third way is more algebraic in nature. The earliest systematic study of valuations in two dimensions was done in the fundamental work of Zariski in [Za1], [Za2] who, among other things, identified the set $\mathcal{V}_K$ of (not necessarily real-valued) valuations on R, vanishing on $\mathbf{C}^*$ and positive on the maximal ideal $\mathfrak{m} = (x, y)$, with sequences of infinitely near points. The space $\mathcal{V}_K$ carries a natural topology (the Zariski topology) and is known as the *Riemann-Zariski variety.* It is a non-Hausdorff quasi-compact space. The obstruction for $\mathcal{V}_K$ being Hausdorff stems from the fact that divisorial valuations do not define closed points. Namely, their associated valuation rings strictly contain valuation rings associated to exceptional curve valuations. One can then build a quotient space by identifying all valuations in the closure of a single divisorial one. This produces a compact Hausdorff space. Theorem 5.24 states that this space is precisely $\mathcal{V}$ (endowed with the topology of pointwise convergence).

The last way uses Zariski's identification of valuations with sequences of infinitely near points. We let Γ_π be the dual graph of a finite composition of blowups π. It is a simplicial tree whose set of vertices defines a poset Γ_π^*. When one sequence π contains another π', the poset Γ_π^* naturally contains $\Gamma_{\pi'}^*$. These posets therefore form an injective system whose injective limit (or, informally, union) Γ^* is a poset with a natural tree structure modeled on the rational numbers. By filling in the irrational points and adding all the ends to the tree we obtain a nonmetric tree Γ, the *universal dual graph.* This nonmetric tree can in fact be equipped with a parameterization and multiplicity function. These both derive from a combinatorial procedure that to each element in Γ^* attaches a vector $(a, b) \in (\mathbf{N}^*)^2$, the *Farey weight.*[5] A fundamental result (Theorem 6.22) asserts that the universal dual graph equipped with the Farey parameterization is canonically isomorphic to the valuative tree with the thinness parameterization, and that this isomorphism preserves multiplicity.

As we mentioned above, singularities can be understood through *functions* on the valuative tree. It is a remarkable fact that the information carried by these functions can also be described in terms of *complex measures* on $\mathcal{V}$. Let us be more precise. In the case of an ideal $I \subset R$, the function on $\mathcal{V}$ is given by $g_I(\nu) := \nu(I)$, and the measure ρ_I is a positive atomic measure supported on the Rees valuations of I. This decomposition into atoms of ρ_I corresponds exactly to the Zariski decomposition of I into simple complete ideals.

In [FJ1], we shall show that a plurisubharmonic function u also determines a function g_u on $\mathcal{V}$. The corresponding measure ρ_u, which is still positive but not necessarily atomic, captures essential information on the singularity of u. In particular, we shall show in [FJ2] that ρ_u determines the multiplier ideals of all multiples of u.

The identifications $g_I \leftrightarrow \rho_I$ and $g_u \leftrightarrow \rho_u$ are particular instances of a general correspondence between measures on $\mathcal{V}$ and certain functions on $\mathcal{V}_{\mathrm{qm}}$. In fact, this correspondence, being purely tree-theoretic in nature, is even more general, and extends the equivalence between positive measures and suitably normalized concave functions on the real line. By analogy, we thus write $\rho_I = \Delta g_I$, $\rho_u = \Delta g_u$ and speak about the *Laplace operator* Δ on the valuative tree.

There is a second instance where complex measures on $\mathcal{V}$ naturally appear, namely when we study the sheaf cohomology of the *voûte étoilée* $\mathfrak{X}$. In our setting, $\mathfrak{X}$ can be viewed as the total space of the set of all blowups above the origin. Elements of $H^2(\mathfrak{X}, \mathbf{C})$ naturally define functions on $\mathcal{V}_{\mathrm{qm}}$ whose Laplacians are atomic measures supported on divisorial valuations. The cup product on cohomology has a natural interpretation as an inner product on measures. This inner product is a bilinear extension of an inner product on the valuative tree itself, and ultimately derives from intersections of curves.

[5] We follow the terminology used in [HP].

We have tried to write this monograph with an eye towards applications, such as the study of singularities of plurisubharmonic functions and dynamics of fixed point germs. Our hope is that people who are new to valuation theory will be able to follow the exposition, which we have tried to make self-contained and elementary.

Experts on valuation theory will undoubtedly notice that we reproduce many known results and that we do not work in the most general setting possible. Indeed, the assumption that R be the ring of formal power series in two complex variables is unnecessarily restrictive. While we refer to Appendix E for a precise discussion, we mention here that our analysis goes through in two important cases: the ring of holomorphic germs at the origin in $\mathbf{C}^2$, and the local ring at a smooth (closed) point of an algebraic surface over an algebraically closed field. We decided to work in the concrete setting of formal power series; readers with a good background in algebra may easily adapt the arguments to more general situations.

It would certainly be interesting to investigate generalizations to the case when R is the local ring at a normal surface singularity, or any local ring of dimension at least three. However, not only would this require the introduction of a substantial amount of new material, but the corresponding valuation space would no longer be a tree in general. Thus we shall not consider these more general situations here.

We remark that a fair amount of the structure of the valuative tree is implicitly contained in the analysis by Spivakovsky [Sp]. In particular, his description of the dual graph of a valuation is closely related to the construction of the valuative tree as the universal dual graph. However, our approach is quite different from his; in particular we do not use continued fractions. A tree structure was described in a context similar to ours in [AA], but without any explicit reference to valuations.

The main applications of the tree structure on the valuative tree to analysis and dynamics will be explored in forthcoming papers: see [FJ1], [FJ2], and [FJ3].

We end this introduction by indicating the organization of the monograph, which is divided into eight chapters and an appendix.

In the first chapter we give basic definitions, examples and results on valuations. In particular we describe the relationship between valuations and sequences of infinitely near points (in dimension two).

Chapter 2 is technical in nature. We encode valuations by finite or countable pieces of data that we call sequences of key polynomials, or *SKP's*. This encoding is an adaptation of a method by MacLane [Ma], the possibility of which was indicated to us by B. Teissier. An SKP (or at least a subsequence of it) corresponds to generating polynomials and approximate roots in the language of Spivakovsky [Sp] and Abhyankar-Moh [AM], respectively. We are thus able to classify valuations on R. This classification is well-known to specialists (see [ZS2, Sp] for instance) but we feel that our concrete approach is of independent interest. The representation of valuations by SKP's is the key to the tree structure on the valuative tree.

The third chapter concerns trees. Our main goal is to visualize the encoding by SKP's in an elegant and coordinate free way. We first discuss different definitions of trees and the relations between them. Using SKP's we then show that valuation space $\mathcal{V}$ does carry an intricate tree structure that we later in the chapter analyze in detail.

As an alternative to SKP's, Chapter 4 contains an approach to the tree structure on $\mathcal{V}$ through Puiseux series. The results can be interpreted in the language of Berkovich. Specifically we indicate how the valuative tree embeds inside the Berkovich projective line and Bruhat-Tits building of PGL_2 over the local field of Laurent series in one variable. In fact, most of these results are at least implicitly contained in [Be] but we felt it was worthwhile to write down the details.

In Chapter 5 we analyze and compare different topologies on the valuative tree. The definition of a valuation as a function on R, as well as the tree structure on $\mathcal{V}$, gives rise to three types of topologies: the weak, the strong and the thin topology. In addition, we have two topologies on $\mathcal{V}_K$: the Zariski topology and the Hausdorff-Zariski (or HZ) topology. As mentioned above, the former gives rise to the weak topology on $\mathcal{V}$ through the quotient construction. The HZ topology is in fact equivalent to the weak tree topology induced by a natural *discrete* tree structure on $\mathcal{V}_K$.

In Chapter 6 we build the universal dual graph described above, and show how to identify it with the valuative tree. The fact that valuations can be simultaneously viewed algebraically as functions on R and geometrically in terms of blowups is extremely powerful and we spend a fair amount of time detailing some of the connections and implications. In particular, we show that the valuative tree has a natural self-similar, or fractal, structure, see Figure 6.12.

Chapter 7 is concerned with the relationship between measures on $\mathcal{V}$ and certain classes of functions on $\mathcal{V}_{\mathrm{qm}}$. The analysis is purely tree-theoretic and gives a connection between (complex) measures on a parameterized tree and functions on the (interior of the) tree satisfying certain regularity properties. Apart from being of independent interest, this analysis is fundamental to many applications.

In Chapter 8, we describe two instances where these measures appear naturally. First we reinterpret in our context Zariski's theory of simple complete ideals as explained in [ZS2]. This gives a new point of view on the decomposition of any integrally closed ideal as a product of simple ideals. We then construct Hironaka's "voûte étoilée" $\mathfrak{X}$ as the projective limit of the total spaces of all sequences of blowups above the origin. We use measures on $\mathcal{V}$ to understand the structure of the sheaf cohomology group $H^2(\mathfrak{X}, \mathbf{C})$.

Finally we conclude this monograph by an appendix containing a few results and discussions that did not find a natural home elsewhere in the monograph. Specifically, we discuss infinitely singular valuations; analyze the tangent space at a divisorial valuation; give tables summarizing the classification of valuations on R from different points of view; present a short dictionary between our terminology and the more standard terminology from the theory of singular plane curves; and finally discuss what the essential assumptions are on the ring R.

Different interpretations of valuations

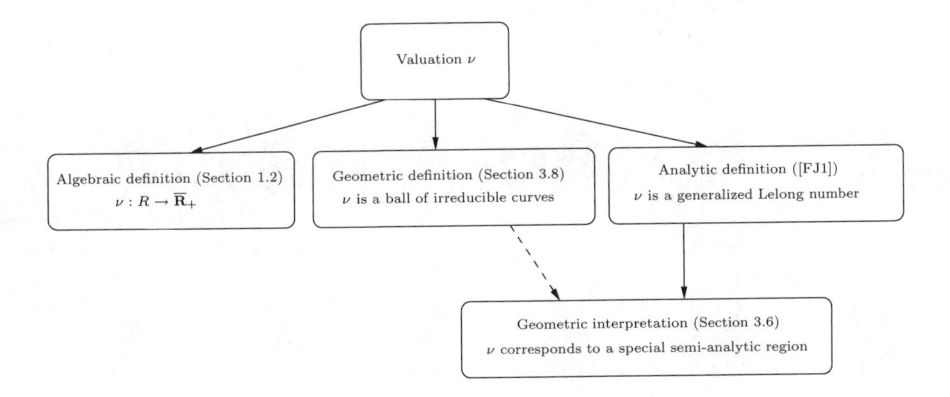

Different approaches to the valuative tree

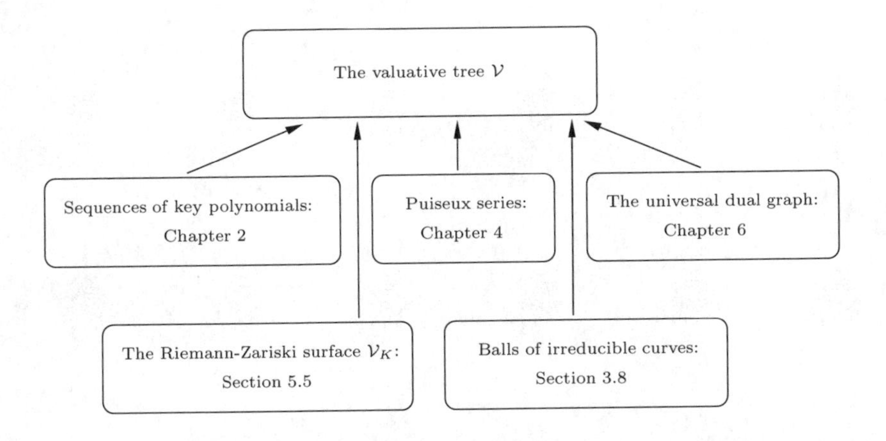

1

Generalities

In this chapter we give basic results on valuations. For definiteness we mostly restrict our attention to the ring of formal power series in two complex variables.

We distinguish between *valuations* and *Krull valuations*[1]. Valuations will always be $\mathbf{R}\cup\{\infty\}$ valued, whereas Krull valuations take values in an abstract totally ordered group. This distinction may seem unnatural but is useful for our purposes.

We start in Sections 1.1, 1.2 and 1.3 by giving precise definitions and noting basic facts on valuations and Krull valuations. Most of this is standard and can be found in our main references which are [ZS2] and [Va]; see also [Sp]. However, we avoid using any deep facts from valuation theory and we have tried to make the exposition as self-contained as possible.

As local formal curves play an important role in our study, we spend some time discussing them in Section 1.4.

Then we survey the valuative landscape before embarking on the journey through it in subsequent chapters. More precisely, we discuss in Section 1.5 six classes of Krull valuations that we shall later show form a complete list. The purpose of the discussion is to give the non-expert reader a feeling for the different valuations that will later appear in the monograph in various disguises.

We also try to give analytic interpretations of valuations whenever possible. For example, when applied to holomorphic germs, valuations often give the the order of vanishing at the origin along a particular approach. When applied to plurisubharmonic functions, many valuations can be viewed as Lelong numbers, (pushforward of) Kiselman numbers, or generalized Lelong numbers. While these remarks are for the most part not necessary for the purposes of the current monograph, they are important for the applications.

---

[1]This is not standard terminology but will be used throughout the monograph.

In Section 1.6 we make a precise comparison between valuations and Krull valuations. Finally, Section 1.7 contains a discussion of Zariski's correspondence between Krull valuations and sequences of infinitely near points.

1.1 Setup

Unless otherwise stated, R will denote the ring of formal power series in two complex variables. This assumption on R can be relaxed somewhat for the analysis in this monograph; see Appendix E for a discussion. Suffice it to say here that our method works for the ring of holomorphic germs at the origin in $\mathbf{C}^2$, and for the local ring at a smooth (closed) point of an algebraic surface over an algebraically closed field. Certainly, most of the background results presented in this chapter hold in a quite general context.

As is both convenient and customary (see [Te3]) we sometimes "ignore" the fact that the power series in R are not necessarily convergent. Thus, rather than talking about Spec R we most often write $(\mathbf{C}^2, 0)$ and view the elements of R as functions on the latter space. In the same vein, we talk about local coordinates (x, y) (rather than regular parameters) and local formal mappings $f : (\mathbf{C}^2, 0) \to (\mathbf{C}^2, 0)$ (rather than endomorphisms of R).

The ring R is local, with unique maximal ideal $\mathfrak{m}$. If (x, y) are local coordinates, then $R = \mathbf{C}[[x, y]]$ and $\mathfrak{m} = xR + yR$. We let K denote the fraction field of R and set $K^* = K \setminus \{0\}$ and $R^* = R \setminus \{0\}$.

1.2 Valuations

Denote the set of nonnegative real numbers by $\mathbf{R}_+$. Set $\overline{\mathbf{R}}_+ := \mathbf{R}_+ \cup \{\infty\}$, extending the addition, multiplication and order on $\mathbf{R}_+$ to $\overline{\mathbf{R}}_+$ in the usual way.

Definition 1.1. *A* valuation *on R is a nonconstant function $\nu : R \to \overline{\mathbf{R}}_+$ with:*

(V1) $\nu(\phi\psi) = \nu(\phi) + \nu(\psi)$ for all $\phi, \psi \in R$;
(V2) $\nu(\phi + \psi) \geq \min\{\nu(\phi), \nu(\psi)\}$ for all $\phi, \psi \in R$;
(V3) $\nu(1) = 0$.

Remark 1.2. For applications to analysis and dynamics it is useful to think of a valuation as an *order of vanishing* at the origin along a particular approach direction. See the examples in Section 1.5 below.

From the definition we see that $\nu(0) = \infty$ and $\nu|_{\mathbf{C}^*} = 0$. The set $\mathfrak{p} := \{\phi \in R\ ;\ \nu(\phi) = \infty\}$ is a prime ideal in R; we say ν is *proper* if $\mathfrak{p} \subsetneq \mathfrak{m}$. It is *centered* if it is proper and if $\nu(\mathfrak{m}) := \min\{\nu(\phi)\ ;\ \phi \in \mathfrak{m}\} > 0$. Two centered valuations ν_1, ν_2 are *equivalent*, $\nu_1 \sim \nu_2$, if $\nu_1 = C\nu_2$ for some constant $C > 0$.

We denote by $\tilde{\mathcal{V}}$ the set of all centered valuations on R. The primary object of study in this monograph will be quotient of $\tilde{\mathcal{V}}$ by the equivalence relation $\sim$. We call it *valuation space* for the time being, although we shall later give it the name *the valuative tree* as it possesses a tree structure.

Notice that $\tilde{\mathcal{V}}$ is naturally a sort of bundle over $\tilde{\mathcal{V}}/\sim$ with fibers isomorphic to the real interval $(0,\infty)$. For our purposes it will be important to pick a section of this bundle, or, what amounts to the same thing, view $\tilde{\mathcal{V}}/\sim$ as a subset of $\tilde{\mathcal{V}}$ singled out by a normalizing condition. There are several possible normalizations. One is to pick any element $x \in \mathfrak{m}$ that defines a smooth curve $\{x = 0\}$ and demand that $\nu(x) = 1$. This is useful in cases where the curve $\{x = 0\}$ is naturally given, e.g. when it is an irreducible component of the exceptional divisor of some composition of blowups. We thus define $\mathcal{V}_x$ to be the set of centered valuations ν on R satisfying $\nu(x) = 1$.[2] We shall explore this space further in Section 3.9.

However, the most important normalization in this monograph will be $\nu(\mathfrak{m}) = 1$, i.e. $\min\{\nu(x), \nu(y)\} = 1$ for any local coordinates (x, y). We will then say that ν is *normalized*. This normalization has the advantage of being coordinate independent. We define $\mathcal{V}$ as the set of centered valuations ν normalized by $\nu(\mathfrak{m}) = 1$.

We endow $\tilde{\mathcal{V}}$, $\mathcal{V}$ and $\mathcal{V}_x$ with the weak topology: if ν_k, ν are valuations in $\tilde{\mathcal{V}}$ ($\mathcal{V}$, $\mathcal{V}_x$), then $\nu_k \to \nu$ in $\tilde{\mathcal{V}}$ ($\mathcal{V}$, $\mathcal{V}_x$) iff $\nu_k(\phi) \to \nu(\phi)$ for all $\phi \in R$. As we will see later, the weak topologies on $\mathcal{V}$ and $\mathcal{V}_x$ are induced by tree structures.

Finally $\tilde{\mathcal{V}}$, $\mathcal{V}$ and $\mathcal{V}_x$ come with natural partial orderings: $\nu_1 \leq \nu_2$ iff $\nu_1(\phi) \leq \nu_2(\phi)$ for all $\phi \in R$. Again, the partial orderings on $\mathcal{V}$ and $\mathcal{V}_x$ arise from tree structures.

1.3 Krull Valuations

We now introduce Krull valuations.[3] They are defined in the same way as valuations, except that we replace $\overline{\mathbf{R}}_+$ with a general totally ordered group. We shall analyze to what extent this really makes a difference in Section 1.6.

Write K for the fraction field of R and set $R^* := R \setminus \{0\}$, $K^* := K \setminus \{0\}$.

Definition 1.3. *Let Γ be a totally ordered abelian group. A* Krull valuation *on R is a function $\nu : R^* \to \Gamma$ satisfying (V1)-(V3) above.*

A Krull valuation is *centered* if $\nu \geq 0$ on R^* and $\nu > 0$ on $\mathfrak{m}$. Two Krull valuations $\nu_1 : R^* \to \Gamma_1$, $\nu_2 : R^* \to \Gamma_2$ are *equivalent* if $h \circ \nu_1 = \nu_2$ for some strictly increasing homomorphism $h : \Gamma_1 \to \Gamma_2$. Any Krull valuation extends naturally to K^* by $\nu(\phi/\psi) = \nu(\phi) - \nu(\psi)$.

---

[2]In fact we shall later add to this space one valuation which is not centered: see Section 3.9.

[3]The reader mainly interested in the analytic and dynamic applications of valuations may skip this section on a first reading.

A *valuation ring* S is a local ring with fraction field K such that $x \in K^*$ implies $x \in S$ or $x^{-1} \in S$. Say that $x, y \in S$ are equivalent iff xy^{-1} is a unit in S. The quotient Γ_S is endowed with a natural total order given by $x > y$ iff $xy^{-1} \in \mathfrak{m}_S$, the maximal ideal of S. The projection $S \to \Gamma_S$ then extends to a Krull valuation $\nu_S : K^* \to \Gamma_S$.

Conversely, if ν is a centered Krull valuation, then $R_\nu := \{0\} \cup \{\phi \in K^* \,;\, \nu(\phi) \geq 0\}$ is a valuation ring with maximal ideal $\mathfrak{m}_\nu := \{0\} \cup \{\phi \,;\, \nu(\phi) > 0\}$. One can show that ν is equivalent to ν_{R_ν} defined above. In particular, two Krull valuations ν, ν' are equivalent iff $R_\nu = R_{\nu'}$.

We let $\tilde{\mathcal{V}}_K$ be the set of all centered Krull valuations, and $\mathcal{V}_K$ be the quotient of $\tilde{\mathcal{V}}_K$ by the equivalence relation. Equivalently, $\mathcal{V}_K$ is the set of valuation rings S in K whose maximal ideal $\mathfrak{m}_S$ satisfies $R \cap \mathfrak{m}_S = \mathfrak{m}$.

The group $\nu(K^*)$ is called the *value group* of ν; similarly $\nu(R^*)$ is the *value semigroup*.

A classical way to analyze a Krull valuation is through its *numerical invariants*.

Definition 1.4. *Let $\nu : R^* \to \Gamma$ be a centered Krull valuation.*

- *The* rank $\mathrm{rk}(\nu)$ *of ν is the Krull dimension of the ring R_ν.*
- *The* rational rank *of ν is defined by* $\mathrm{rat.rk}(\nu) := \dim_{\mathbf{Q}}(\nu(K^*) \otimes_{\mathbf{Z}} \mathbf{Q})$.
- *The* transcendence degree $\mathrm{tr.deg}(\nu)$ *of ν is defined as follows. Since ν is centered, we have a natural inclusion $\mathbf{C} = R/\mathfrak{m} \subset k_\nu := R_\nu/\mathfrak{m}_\nu$. The field k_ν is called the* residue field *of ν. We let* $\mathrm{tr.deg}(\nu)$ *be the transcendence degree of this field extension.*

One can show that $\mathrm{rk}(\nu)$ is the least integer l so that $\nu(K^*)$ can be embedded as an ordered group into $(\mathbf{R}^l, +)$ endowed with the lexicographic order. Hence both $\mathrm{rk}(\nu)$ and $\mathrm{rat.rk}(\nu)$ depend only on the value group $\nu(K^*)$ of the valuation. We will later give a geometric interpretation of $\mathrm{tr.deg}(\nu)$ (see Remark 1.11).

The main relation between these numerical invariants is given by Abhyankar's inequalities,[4] which assert that:

$$\mathrm{rk}(\nu) + \mathrm{tr.deg}(\nu) \leq \mathrm{rat.rk}(\nu) + \mathrm{tr.deg}(\nu) \leq \dim R = 2. \tag{1.1}$$

Moreover, if $\mathrm{rat.rk}(\nu) + \mathrm{tr.deg}(\nu) = 2$, then $\nu(K^*)$ is isomorphic (as a group) to $\mathbf{Z}^\varepsilon$ with $\varepsilon = \mathrm{rat.rk}(\nu)$. When $\mathrm{rk}(\nu) + \mathrm{tr.deg}(\nu) = 2$, $\nu(K^*)$ is isomorphic as an *ordered* group to $\mathbf{Z}^\varepsilon$ endowed with the lexicographic order.

To ν we associate a *graded ring* $\mathrm{gr}_\nu R = \oplus_{r \in \Gamma} \{\nu \geq r\}/\{\nu > r\}$. It is in bijection with equivalence classes of R under the equivalence relation $\phi = \psi$ *modulo* ν iff $\nu(\phi - \psi) > \nu(\phi)$. We have $\overline{\phi \cdot \psi} = \overline{\phi} \cdot \overline{\psi}$ in $\mathrm{gr}_\nu R$ for any $\phi, \psi \in R$.

Any formal mapping $f : (\mathbf{C}^2, 0) \to (\mathbf{C}^2, 0)$ induces a ring homomorphism $f^* : R \to R$. The latter in turn induces actions on $\tilde{\mathcal{V}}$ and $\tilde{\mathcal{V}}_K$ given by

[4] Although we shall not prove this formula, one may use results from Chapter 2 to do so (see Theorems 2.28, 2.29).

$f_*\nu(\phi) = \nu(f^*\phi)$. When f is invertible f_* is a bijection and preserves all three invariants rk, rat. rk and tr. deg defined above. It also restricts to a bijection on $\mathcal{V}$ preserving the natural partial order: $f_*\nu \leq f_*\nu'$ as soon as $\nu \leq \nu'$.

1.4 Plane Curves

Curves play a key role in our approach to valuations. Indeed, we shall later see that "most" valuations can be seen as balls of irreducible curves in a particular metric. Here we recall a few classical results regarding plane curves: see [Te3] for details.

Recall that R is the ring of formal power series in two complex variables. Thus a *curve* C for us is simply an equivalence class of elements $\phi \in \mathfrak{m}$, where two elements are equivalent if they differ by multiplication by a unit in R, i.e. by an element of $R \setminus \mathfrak{m}$. We write $C = \{\phi = 0\}$ even though ϕ is not necessarily a convergent power series (see Section 1.1) and say that the curve is *represented* by ϕ.

The *multiplicity* $m(C)$ of a curve $C = \{\phi = 0\}$ is defined by $m(C) = m(\phi) = \max\{n \; ; \; \phi \in \mathfrak{m}^n\}$; this does not depend on the choice of ϕ. Recall that R is a unique factorization domain. A curve is *reduced* if it is represented by an element of $\mathfrak{m}$ without repeated irreducible factors. It is *irreducible* if it is represented by an irreducible element of $\mathfrak{m}$.

Any irreducible curve C admits a *parameterization* as follows. Pick local coordinates (x, y) such that C is transverse to the curve $\{x = 0\}$, i.e. if $C = \{\phi(x, y) = 0\}$, then $\phi(0, y)$ is a formal power series in y whose lowest order term has degree $m = m(C)$. Then there exists a formal power series $y(t) \in \mathbf{C}[[t]]$ such that $\phi(t^m, y(t)) = 0$.

The *intersection multiplicity* of two curves $C = \{\phi = 0\}$ and $D = \{\psi = 0\}$ is defined by $C \cdot D = \dim_{\mathbf{C}} R/\langle\phi, \psi\rangle$, where $\langle\phi, \psi\rangle$ denotes the ideal of R generated by ϕ and ψ; this intersection multiplicity does not depend on the choice of ϕ and ψ. We sometimes write $\phi \cdot \psi$ for $\{\phi = 0\} \cdot \{\psi = 0\}$. Then $\phi \cdot \psi = \infty$ iff ϕ and ψ have a common irreducible factor.

The intersection multiplicity can also be computed in terms of parameterizations. Suppose C is an irreducible curve, (x, y) are coordinates such that C is transverse to $\{x = 0\}$ and $(x, y) = (t^m, y(t))$ is a parameterization of C. Then for any curve $D = \{\psi = 0\}$ we have that $C \cdot D$ is the lowest order term in the formal power series $\psi(t^m, y(t)) \in \mathbf{C}[[t]]$. In particular when D is smooth and transverse to C then $C \cdot D = m(C)$.

1.5 Examples of Valuations

It is now time to present examples of valuations and Krull valuations. In fact, the list that we give below is complete, but that will only be proved in Chapter 2, see also Appendix C. The purpose of presenting the list below is not so much to give precise definitions, but rather to introduce some notation and give a road map to the valuative landscape that lies ahead. In particular we do not hesitate to mention features that are not obvious from the definitions.

We also give some hints to how the valuations can be interpreted analytically when applied to holomorphic germs or even to plurisubharmonic functions. This analytic point of view is not strictly necessary for anything that we do in this monograph, but hopefully serves to add to the reader's intuition and explain how valuations can be used in applications such as in [FJ1] [FJ2].

1.5.1 The Multiplicity Valuation

The function

$$\nu_{\mathfrak{m}}(\phi) := \max\{k \ ; \ \phi \in \mathfrak{m}^k\},$$

defines both a normalized valuation and a Krull valuation. We call it the *multiplicity valuation*. Notice that $\nu_{\mathfrak{m}}(\phi) = m(\phi)$ is the multiplicity of the curve $\{\phi = 0\}$ as defined in Section 1.4.

The multiplicity valuation is the root of the valuative tree in the sense that it is the minimal element under the partial ordering defined above. See Proposition 3.20.

The numerical invariants are easy to compute (alternatively see Theorem 2.28). Clearly $\nu_{\mathfrak{m}}(K^*) = \mathbf{Z}$ so $\mathrm{rk}(\nu_{\mathfrak{m}}) = \mathrm{rat.rk}(\nu_{\mathfrak{m}}) = 1$. It is not hard to see that the residue field is given by $k_\nu \simeq \mathbf{C}(y/x)$, where (x, y) are local coordinates. Hence $\mathrm{tr.deg}(\nu_{\mathfrak{m}}) = 1$.

Analytically, the multiplicity valuation can be interpreted as the order of vanishing at the origin: if ϕ is a holomorphic germ, then

$$\nu_{\mathfrak{m}}(\phi) = \lim_{r\to 0} \frac{1}{\log r} \sup_{B(r)} \log|\phi|, \tag{1.2}$$

where $B(r)$ is a ball of radius r centered at the origin. In fact, this equation makes sense even when $\log|\phi|$ is replaced by a plurisubharmonic function u and recovers the *Lelong number* [De] of u at the origin.

1.5.2 Monomial Valuations

Fix local coordinates (x, y) and $\alpha \geq 1$. Define a valuation $\nu_{y,\alpha}$ as follows:[5] $\nu_{y,\alpha}(x) = 1$, $\nu_{y,\alpha}(y) = \alpha$ and

$$\nu_{y,\alpha}(\phi) := \min\{i + \alpha j \ ; \ a_{ij} \neq 0\},$$

for $\phi = \sum a_{ij} x^i y^j$. Then $\nu_{y,\alpha}$ is a valuation and a Krull valuation: we say that $\nu_{y,\alpha}$ is *monomial* in the coordinates (x, y).

[5] We will see later that $\nu_{y,\alpha}$ does not depend on the choice of the other coordinate x.

In general, a valuation ν is monomial if it is monomial in some local coordinates (x, y). Notice that since $\alpha \geq 1$, $\nu_{y,\alpha}$ is normalized in the sense that $\nu_{y,\alpha}(\mathfrak{m}) = 1$. If $\alpha = 1$, then $\nu_{y,1} = \nu_{\mathfrak{m}}$ is the multiplicity valuation.

A monomial valuation has numerical invariants given as follows (see Theorem 2.28). We have $\mathrm{rk}(\nu_{y,\alpha}) = 1$ for any α. When $\alpha \in \mathbf{Q}$, $\mathrm{tr.deg}(\nu_{y,\alpha}) = 1$ and $\mathrm{rat.rk}(\nu_{y,\alpha}) = 1$; when $\alpha \notin \mathbf{Q}$, $\mathrm{tr.deg}(\nu_{y,\alpha}) = 0$ and $\mathrm{rat.rk}(\nu_{y,\alpha}) = 2$.

Monomial valuations have the following analytic interpretation. Assume (x, y) are holomorphic coordinates (this is in fact not a restriction as we may perturb x and y slightly without changing ν). If ϕ is a holomorphic germ then

$$\nu_{y,\alpha}(\phi) = \lim_{r \to 0} \frac{1}{\log r} \sup_{|x|<r, |y|<r^\alpha} \log |\phi|, \tag{1.3}$$

As before, this equation makes sense even when $\log |\phi|$ is replaced by a plurisubharmonic function u and recovers the *Kiselman number* [De] of u (with weights $(\alpha, 1)$) at the origin.[6]

1.5.3 Divisorial Valuations

If $C = \{\psi = 0\}$ is an irreducible curve, then

$$\mathrm{div}_C(\phi) := \max\{k \; ; \; \psi^k \mid \phi\}$$

defines a valuation on R. In other words, div_C is the order of vanishing along C. Notice that such a valuation is not centered, but it is the basis for a whole class of centered valuations defined as follows.

Consider a birational morphism $\pi : X \to (\mathbf{C}^2, 0)$. This means that X is a smooth complex surface, π is holomorphic and proper, and π is a biholomorphism outside the exceptional divisor $\pi^{-1}(0)$. It is a classical theorem of Zariski that such a π is a composition of point blowups. Let E be an irreducible component of the exceptional divisor. In the sequel, we sometimes refer to such a divisor as an *exceptional component*. There is then a unique integer $b = b_E > 0$ such that

$$\nu_E(\phi) := b^{-1} \, \mathrm{div}_E(\pi^*\phi)$$

defines a normalized valuation (and a Krull valuation) on R, called a *divisorial* valuation. Thus $b_E \, \nu_E(\phi)$ is the order of vanishing of the pullback $\pi^*\phi$ along E.

The multiplicity valuation $\nu_{\mathfrak{m}}$ is divisorial, with $\pi : X \to (\mathbf{C}^2, 0)$ being a single blowup of the origin, $E = \pi^{-1}(0)$ and $b_E = 1$.

The numerical invariants of ν_E are given as follows: $\mathrm{rk}(\nu_E) = \mathrm{rat.rk}(\nu_E) = 1$ and $\mathrm{tr.deg}(\nu_E) = 1$. In fact the residue field of ν is canonically isomorphic to the field of rational functions on E: to the class of $\phi \in K$ is associated the rational function $\pi^*\phi|_E$.

---

[6] Kiselman numbers are also sometimes called *directional Lelong numbers*.

Any valuation with the invariants above is in fact divisorial (see [ZS2, p.89]). We shall give an argument below: see Proposition 1.12. In particular a monomial valuation $\nu_{y,\alpha}$ is divisorial iff $\alpha \in \mathbf{Q}$.

As we will show (Proposition 3.20), a divisorial valuation ν_E constitutes a branch point in the valuative tree. The tree tangent space at ν_E, i.e. the set of branches emanating from ν_E, is naturally in bijection with the set of points on E (Theorem B.1).

Analytically, divisorial valuations can be interpreted as follows. If C is a holomorphic curve and ϕ a holomorphic germ, then $\operatorname{div}_C(\phi)$ is the order of vanishing of ϕ at a generic point of C. Similarly, $b_E\,\nu_E(\phi)$ the order of vanishing of $\pi^*\phi$ at a generic point of E.

These interpretations generalize to plurisubharmonic functions u (an example is $u = \log|\phi|$ for a holomorphic germ ϕ). If C is a holomorphic curve, then $\operatorname{div}_C(u)$ is the Lelong number of u at a generic point of C, or equivalently the (normalized) mass of the current $dd^c u$ on C. Similarly, $b_E\,\nu_E(u)$ is both the Lelong number of $\pi^* u$ at a generic point of E and the normalized mass of the pullback $\pi^* dd^c u$ along E.

Divisorial valuations are special cases of quasimonomial valuations, to be discussed next. This gives yet another analytic interpretation.

1.5.4 Quasimonomial Valuations

As in the definition of a divisorial valuation above, consider a birational morphism $\pi : X \to (\mathbf{C}^2, 0)$ and an exceptional component $E \subset \pi^{-1}(0)$. Pick a point $p \in E$ and let μ be a monomial valuation at p. The image $\nu = \pi_*\mu$ defined by $\nu(\phi) = \mu(\pi^*\phi)$ is a *quasimonomial* valuation. Its numerical invariants are the same as those of μ (see Theorem 2.28). In particular, quasimonomial valuations for which the the second inequality in (1.1) above is an equality. Such valuations were called rank one Abhyankar valuations in [ELS]. Conversely, any rank one Abhyankar valuation is a quasimonomial valuation, see Theorem 2.28.

Beware that we will use an alternative definition of quasimonomial valuations in Section 2.2 and it will take until Chapter 6 before we see that the two definitions are equivalent.

Any divisorial valuation is quasimonomial. A quasimonomial valuation that is not divisorial is called *irrational*. Such a valuation has numerical invariants $\operatorname{rk}(\nu) = 1$, $\operatorname{rat.rk}(\nu) = 2$ and $\operatorname{tr.deg}(\nu) = 0$, see Theorem 2.28.

We already observed that divisorial valuations are branch points in the valuative tree. The irrational valuations are regular points in the sense that there are exactly two branches emanating from each such point. See Proposition 3.20. In fact, the quasimonomial valuations are exactly the points in the valuative tree that are not ends. The set of quasimonomial valuations therefore constitute a subtree $\mathcal{V}_{\mathrm{qm}}$ of $\mathcal{V}$.

Quasimonomial valuations are useful for applications as they can be interpreted as the order of vanishing at the origin along a particular approach direction. More precisely, we can associate to any quasimonomial valuation ν a *characteristic region* of the form

$$\Omega(r) = \left\{(x, y) \in \mathbf{C}^2 \ ; \ |x| < r, \ |\phi(x, y)| < |x|^{tm}\right\} \subset \mathbf{C}^2.$$

Here $\phi \in \mathfrak{m}$ is an irreducible *holomorphic* germ, (x, y) are local (holomorphic) coordinates such that the curves $\{x = 0\}$ and $\{\phi = 0\}$ are transverse, $m = m(\phi)$ is the multiplicity of ϕ and $t > 1$. For $r > 0$ small, the region $\Omega(r)$ is a small neighborhood of the curve $\{\phi = 0\}$ with the origin removed. See Figure 3.7 on page 69. Then $\nu(\psi)$ is given by

$$\nu(\psi) = \lim_{r \to 0} \frac{1}{\log r} \sup_{\Omega(r)} \log |\psi|. \tag{1.4}$$

In fact, this limit exists even when $\log |\psi|$ is replaced by a general plurisubharmonic function, and this is the basis for the valuative study of singularities of plurisubharmonic functions. The fact that a quasimonomial valuation is the pushforward of a monomial valuation means that, analytically, a quasimonomial valuation is the pushforward of a Kiselman number.

1.5.5 Curve Valuations

Any irreducible curve defines a valuation by intersection. More precisely, let C be an irreducible curve of multiplicity $m(C)$ and define

$$\nu_C(\psi) = \frac{C \cdot \{\psi = 0\}}{m(C)},$$

using the intersection multiplicity between curves as defined in Section 1.4. Then ν_C is a centered valuation on R and normalized in the sense that $\nu_C(\mathfrak{m}) = 1$. We call ν_C a *curve valuation*. If $C = \{\phi = 0\}$ then we also write $\nu_C = \nu_\phi$. Notice that $\nu_\phi(\psi) = \infty$ iff ϕ divides ψ.

Curve valuations will play a very important role in our approach. Indeed, a quasimonomial valuation can be accurately understood in terms of the curve valuations that dominate it. This explains Spivakovsky's observation [Sp] that the classification of valuations and curves is essentially the same, see Section 3.8.

If the curve C is holomorphic, then the value of ν_C on a holomorphic germ ψ can be interpreted as the (normalized) order of vanishing of ψ at the origin of the restriction $\psi|_C$; compare (1.4).

As a curve valuation $\nu_C = \nu_\phi$ can take infinite values, it is not strictly speaking a Krull valuation, but it can be turned into one as follows. For $\psi \in R^*$, write $\psi = \phi^k \hat{\psi}$ with $k \in \mathbf{N}$, $\hat{\psi}$ prime with ϕ, and define

$$\nu'_C(\psi) := (k, \nu_\phi(\hat{\psi})) \in \mathbf{Z} \times \mathbf{Q}.$$

Then $\nu'_C = \nu'_\phi$ is a centered Krull valuation, where $\mathbf{Z} \times \mathbf{Q}$ is lexicographically ordered. The numerical invariants of ν'_C are $\mathrm{rk}(\nu'_C) = \mathrm{rat.\,rk}(\nu'_C) = 2$ and $\mathrm{tr.\,deg}(\nu'_C) = 0$ (use (1.1) or see Theorem 2.28).

1.5.6 Exceptional Curve Valuations

Consider a birational morphism $\pi : X \to (\mathbf{C}^2, 0)$ as above and a curve valuation μ at a point p on the exceptional divisor $\pi^{-1}(0)$. If the corresponding curve is not contained in the exceptional divisor, then the pushforward of μ under π is proportional to the curve valuation at the image curve.

If the curve is instead an irreducible component E of the exceptional divisor, then the pushforward of the curve valuation is not proper. However, the pushforward of the corresponding Krull valuation gives a new type of Krull valuation.

Let div_E denote the order of vanishing along E and $\nu_E \in \mathcal{V}$ the normalized divisorial valuation as in Section 1.5.3. Pick local coordinates (z, w) at p such that $E = \{z = 0\}$. For $\phi \in R^*$ write $\pi^*\phi = z^{\operatorname{div}_E(\pi^*\phi)}\psi$, where $\psi \in \mathbf{C}[[z, w]]$. Thus $\psi(0, w) = w^k\hat{\psi}(w)$, where $1 \leq k < \infty$ and $w \nmid \hat{\psi}$. Then

$$\nu_{E,p}(\phi) := (\nu_E(\phi), k) \in \mathbf{Z} \times \mathbf{Z}$$

defines a centered Krull valuation called an *exceptional curve valuation.* Its numerical invariants are the same as those of a curve valuation, i.e. $\operatorname{rk}(\nu_{E,p}) = \operatorname{rat.rk}(\nu_{E,p}) = 2$ and $\operatorname{tr.deg}(\nu_{E,p}) = 0$.

As we shall see in Lemma 1.5, the exceptional curve valuations are exactly the Krull valuations that are not valuations. They are therefore not points in the valuative tree, but can be interpreted as tangent vectors at divisorial valuations. See Appendix B.

1.5.7 Infinitely Singular Valuations

The remaining centered valuations on R are *infinitely singular* valuations. We shall define them in Section 2.2 but they can be characterized in a number of equivalent ways. One way is through their numerical invariants: $\operatorname{rat.rk} = \operatorname{rk} = 1$ and $\operatorname{tr.deg} = 0$ (see Theorem 2.28). Another way is to say that their value (semi)groups are not finitely generated. Examples of valuations with this latter property are given in [ZS2, p.102-104].

Yet another, perhaps more illustrative, way is through Puiseux series. Namely, any (normalized) infinitely singular valuation ν is represented by a (generalized) Puiseux series $\hat{\phi}$ as follows. Pick local coordinates (x, y) such that $\nu(x) = 1$, $\nu(y) \geq 1$. Then $\hat{\phi}$ is a series of the form $\hat{\phi} = \sum_{j=1}^{\infty} a_j x^{\hat{\beta}_j}$, where $a_j \in \mathbf{C}^*$ and $(\hat{\beta}_j)_1^{\infty}$ is a strictly increasing sequence of positive rational numbers with unbounded denominators. If $\psi = \psi(x, y) \in \mathfrak{m}$, then $\psi(x, \hat{\phi})$ defines a series of the same form and $\nu(\psi)$ is the minimum order of the terms in this series.

See Chapter 4 for details on the Puiseux point of view and Appendix A for even more characterizations of infinitely singular valuations. In terms of the valuative tree, the infinitely singular valuations are ends, sharing the latter property with the curve valuations. In a way, the infinitely singular valuations can be viewed as curve valuations associated to "curves" of infinite multiplicity.[7]

Infinitely singular valuations are nontrivial to interpret analytically, but they can sometimes be understood as a generalized Lelong number in the sense of Demailly, with respect to a plurisubharmonic weight [De].

1.6 Valuations Versus Krull Valuations

As we have made a point of distinguishing between valuations and Krull valuations, it is reasonable to ask what the relation is between the two concepts.[8]

In order to understand the situation we will make use of the following result, whose proof is postponed until the end of the section.

Lemma 1.5. *Let μ, ν be non-trivial Krull valuations on R with $R \subset R_\mu \subsetneq R_\nu$. Then either $\nu \sim \nu_E$ is divisorial and $\mu \sim \nu_{E,p}$ is an exceptional curve valuation; or there exists a curve C such that $\mu \sim \nu'_C$ is a curve valuation and $\nu \sim \mathrm{div}_C$.*

First consider a centered valuation $\nu : R \to \overline{\mathbf{R}}_+$. In order to see whether or not it defines a Krull valuation we consider the prime ideal $\mathfrak{p} = \{\nu = \infty\}$. As ν is proper, $\mathfrak{p} \subsetneq \mathfrak{m}$, so either $\mathfrak{p} = (0)$ or $\mathfrak{p} = (\phi)$ for an irreducible $\phi \in \mathfrak{m}$.

When $\mathfrak{p} = (0)$, ν defines a Krull valuation whose value group is included in $\mathbf{R}$. In particular $\mathrm{rk}(\nu) = 1$. When $\mathfrak{p} = (\phi)$ is nontrivial, we may define a Krull valuation $\mathrm{krull}[\nu] : R^* \to \mathbf{Z} \times \mathbf{R}$ as in Section 1.5.5. Namely, for $\psi \in R^*$, write $\psi = \phi^k \hat{\psi}$ with $k \in \mathbf{N}$, $\hat{\psi}$ prime with ϕ, and set $\mathrm{krull}[\nu](\psi) = (k, \nu(\hat{\psi}))$. Then Lemma 1.5 applies to the pair $\mathrm{krull}[\nu]$, div_ϕ. This shows that $\mathrm{krull}[\nu]$ is the associated Krull valuation ν'_ϕ. As a consequence, ν is equivalent to the curve valuation ν_ϕ.

It is straightforward to check that if two valuations give rise to equivalent Krull valuations, then the valuations are themselves equivalent. In particular we may define the numerical invariants rk, rat. rk and tr. deg for all valuations.

Conversely, pick a Krull valuation ν. Let us see if it arises from a valuation by the preceding procedure.

When $\mathrm{rk}(\nu) = 1$ i.e. when $\mathfrak{p} := \{\nu = \infty\} = (0)$, the value group $\nu(K^*)$ of ν can be embedded (as an ordered group) in $(\mathbf{R}, +)$ so that ν defines a valuation. When $\mathrm{rk}(\nu) = 2$, two cases may appear.

(1) There are non-units $\phi_1, \phi_\infty \in R^*$ with $n\nu(\phi_1) < \nu(\phi_\infty)$ for all $n \geq 0$. Then

$$\mathfrak{p} := \{\phi \in R^* \ ; \ n\nu(\phi_1) < \nu(\phi) \text{ for all } n \geq 0\} \sup\{0\}$$

is a proper prime ideal generated by an irreducible element, which we again denote by ϕ_∞. We define a valuation $\nu_0 : R \to \overline{\mathbf{R}}_+$ as follows. First set $\nu_0(\phi_1) = 1$, and $\nu_0(\phi) = \infty$ for $\phi \in \mathfrak{p}$. For any $\phi \in R \setminus \mathfrak{p}$, let

[7] At least this is the reason for their names.

[8] The reader mainly interested in the analytic applications of valuations may skip this section on a first reading.

$$n_k := \max\{n \in \mathbf{N} \; ; \; n\nu(\phi_1) \leq \nu(\phi^k)\}.$$

One checks that $k^{-1}n_k \leq (kl)^{-1}n_{kl} \leq k^{-1}n_k + k^{-1} - (kl)^{-1}$ for all $k, l \geq 0$. This implies that the sequence n_k/k converges towards a real number we define to be $\nu_0(\phi)$. The function ν_0 is a valuation, and $\mathrm{krull}[\nu_0]$ is isomorphic to ν.

(2) For all non-units $\phi, \psi \in R^*$ there exists $n \geq 0$ so that $n\nu(\psi) > \nu(\phi)$. In this case, there can be no valuation $\nu_0 : R \to \overline{\mathbf{R}}_+$ so that $\mathrm{krull}[\nu_0] \sim \nu$. Let us show that ν is an exceptional curve valuation. As $\mathrm{rk}\,\nu = 2$, we may assume $\nu(R^*) \subset \mathbf{R} \times \mathbf{R}$, and $\nu = (\nu_0, \nu_1)$. The function $\nu_0 : R^* \to \mathbf{R}$ is also a Krull valuation, as ν is and $\mathbf{R} \times \mathbf{R}$ is endowed with the lexicographic order. Since $R_\nu \subsetneq R_{\nu_0}$, Lemma 1.5 shows that ν_0 is divisorial and ν an exceptional curve valuation.

We have hence proved

Proposition 1.6. *To any Krull valuation ν which is not an exceptional curve valuation is associated a unique (up to equivalence) valuation $\nu_0 : R \to \overline{\mathbf{R}}_+$ with* $\mathrm{krull}[\nu_0] \sim \nu$.

Remark 1.7. In Section 5.5 we shall strengthen this result and show that $\mathcal{V}$, endowed with the weak topology, is homeomorphic to the quotient space resulting from $\mathcal{V}_K$ with the Zariski topology after having identified all exceptional curve valuations with their associated divisorial valuations. See Theorem 5.24.

Proof (Lemma 1.5). Whereas we have tried to keep the monograph as self-contained as possible, the following proof does make use of some (well-known) results from the literature. Lemma 1.5 is only used to explain the connections between Krull valuations and valuations and is not necessary for obtaining most of the structure on the valuative tree.

The proof goes as follows. As $R_\mu \subsetneq R_\nu$, the ideal $\mathfrak{q} := \mathfrak{m}_\nu \cap R_\mu$ is prime, strictly included in R_μ, and the quotient ring $R_\mu/\mathfrak{q}$ is a non-trivial valuation ring of the residue field $k_\nu = R_\nu/\mathfrak{m}_\nu$. Conversely, a valuation κ on the residue field k_ν determines a unique valuation μ' on K with $R_{\mu'} \subsetneq R_\nu$, and $R_{\mu'}/\mathfrak{m}_\nu \cap R_{\mu'} = R_\kappa$. One says that μ' is the composite valuation of ν and κ. When the value group of ν is isomorphic to $\mathbf{Z}$, we have $\mu'(\psi) = (\nu(\psi), \kappa(\overline{\psi}))$, where $\overline{\psi}$ is the class of ψ in k_ν (see e.g. [Va, p.554]).

Suppose ν is centered. By Abhyankar's inequality (1.1), $\mathrm{tr.deg}(\nu)$ is either 0 or 1. In the first case, k_ν is isomorphic to $\mathbf{C}$, which admits no nontrivial nonnegative valuation. Hence $\mathrm{tr.deg}(\nu) = 1$, $\nu \sim \nu_E$ is divisorial and the valuation κ defined above has rank 1. We may hence assume that $\kappa(K^*) \subset \mathbf{R}$ and that $\mu = (\nu, \kappa)$ takes values in $\mathbf{R} \times \mathbf{R}$.

Performing a suitable sequence of blowups and lifting the situation, one can suppose $\nu = \nu_\mathfrak{m}$, the multiplicity valuation. Assume $\nu(y) \geq \nu(x)$, or else switch the roles of x and y. Since $\mu \neq \nu_\mathfrak{m}$ there exists a unique $\theta \in \mathbf{C}$ for which $\mu(y - \theta x) > \mu(x)$. Then μ is equivalent to the unique valuation sending x to $(1, 0)$ and $y - \theta x$ to $(1, 1)$. This is the exceptional curve valuation attached to the exceptional curve obtained by blowing-up the origin once and at the point of intersection with the strict transform of $\{y - \theta x = 0\}$.

When ν is not centered, the prime ideal $R \cap \mathfrak{m}_\nu$ is generated by an irreducible element $\phi \in \mathfrak{m}$. Then $\nu \sim \mathrm{div}_C$, where $C = \{\phi = 0\}$. The residue field k_ν is isomorphic to $\mathbf{C}((t))$, on which there exists a unique (up to equivalence) nontrivial valuation vanishing on $\mathbf{C}^*$. Therefore μ is equivalent to the (Krull) curve valuation ν'_C. □

Remark 1.8. Let ν be a divisorial valuation on R, and pick two Krull valuations μ_1, μ_2 such that $R_{\mu_i} \subsetneq R_\nu$ for $i = 1, 2$. Then $R_{\mu_1} = R_{\mu_2}$ iff one can find an element $\psi \in \mathfrak{m}_{\mu_1} \cap \mathfrak{m}_{\mu_2}$ such that $\nu(\psi) = 0$.

To see this, first note that the proof implies that $R_{\mu_1} = R_{\mu_2}$ iff the valuation rings $S_i := R_{\mu_i}/(\mathfrak{m}_\nu \cap R_{\mu_i})$ coincide for $i = 1, 2$ inside k_ν. But k_ν is isomorphic to $\mathbf{C}(T)$ so the two valuations rings coincide iff their maximal ideals intersect. This happens iff there exists $\psi \in \mathfrak{m}_{\mu_1} \cap \mathfrak{m}_{\mu_2}$ such that $\nu(\psi) = 0$.

1.7 Sequences of Blowups and Krull Valuations

So far we have defined (Krull) valuations as functions on the ring R satisfying certain conditions. Let us now describe the geometric point of view of valuations developed by Zariski (see [ZS2, p.122], [Va]). This point of view will be exploited systematically in Chapter 6 as a geometric approach to the valuative tree.

We start by a naive approach. Let $\nu \in \tilde{\mathcal{V}}_K$ be a centered Krull valuation on R. Fix local coordinates (x, y) so that $R = \mathbf{C}[[x, y]]$. Assume $\nu(y) \geq \nu(x)$. Then $\nu \geq 0$ on the larger ring $R[y/x] \subset K$ and $Z_\nu := \{\phi \in R[y/x]\ ;\ \nu(\phi) > 0\}$ is a prime ideal containing x.

When $Z_\nu = (x)$, the value $\nu(\phi)$ for $\phi \in R[y/x]$ is proportional to the order of vanishing of ϕ along x. Thus $\nu = \nu_\mathfrak{m}$ is the multiplicity valuation. Otherwise Z_ν is of the form $Z_\nu = (y/x - \theta, x)$ where $\theta \in \mathbf{C}$. Define $x_1 = x$ and $y_1 = y/x - \theta$. Then ν is a centered Krull valuation on the local ring $\mathbf{C}[[x_1, y_1]] \supset R$. We have assumed $\nu(y) \geq \nu(x)$, but when $\nu(y) < \nu(x)$ we get the same conclusion by exchanging the roles of x and y. The idea is now to iterate the procedure and organize the resulting information.

A more geometric formulation of the construction above requires some terminology and is usually formulated in the language of schemes. If $(R_1, \mathfrak{m}_1)$, $(R_2, \mathfrak{m}_2)$ are local rings with common fraction field K, then we say that $(R_1, \mathfrak{m}_1)$ *dominates* $(R_2, \mathfrak{m}_2)$ if $R_1 \supset R_2$ and $\mathfrak{m}_2 = R_2 \cap \mathfrak{m}_1$.

Definition 1.9. *Let X be a (complete) scheme whose field of rational functions is K and let ν be a Krull valuation on K with valuation ring R_ν. The* center *of ν in X is the unique (not necessarily closed) point $x \in X$ whose local ring $\mathcal{O}_{X,x} \subset K$ is dominated by R_ν.*

In our setting, the situation is quite concrete. The space X is the total space of a finite composition $\pi : X \to (\mathbf{C}^2, 0)$ of (point) blowups above the origin. There are then two cases:

(i) the center of ν in X is an irreducible component E of the exceptional divisor $\pi^{-1}(0)$; in this case ν is a divisorial valuation as in Section 1.5.3: $\nu(\phi)$ is proportional to the order of vanishing of $\pi^*\phi$ along E for any $\phi \in R$;
(ii) the center of ν in X is a (closed) point p on the exceptional divisor $\pi^{-1}(0)$; in this case the point p is characterized by the property that there exists a centered valuation μ on the local ring $\mathcal{O}_p$ at p such that $\nu = \pi_*\mu$; in other words, there exist local coordinates (z, w) at p and a valuation μ on the ring $\mathbf{C}[[z, w]]$ such that $\mu(z), \mu(w) > 0$ and $\nu(\phi) = \mu(\pi^*\phi)$ for every $\phi \in R$.

We shall take this more concrete point of view in the sequel.

Again consider a centered Krull valuation $\nu \in \tilde{\mathcal{V}}_K$. The center of ν in $(\mathbf{C}^2, 0)$ is the origin, which we denote by p_0 (in the language of schemes, the center of ν in $\operatorname{Spec} R$ is the maximal ideal $\mathfrak{m}$). blowup the origin: $\tilde{\pi}_0 : X_0 \to (\mathbf{C}^2, 0)$, and let $E_0 = \tilde{\pi}_0^{-1}(0)$ be the exceptional divisor. Consider the center of ν in X_1. Either this is the exceptional component E_0, in which case ν is equivalent to the multiplicity valuation, $\nu \sim \nu_{\mathfrak{m}}$, or the center is a unique point $p_1 \in E_0$.

In the latter case, ν restricts to a centered Krull valuation on the local ring $\mathcal{O}_{p_1}$ at p_1 (if we were to follow (ii), then we should write $\nu = \pi_*\mu$, where μ is a centered valuation on $\mathcal{O}_{p_1}$, but we shall not do this here). Concretely, $\mathcal{O}_{p_1}$ is of the form $\mathbf{C}[[x_1, y_1]]$ above. The fraction field of $\mathcal{O}_{p_1}$ is (isomorphic to) K, hence we can repeat the procedure. That is, we blowup p_1: $\tilde{\pi}_1 : X_1 \to X_0$, and consider the center of ν in X_1. This is either the exceptional divisor $E_1 = \tilde{\pi}_1^{-1}(p_1)$, in which case $\nu \sim \nu_{E_1}$ is divisorial, or a unique closed point $p_2 \in E_1$. Iterating this, we obtain a sequence $\Pi[\nu] = (p_j)_0^n$, $0 \leq n \leq \infty$, of *infinitely near points* above the origin p_0.

Conversely, let $\bar{p} = (p_j)_0^n$, $0 \leq n \leq \infty$, be a sequence of infinitely near points above the origin. By definition, this means p_0 is the origin ; $\tilde{\pi}_0 : X_0 \to (\mathbf{C}^2, 0)$ is the blowup of the origin with exceptional divisor E_0, and inductively, for $j < n$, $\tilde{\pi}_{j+1} : X_{j+1} \to X_j$ is the blowup of a point p_{j+1} on the exceptional divisor $E_j = \tilde{\pi}_j^{-1}(p_j)$ of $\tilde{\pi}$. Write $\pi_j := \tilde{\pi}_0 \circ \cdots \circ \tilde{\pi}_j$.

Let us show how to associate a Krull valuation to $\bar{p}$. In fact, rather than defining a Krull valuation directly, we shall define a valuation ring $R_{\bar{p}} \subset K$.

When $n < \infty$, we declare $\phi \in R_{\bar{p}}$ iff $\pi_n^*\phi$ is a regular function at a generic point on E_n. If $n = \infty$, then $\phi \in R_{\bar{p}}$ iff there exists $j \geq 1$ such that $\pi_j^*\phi$ is regular at p_{j+1} (which implies that $\pi_k^*\phi$ is regular at p_{k+1} for $k \geq j$). A direct check shows that $R_{\bar{p}}$ is a ring, and since any curve can be desingularized by a finite sequence of blowups, we get either $\phi \in R_{\bar{p}}$ or $\phi^{-1} \in R_{\bar{p}}$ for any $\phi \in K^*$. Thus $R_{\bar{p}}$ is a valuation ring, and we denote by $\operatorname{val}[\bar{p}]$ its associated Krull valuation.

It is clear that the map $\bar{p} \to \mathrm{val}[\bar{p}]$ is injective. Let us indicate how to show that $R_{\Pi[\nu]} = R_\nu$. This will prove that the maps $\bar{p} \to \mathrm{val}[\bar{p}]$ and $\nu \to \Pi[\nu]$ are inverse one to another. For sake of simplicity we suppose that $\Pi[\nu] = (p_j)$ is an infinite sequence. Pick $\phi \in R_\nu$, i.e. $\nu(\phi) \geq 0$. Write $\phi = \phi_1/\phi_2$ with ϕ_ε regular functions. For j large enough, the strict transform of $\{\phi_\varepsilon = 0\}$, $\varepsilon = 1, 2$, is smooth at p_{j+1} and transverse to the exceptional divisor, or does not contain p_{j+1}. In local coordinates (z, w) at p_{j+1}, we therefore have $\pi_j^*\phi = z^a w^b \times$ unit for large j. When $a, b \geq 0$, $\pi_j^*\phi$ is regular at p_{j+1} and we are done. When $a, b \leq 0$ and $ab \neq 0$, ϕ^{-1} lies in the maximal ideal of the ring $\mathcal{O}_{p_j}$. By construction, this gives $\nu(\phi) = -\nu(\phi^{-1}) < 0$, a contradiction. Finally when $a > 0$ and $b < 0$, one goes one step further considering the lift of ϕ to p_{j+2}. The coefficients (a, b) are replaced by $(a, a+b)$ or $(b, a+b)$. After finitely many steps, we are hence in one of the two preceding situations. This proves that $\pi_j^*\phi$ is regular at p_{j+1} for j sufficiently large, hence $\phi \in R_{\Pi[\nu]}$. Conversely, pick $\phi \in R_{\Pi[\nu]}$. For j large enough, $\pi_j^*\phi$ is a regular function at p_{j+1}. By definition, the valuation ring associated to ν at p_{j+1} contains $\mathcal{O}_{p_{j+1}}$. Hence $\phi \in R_\nu$, which concludes the proof that $R_{\Pi[\nu]} = R_\nu$.

Summing up, we have

Theorem 1.10. *The set $\mathcal{V}_K$ of centered Krull valuations modulo equivalence is in bijection with the set of sequences of infinitely near points. The bijection is given as follows: $\nu \in \mathcal{V}_K$ corresponds to (p_j) iff R_ν dominates $\mathcal{O}_{p_j}$ for all j.*

Remark 1.11. The correspondence above is a particular case of a general result: see [ZS2, p.122]. Moreover, the residue field $k_\nu \cong R_\nu/\mathfrak{m}_\nu$ of a valuation is isomorphic to the field of rational functions of the last exceptional divisor of the sequence $\Pi[\nu]$. In our case, either ν is non-divisorial and $k_\nu \cong \mathbf{C}$; or ν is divisorial and $k_\nu \cong \mathbf{C}(T)$ for some independent variable T.

Theorem 1.10 yields the following characterization of divisorial valuations.

Proposition 1.12. *For any centered Krull valuation $\nu \in \mathcal{V}_K$, the following three conditions are equivalent:*

- *ν is divisorial in the sense of Section 1.5.3;*
- *its sequence $\Pi[\nu]$ of infinitely near points is finite;*
- $\mathrm{tr.deg}\,\nu = 1$.

Proof. It is clear that if the sequence of infinitely near points associated to a Krull valuation is finite, then the valuation is divisorial. We saw in Section 1.5.3 that any divisorial valuation has transcendence degree 1. Finally, suppose the sequence $\Pi[\nu] = (p_j)_0^\infty$ of infinitely near points associated to ν is infinite. We follow the notation of the proof of Theorem 1.10. Pick $\phi \in R_\nu \setminus \mathfrak{m}_\nu$. Then $\pi_j^*\phi$ is a regular function near p_{j+1} for j large enough. As $\phi \notin \mathfrak{m}_\nu$, it is invertible in R_ν, so we may suppose that $\pi_j^*\phi^{-1}$ is also regular at p_{j+1} by increasing j. This implies that ϕ does not vanish at p_j, so $\phi = \phi(0)$ inside $k_\nu := R_\nu/\mathfrak{m}_\nu$. The residue field of ν is thus isomorphic to $\mathbf{C}$, and we have $\mathrm{tr.deg}\,\nu = 0$. □

In several cases, we may interpret geometrically the sequence of infinitely near points associated to a Krull valuation.

Example 1.13. When $\nu = \nu_C$ is a curve valuation, then $\Pi[\nu] = (p_j)_0^\infty$ is the sequence of infinitely near points associated to C. This sequence can be defined as follows: p_0 is the origin and, inductively, p_{j+1} is the intersection point of E_j and the strict transform of C under π_j. See Section 6.2, and Table C.3 in Appendix C for more information on this sequence of infinitely near points.

Example 1.14. If $\nu = \nu_{E,p}$ is an exceptional curve valuation, then $\Pi[\nu] = (p_j)_0^\infty$, where $\Pi[\nu_E] = (p_j)_0^{j_0}$, $p_{j_0+1} = p$ and, inductively for $j > j_0$, p_{j+1} is the intersection point of E_j and the strict transform of $E = E_{j_0}$. See Appendix B.

Example 1.15. Let $s \geq 1$, and $\nu = \nu_s$ the monomial valuation with $\nu(x) = 1$, $\nu(y) = s$. Then $\Pi[\nu]$ is determined by the *continued fractions expansion*

$$s = a_1 + \frac{1}{a_2 + \ldots} = [[a_1, a_2, \ldots]],$$

and in particular $\Pi[\nu]$ is finite iff the continued fractions expansion is finite, which happens iff s is rational. See [Sp] for details.

2

MacLane's Method

We now do the preparatory work for obtaining the tree structure on valuation space. Namely, we show how to represent a valuation conveniently by a finite or countable sequence of polynomials and real numbers. In the next chapter we shall see how to visualize this encoding through the language of trees.

Our approach is an adaptation of MacLane's method [Ma]. In his original article, MacLane gives a description of all extensions of a discrete valuation of rank 1 defined on an arbitrary field, in two different contexts: to a finite extension or to a purely transcendental extension of degree 1. The situation we have to deal with here is slightly different as we are working on a ring, namely $\mathbf{C}[[x, y]]$, and there are some adaptations to be made.

The method can be outlined as follows. Fix local coordinates (x, y). Pick a centered valuation ν on $R = \mathbf{C}[[x, y]]$. It corresponds uniquely to a valuation, still denoted ν, on the Euclidean domain $\mathbf{C}(x)[y]$. Write $U_0 = x$ and $U_1 = y$. The value $\tilde{\beta}_0 = \nu(U_0)$ determines ν on $\mathbf{C}(x)$ and the value $\tilde{\beta}_1 = \nu(U_1)$ determines ν on many polynomials in y. The idea is now to find a polynomial $U_2(x, y)$ of minimal degree in y such that $\tilde{\beta}_2 = \nu(U_2)$ is not determined by the previous data. Inductively we construct polynomials U_j and numbers $\tilde{\beta}_j = \nu(U_j)$ which, as it turns out, represent ν completely.

More precisely we proceed as follows. First we define what data $[(U_j); (\tilde{\beta}_j)]$ to use, namely sequences of key polynomials, or SKP's. These are introduced in Section 2.1 where we also show how to associate a canonical valuation to an SKP. In Section 2.3 we study such a valuation in detail, describing its graded ring and computing its numerical invariants. Then in Section 2.4 we show that in fact any valuation is associated to an SKP. Finally, in Section 2.5 we compute $\nu(\phi)$ for $\phi \in \mathfrak{m}$ irreducible in terms of the SKP's associated to ν and the curve valuation ν_ϕ. This computation is the key to parameterization of valuation space (by skewness) as described in Section 3.3.

We formulate almost all results for valuations rather than Krull valuations, but the method works for both classes. We leave it to the reader to make the suitable adaptation in the notation and statements.

This chapter is quite technical in nature. A number of proofs require a fine combinatorial analysis and elaborate inductions. We shall later exploit two other approaches to the valuative tree: one based on Puiseux expansions (Chapter 4) and one based on the geometrical interpretation of valuations (Chapter 6). The method by SKP's has, however, two advantages. First, it is purely algebraic and may thus be generalized to more general contexts, especially when $\mathbf{C}$ is replaced by a field of positive characteristic. Second, it is well adapted to global problems, for instance when analyzing singularities of plane curves at infinity in $\mathbf{C}^2$. These two points of view already appear in the work of Abhyankar-Moh on approximate roots, which are very closely related to SKP's.

2.1 Sequences of Key Polynomials

2.1.1 Key Polynomials

Let us define the data we use to represent valuations. Fix local coordinates (x, y) for the rest of this chapter.

Definition 2.1. *A sequence of polynomials* $(U_j)_{j=0}^k$, $1 \leq k \leq \infty$, *in* $\mathbf{C}[x, y]$ *is called a* sequence of key-polynomials *(SKP) if it satisfies:*

(P0) $U_0 = x$ *and* $U_1 = y$;

(P1) to each polynomial U_j *is attached a number* $\tilde{\beta}_j \in \overline{\mathbf{R}}_+$ *(not all* ∞*) with*

$$\tilde{\beta}_{j+1} > n_j \tilde{\beta}_j = \sum_{l=0}^{j-1} m_{j,l} \tilde{\beta}_l \quad \textit{for } 1 \leq j < k, \tag{2.1}$$

where $n_j \in \mathbf{N}^*$ *and* $m_{j,l} \in \mathbf{N}$ *satisfy, for* $j > 1$ *and* $1 \leq l < j$,

$$n_j = \min\{l \in \mathbf{Z} \ ; \ l\,\tilde{\beta}_j \in \mathbf{Z}\tilde{\beta}_0 + \cdots + \mathbf{Z}\tilde{\beta}_{j-1}\} \quad \textit{and} \quad 0 \leq m_{j,l} < n_l; \tag{2.2}$$

(P2) for $1 \leq j < k$ *there exists* $\theta_j \in \mathbf{C}^*$ *such that*

$$U_{j+1} = U_j^{n_j} - \theta_j \cdot U_0^{m_{j,0}} \cdots U_{j-1}^{m_{j,j-1}} \tag{2.3}$$

In the sequel, we call k the *length* of the SKP.

Remark 2.2. If an abstract semi-group Γ is given, a sequence of polynomials satisfying (P0)-(P2) with $\tilde{\beta}_j \in \Gamma$ will be called a Γ-SKP.

Remark 2.3. If $(U_j)_0^k$ is an SKP of length $k \geq 2$, then $(\tilde{\beta}_j)_1^{k-1}$ are determined by $\tilde{\beta}_0$ and $(U_j)_0^k$. Conversely, given a sequence $(\tilde{\beta}_j)_1^{k+1}$ satisfying (P2) and any sequence $(\theta_j)_1^k$ in $\mathbf{C}^*$ there exists a unique associated SKP. Despite this redundancy we will typically write an SKP as $[(U_j); (\tilde{\beta}_j)]$.

Lemma 2.4. *For $1 \leq j \leq k$, the polynomial U_j is irreducible and of Weierstrass form $U_j = y^{d_j} + a_1(x)y^{d_j-1} + \cdots + a_{d_j}(x)$ with $a_l(0) = 0$ for all l. Moreover, $d_{j+1} = n_j d_j$ for $1 \leq j < k$.*

Proof. The proof of all assertions is by induction. The fact that U_j is irreducible is not obvious and will follow from the proof of Theorem 2.8. We included this fact here for clarity. Let us show by induction on k that $d_{j+1} = n_j d_j$ for $1 \leq j < k$.

By (2.2) and the induction hypothesis, we have $m_{j,l} < n_l = d_{l+1}/d_l$, hence $m_{j,l} + 1 \leq d_{l+1}/d_l$ as both sides are integers. We infer

$$\sum_{l=1}^{l-1} m_{j,l} d_l \leq \sum_{l=1}^{j-1} (\frac{d_{l+1}}{d_l} - 1) d_l = d_j - 1 < n_j d_j,$$

hence $\deg_y(U_{j+1}) = n_j d_j$. □

Although we shall not need the following lemma in this chapter, we include it here for convenience. Recall that $m(\phi)$ denotes the multiplicity of $\phi \in R$.

Lemma 2.5. *When $\tilde{\beta}_1 \geq \tilde{\beta}_0$, $m(U_j) = d_j$ for $j > 0$.*

Proof. The proof goes by induction on j. For $j = 1$ the assertion is obvious as $U_1 = y$ and $d_1 = 1$. For $j = 2$, $U_2 = y^{n_1} - x^{m_{1,0}}$ with $n_1\tilde{\beta}_1 = m_{1,0}\tilde{\beta}_0$. As $\tilde{\beta}_1 \geq \tilde{\beta}_0$, $m(U_2) = n_1 = d_1$.

Now assume $j \geq 3$. To prove $m(U_j) = n_j m(U_{j-1})$ it is sufficient to prove that $n_j d_j < \sum_0^{j-1} m_{j,l} d_l$. We have $d_j = n_{j-1} d_{j-1}$ and $\tilde{\beta}_j > n_{j-1}\tilde{\beta}_{j-1}$ hence $n_j d_j = n_j n_{j-1} d_{j-1} < n_j \tilde{\beta}_j d_{j-1}/\tilde{\beta}_{j-1}$. Now $n_j\tilde{\beta}_j = \sum_0^{j-1} m_{j,l}\tilde{\beta}_l$ and $\tilde{\beta}_{j-1} > (n_{j-2} \ldots n_l)\tilde{\beta}_l$ for any $l < j$. Hence

$$n_j d_j < \sum_{l=0}^{j-1} m_{j,l} \tilde{\beta}_l \frac{d_{j-1}}{\tilde{\beta}_{j-1}} < \sum_{l=0}^{j-1} m_{j,l} d_l.$$

This concludes the proof. □

Remark 2.6. In the notation of Zariski and Spivakovsky the sequence $(\overline{\beta}_i)$ can be extracted from $(\tilde{\beta}_j)$ as follows. One has $\overline{\beta}_i = \tilde{\beta}_{k_i}$ where k_i is defined inductively to be the smallest integer $k_i > k_{i-1}$ so that $n_{k_i} \geq 2$. In the Abhyankar-Moh terminology [AM], the U_{k_i} are the approximate roots of U_k. See Sections 3.5 and 3.7 and Appendix C for more details.

We include a result of arithmetic nature which will be used repeatedly.

Lemma 2.7. *Assume $\tilde{\beta}_0, \ldots, \tilde{\beta}_{k+1}$ are given so that for all $j = 1, \ldots, k$ one has $\tilde{\beta}_{j+1} > n_j\tilde{\beta}_j$ with $n_j := \min\{n \in \mathbf{N}^* \; ; \; n\tilde{\beta}_j \in \sum_0^{j-1} \mathbf{Z}\tilde{\beta}_i\}$. Then for any $j = 1, \ldots, k$ there exists a unique decomposition $n_j\tilde{\beta}_j = \sum_0^{j-1} m_{j,l}\tilde{\beta}_l$ where $0 \leq m_{j,l} < n_l$ for $l = 1, \ldots, j-1$.*

Proof. By assumption there exist $m_l \in \mathbf{Z}$ with $n_j\tilde{\beta}_j = \sum_0^{j-1} m_l\tilde{\beta}_l$. By Euclidean division, $m_{j-1} = q_{j-1}n_{j-1}+r_{j-1}$ with $0 \leq r_{j-1} < n_{j-1}$ so by the fact that $n_{j-1}\tilde{\beta}_{j-1} \in \sum_0^{j-2} \mathbf{Z}\tilde{\beta}_i$ we can suppose that $0 \leq m_{j-1} < n_{j-1}$. Inductively we get $0 \leq m_l < n_i$ for $1 \leq l < j$. It remains to prove that $m_0 \geq 0$. Set $S = \sum_1^{j-1} m_l\tilde{\beta}_l$. We have $m_1\tilde{\beta}_1 < n_1\tilde{\beta}_1 < \tilde{\beta}_2$ and $m_2 < n_2$ so $m_2 + 1 \leq n_2$ and

$$m_1\tilde{\beta}_1 + m_2\tilde{\beta}_2 < (1 + m_2)\tilde{\beta}_2 \leq n_2\tilde{\beta}_2 < \tilde{\beta}_3.$$

We infer that $S < (1+m_{j-1})\tilde{\beta}_{j-1} \leq n_{j-1}\tilde{\beta}_{j-1} < \tilde{\beta}_j$ so that $m_0\tilde{\beta}_0 \geq \tilde{\beta}_j - S \geq 0$. □

2.1.2 From SKP's to Valuations I

We show how to associate a valuation ν to any finite SKP in a canonical way.

Theorem 2.8. *Let $[(U_j)_0^k; (\tilde{\beta}_j)_0^k]$ be an SKP of length $1 \leq k < \infty$. Then there exists a unique centered valuation $\nu_k \in \tilde{\mathcal{V}}$ satisfying*

(Q1) $\nu_k(U_j) = \tilde{\beta}_j$ for $0 \leq j \leq k$;
(Q2) $\nu_k \leq \nu$ for any $\nu \in \tilde{\mathcal{V}}$ satisfying (Q1).

Further, if $k > 1$ and ν_{k-1} is the valuation associated to $[(U_j)_0^{k-1}; (\tilde{\beta}_j)_0^{k-1}]$, then

(Q3) $\nu_{k-1} \leq \nu_k$;
(Q4) $\nu_{k-1}(\phi) < \nu_k(\phi)$ iff U_k divides ϕ in $\mathrm{gr}_{\nu_k} \mathbf{C}(x)[y]$.

Remark 2.9. The valuation ν_k is normalized in the sense that $\nu_k(\mathfrak{m}) = 1$ iff $\min\{\tilde{\beta}_0, \tilde{\beta}_1\} = 1$. It is normalized in the sense that $\nu_k(x) = 1$ iff $\tilde{\beta}_0 = 1$.

The proof of Theorem 2.8 occupies all of Section 2.1.3. It is a subtle induction on k using divisibility properties in the graded ring $\mathrm{gr}_\nu \mathbf{C}(x)[y]$.

First, ν_1 is defined to be the monomial valuation with $\nu_1(x) = \tilde{\beta}_0, \nu_1(y) = \tilde{\beta}_1$, i.e.

$$\nu_1(\sum a_{ij}x^iy^j) = \min\{i\tilde{\beta}_0 + j\tilde{\beta}_1 \ ; \ a_{ij} \neq 0\}. \tag{2.4}$$

Property (Q2) clearly holds.

Now assume $k > 1$, that the SKP $[(U_j)_0^k; (\tilde{\beta}_j)_0^k]$ is given and that $\nu_1, \ldots, \nu_{k-1}$ have been defined. First consider a polynomial $\phi \in \mathbf{C}[x, y]$. As U_k is unitary in y, we can divide ϕ by U_k in $\mathbf{C}[x][y]$: $\phi = \phi_0 + U_k\psi$ with $\deg_y(\phi_0) < d_k = \deg_y(U_k)$ and $\psi \in \mathbf{C}[x, y]$. Iterating the procedure we get a unique decomposition

$$\phi = \sum_j \phi_j U_k^j \tag{2.5}$$

with $\phi_j \in \mathbf{C}[x, y]$ and $\deg_y(\phi_j) < d_k$. Define

$$\nu_k(\phi) := \min_j \nu_k(\phi_j U_k^j) := \min_j\{\nu_{k-1}(\phi_j) + j\tilde{\beta}_k\}. \tag{2.6}$$

The function ν_k is easily seen to satisfy (V2) and (V3). We will show it also satisfies (V1), hence defines a valuation on $\mathbf{C}[x,y]$, which automatically satisfies (Q2). Theorem 2.8 then follows from

Proposition 2.10 ([Sp]). *Any valuation $\nu : \mathbf{C}[x,y] \to \overline{\mathbf{R}}_+$ with $\nu(x), \nu(y) > 0$ has a unique extension to a centered valuation on $R = \mathbf{C}[[x,y]]$. This extension preserves the partial ordering: if $\nu_1(\phi) \leq \nu_2(\phi)$ holds for polynomials ϕ, it also holds for formal power series.*

Proof. We may assume that ν is normalized in the sense that $\nu(\mathfrak{m}) = 1$. Pick $\phi \in R$, and write $\phi = \phi_k + \hat{\phi}_k$ where ϕ_k is the truncation of ϕ at order k, i.e. $m(\hat{\phi}_k) \geq k$. Notice that this implies $\nu(\hat{\phi}_k) \geq k$.

We claim that $\nu(\phi) := \lim_k \nu(\phi_k)$ exists in $\overline{\mathbf{R}}_+$. Indeed, for any k, l, we have

$$\nu(\phi_{k+l}) \geq \min\{\nu(\phi_k), \nu(\phi_{k+l} - \phi_k)\} \geq \min\{\nu(\phi_k), k\}$$

Thus the sequence $\min\{\nu(\phi_k), k\}$ is nondecreasing, hence converges.

Uniqueness can be proved as follows: $\nu(\phi) \geq \min\{\nu(\phi_k), k\}$ for all k. If $\nu(\phi_k)$ is unbounded, then $\nu(\phi) = \infty$. Otherwise $\nu(\phi) = \nu(\phi_k)$ for k large enough. By construction the extension preserves the partial ordering. □

The extension above needs not preserve numerical invariants. For example, if ν is a curve valuation at a non-algebraic curve, then $\mathrm{rk}(\nu|_{\mathbf{C}[x,y]}) = 1$ but $\mathrm{rk}(\nu) = 2$.

Remark 2.11. The expansion (2.5) can be applied for any $\phi \in R = \mathbf{C}[[x,y]]$. Indeed, using Weierstrass' preparation theorem, we can assume that ϕ is a monic polynomial in y. Weierstrass' division theorem then yields (2.5) with $\phi_j \in \mathbf{C}[[x]][y]$. The valuation ν_k can hence be defined by (2.6) for any formal power series.

2.1.3 Proof of Theorem 2.8

By Proposition 2.10, we will restrict our attention to centered valuations defined on the Euclidean ring $\mathbf{C}(x)[y]$. We first make the induction hypothesis precise.

(H_k): ν_k is a valuation satisfying (Q1)-(Q4).
(E_k): the graded ring $\mathrm{gr}_{\nu_k} \mathbf{C}(x)[y]$ is a Euclidean domain.
(I_k): U_k and U_{k+1} are irreducible in $\mathrm{gr}_{\nu_k} \mathbf{C}(x)[y]$.
$(\overline{I}_k)$: U_j is irreducible both in $\mathbf{C}[x,y]$, and $\mathbf{C}(x)[y]$ for $1 \leq j \leq k$.

For the precise definition of a Euclidean domain we refer to [ZS1, p.23]. One immediately checks that H_1 and $\overline{I}_1$ hold. Our strategy is to show successively $\overline{I}_k$ & $H_k \Rightarrow E_k$; $\overline{I}_k$ & $H_k \Rightarrow I_k$; E_k & I_k & $H_k \Rightarrow H_{k+1}$; I_k & $\overline{I}_k$ & $H_{k+1} \Rightarrow \overline{I}_{k+1}$.

Step 1: $\bar{I}_k$ & $H_k \Rightarrow E_k$. We want to show that $\mathrm{gr}_{\nu_k} \mathbf{C}(x)[y]$ is a Euclidean domain. Let $\phi \in R$ and consider the expansion (2.5). We define

$$\delta_k(\phi) := \max\left\{j \ ; \ \nu_k(\phi_j U_k^j) = \nu_k(\phi)\right\}.$$

By convention we let $\delta_k(0) = -\infty$. Note that if $\phi = \phi'$ modulo ν_k, then $\delta_k(\phi) = \delta_k(\phi')$. Hence δ_k is well defined on $\mathrm{gr}_{\nu_k} \mathbf{C}(x)[y]$.

Lemma 2.12. *For all* $\phi, \psi \in \mathbf{C}(x)[y]$, $\delta_k(\phi\psi) = \delta_k(\phi) + \delta_k(\psi)$.

Proof. First assume $\delta_k(\phi) = \delta_k(\psi) = 0$. Then $\phi = \phi_0$, $\psi = \psi_0$ in $\mathrm{gr}_{\nu_k} \mathbf{C}(x)[y]$, so we can assume $\deg_y(\phi), \deg_y(\psi) < d_k$. Thus $\deg_y(\phi\psi) < 2d_k$ so $\phi\psi = \alpha_0 + \alpha_1 U_k$ with $\deg_y(\alpha_0), \deg_y(\alpha_1) < d_k$. Suppose $\nu_k(\alpha_1 U_k) = \nu_k(\phi\psi)$. Then $\nu_{k-1}(\alpha_1 U_k) < \nu_k(\alpha_1 U_k) \leq \nu_k(\alpha_0) = \nu_{k-1}(\alpha_0)$. Thus

$$\nu_{k-1}(\alpha_1 U_k) = \nu_{k-1}(\phi\psi) = \nu_{k-1}(\phi) + \nu_{k-1}(\psi) = \nu_k(\phi) + \nu_k(\psi) = \nu_k(\phi\psi).$$

Hence $\nu_{k-1}(\alpha_1 U_k) = \nu_k(\alpha_1 U_k)$ contradicting (Q4). So $\phi\psi = \alpha_0$ in $\mathrm{gr}_{\nu_k} \mathbf{C}(x)[y]$ and $\delta_k(\phi\psi) = 0$.

In the general case, one has $\phi = \sum \phi_i U_k^i$, $\psi = \sum \psi_j U_k^j$, and $\phi\psi = \sum (\phi_i \psi_j) U_k^{i+j}$. By what precedes $\delta_k(\phi_i\psi_j) = 0$ for all i, j so $\delta_k(\phi\psi) = \delta_k(\phi) + \delta_k(\psi)$. □

Lemma 2.13. *For* $\phi \in \mathbf{C}(x)[y]$, $\delta_k(\phi) = 0$ *iff* ϕ *is a unit in* $\mathrm{gr}_{\nu_k} \mathbf{C}(x)[y]$.

Proof. If $\delta_k(\phi) = 0$, then $\phi = \phi_0$ in $\mathrm{gr}_{\nu_k} \mathbf{C}(x)[y]$. As U_k is irreducible in $\mathbf{C}(x)[y]$ and $\deg_y(U_k) > \deg_y(\phi_0)$, the polynomial U_k is prime with ϕ_0. Hence we can find $A, B \in \mathbf{C}(x)[y]$ with $\deg_y A, \deg_y B < d_k$ so that $A\phi_0 = 1 - BU_k$. Then, comparing with (2.5), we have $\nu_k(A\phi_0) = \nu_k(1) < \nu_k(BU_k)$. Therefore $A\phi_0 = 1$ in $\mathrm{gr}_{\nu_k} \mathbf{C}(x)[y]$ so ϕ_0, and hence ϕ, is a unit in $\mathrm{gr}_{\nu_k} \mathbf{C}(x)[y]$.

Conversely, if ϕ is a unit, say $A\phi = 1$ in $\mathrm{gr}_{\nu_k} \mathbf{C}(x)[y]$ for some $A \in \mathbf{C}(x)[y]$, then $\delta_k(\phi) + \delta_k(A) = \delta_k(1) = 0$, so $\delta_k(\phi) = 0$. □

Lemma 2.14. *If* $\phi, \psi \in \mathbf{C}(x)[y]$, *then there exists* $Q, R \in \mathbf{C}(x)[y]$ *such that* $\phi = Q\psi + R$ *in* $\mathrm{gr}_{\nu_k} \mathbf{C}(x)[y]$ *and* $\delta_k(R) < \delta_k(\psi)$.

Proof. Write $\psi = \sum_j \psi_j U_k^j$. It suffices to prove the lemma when $\psi_j = 0$ for $j > M := \delta_k(\psi)$ and using Lemma 2.13 we may assume $\psi_M = 1$. As $\deg_y(\psi_j) < d_k$ for $j \leq M$ we have $\deg_y(\psi) = Md_k$. Euclidean division in $\mathbf{C}(x)[y]$ yields $Q, R^1 \in \mathbf{C}(x)[y]$ with $\deg_y(R^1) < \deg_y(\psi)$ so that $\phi = Q\psi + R^1$. Write $R^1 = \sum_i R_i U_k^i$ and set $N := \delta_k(R^1)$, $R := \sum_{i \leq N} R_i U_k^i$. Then $\phi = Q\psi + R$ in $\mathrm{gr}_{\nu_k} \mathbf{C}(x)[y]$ and

$$\deg_y(R) = \deg_y(R_N) + Nd_k < Md_k = \deg_y(\psi).$$

Hence $N < M$ and we are done. □

This completes Step 1. The Euclidean property of $\mathrm{gr}_{\nu_k} \mathbf{C}(x)[y]$ makes every ideal principal and supplies us with unique factorization and Gauss' lemma.

Step 2: $\overline{I}_k$ & $H_k \Rightarrow I_k$. We want to show that U_k and U_{k+1} are irreducible in $\mathrm{gr}_{\nu_k} \mathbf{C}(x)[y]$ and proceed in several steps.

Lemma 2.15. *U_k is irreducible in $\mathrm{gr}_{\nu_k} \mathbf{C}(x)[y]$.*

Proof. We trivially have $\delta_k(U_k) = 1$ so if $\phi\psi = U_k$ in $\mathrm{gr}_{\nu_k} \mathbf{C}(x)[y]$, then $\delta_k(\phi) = 0$ or $\delta_k(\psi) = 0$. Hence ϕ or ψ is a unit in $\mathrm{gr}_{\nu_k} \mathbf{C}(x)[y]$ (see Lemma 2.12 and 2.13). □

Lemma 2.16. *If $j < k$ then U_j is a unit in $\mathrm{gr}_{\nu_k} \mathbf{C}(x)[y]$.*

Proof. By Lemma 2.13 it suffices to show that $\delta_k(U_j) = 0$. If $d_j < d_k$ then this is obvious. If $d_j = d_k$, then $U_j = (U_j - U_k) + U_k$, where $\deg_y(U_j - U_k) < d_k$. Now $\nu_k(U_j) = \tilde{\beta}_j < \tilde{\beta}_k = \nu_k(U_k)$, so $\nu_k(U_j - U_k) < \nu_k(U_k)$ and $\delta_k(U_j) = 0$. □

Lemma 2.17. *If $\delta_k(\phi) < n_k$, then $\phi = \phi_i U_k^i$ in $\mathrm{gr}_{\nu_k} \mathbf{C}(x)[y]$ for some $i < n_k$.*

Proof. Suppose $\nu_k(\phi_i U_k^i) = \nu_k(\phi_j U_k^j) = \nu_k(\phi)$, where $i \leq j < n_k$. Then $(j-i)\tilde{\beta}_k = \nu_{k-1}(\phi_i) - \nu_{k-1}(\phi_j) \in \sum_0^{k-1} \mathbf{Z}\tilde{\beta}_j$. By (2.2) n_k divides $j - i$ hence $i = j$. □

Lemma 2.18. *U_{k+1} is irreducible in $\mathrm{gr}_{\nu_k} \mathbf{C}(x)[y]$.*

Proof. We have $U_{k+1} = U_k^{n_k} - \tilde{U}_{k+1}$, where $\tilde{U}_{k+1} = \theta_k \prod_{j=0}^{k-1} U_j^{m_{k,j}}$. Assume $U_{k+1} = \phi\psi$ in $\mathrm{gr}_{\nu_k} \mathbf{C}(x)[y]$ with $0 < \delta_k(\phi), \delta_k(\psi) < n_k$. By Lemma 2.17, we can write $\phi = \phi_i U_k^i$, $\psi = \psi_j U_k^j$. Then $U_{k+1} = \phi_i\psi_j U_k^{n_k}$ so $(1 - \phi_i\psi_j)U_k^{n_k} = \tilde{U}_{k+1}$. As U_k is irreducible and $\tilde{U}_{k+1}$ is a unit, we have $\phi_i\psi_j = 1$ in $\mathrm{gr}_{\nu_k} \mathbf{C}(x)[y]$. But then $\tilde{U}_{k+1} = 0$ in $\mathrm{gr}_{\nu_k} \mathbf{C}(x)[y]$, which is absurd. So we can assume $\delta_k(\phi) = n_k$ and $\delta_k(\psi) = 0$. Hence ψ is a unit, completing the proof. □

Step 3: E_k & I_k & $H_k \Rightarrow H_{k+1}$. We must show $\nu_{k+1}(\phi\psi) = \nu_{k+1}(\phi) + \nu_{k+1}(\psi)$ for $\phi, \psi \in \mathbf{C}[x,y]$ where ν_{k+1} is defined as in (2.6). We first note

(i) $\deg_y(\phi) < d_{k+1}$ implies $\nu_{k+1}(\phi) = \nu_k(\phi)$;
(ii) $\nu_k(U_{k+1}) = n_k\tilde{\beta}_k < \tilde{\beta}_{k+1} = \nu_{k+1}(U_{k+1})$;
(iii) $\nu_{k+1}(U_{k+1}\phi) = \nu_{k+1}(U_{k+1}) + \nu_{k+1}(\phi)$ for all $\phi \in \mathbf{C}[x,y]$.

The second assertion follows from (2.1). We now show

Lemma 2.19. *For $\phi \in \mathbf{C}[x,y]$ we have $\nu_{k+1}(\phi) \geq \nu_k(\phi)$.*

Proof. We argue by induction on $\deg_y(\phi)$. If $\deg_y(\phi) < d_{k+1}$, then we are done by (i). If $\deg_y(\phi) \geq d_{k+1}$, then we write $\phi = \sum_i \phi_i U_{k+1}^i$ as in (2.5). By the induction hypothesis and (ii)-(iii) above we may assume that $\phi_0 \not\equiv 0$. Write $\phi = \phi_0 + \psi$ with $\psi = \sum_{i\geq 1} \phi_i U_{k+1}^i$. When $\nu_k(\phi) = \min\{\nu_k(\phi_0), \nu_k(\psi)\}$, one has

$$\nu_{k+1}(\phi) = \min\{\nu_{k+1}(\phi_0), \nu_{k+1}(\psi)\} \geq \nu_k(\phi),$$

proving the lemma in this case. Otherwise, $\phi_0 + \psi = 0$ in $\mathrm{gr}_{\nu_k} \mathbf{C}(x)[y]$. This implies that U_{k+1} divides ϕ_0 in this ring. But $\deg_y(\phi_0) < d_{k+1}$, hence $\delta_k(\phi_0) < n_k$, so that ϕ_0 is a power of the irreducible U_k times a unit in $\mathrm{gr}_{\nu_k} \mathbf{C}(x)[y]$ by Lemma 2.17. By Lemma 2.18, U_{k+1} is also irreducible, a contradiction. □

Introduce $\mathfrak{p} := \{\nu_{k+1} > \nu_k\} \subset \mathrm{gr}_{\nu_k} \mathbf{C}(x)[y]$. Using the preceding lemma, one easily verifies that $\mathfrak{p}$ is a proper ideal, which contains the irreducible element U_{k+1} by (ii). As $\mathrm{gr}_{\nu_k} \mathbf{C}(x)[y]$ is a Euclidean domain, $\mathfrak{p}$ is generated by U_{k+1}.

Fix $\phi, \psi \in R$. We want to show that $\nu_{k+1}(\phi\psi) = \nu_{k+1}(\phi) + \nu_{k+1}(\psi)$. First assume $\phi, \psi \notin \mathfrak{p}$. Then $\phi\psi \notin \mathfrak{p}$. By H_k, ν_k is a valuation, so

$$\nu_{k+1}(\phi\psi) = \nu_k(\phi\psi) = \nu_k(\phi) + \nu_k(\psi) = \nu_{k+1}(\phi) + \nu_{k+1}(\psi).$$

In the general case, write $\phi = \widehat{\phi} U_{k+1}^n$, $\psi = \widehat{\psi} U_{k+1}^m$ in $\mathrm{gr}_{\nu_k} \mathbf{C}(x)[y]$ where $\widehat{\phi}$, $\widehat{\psi}$ are prime with U_{k+1}. In particular they do not belong to $\mathfrak{p}$ so that

$$\begin{aligned}\nu_{k+1}(\phi\psi) &= \nu_{k+1}(\widehat{\phi}\,\widehat{\psi}\,U_{k+1}^{m+n}) \\ &= \nu_{k+1}(\widehat{\phi}) + \nu_{k+1}(\widehat{\psi}) + (m+n)\nu_{k+1}(U_{k+1}) = \nu_{k+1}(\phi) + \nu_{k+1}(\psi).\end{aligned}$$

This completes Step 3.

Step 4: I_k & $\overline{I}_k$ & $H_{k+1} \Rightarrow \overline{I}_{k+1}$. What we have to prove is

Lemma 2.20. *U_{k+1} is irreducible in $\mathbf{C}[x,y]$ and $\mathbf{C}(x)[y]$.*

Proof. As U_{k+1} is monic in y, irreducibility in $\mathbf{C}[x,y]$ implies irreducibility in $\mathbf{C}(x)[y]$. So suppose $\phi\phi' = U_{k+1}$ for $\phi, \phi' \in \mathbf{C}[x,y]$. As U_{k+1} is irreducible in $\mathrm{gr}_{\nu_k} \mathbf{C}(x)[y]$, we may assume that the image of ϕ in $\mathrm{gr}_{\nu_k} \mathbf{C}(x)[y]$ is a unit i.e. $\delta_k(\phi) = 0$ and $\delta_k(\phi') = n_k$. On the other hand, one has $\delta_k(\psi) d_k \leq \deg_y(\psi)$ for any $\psi \in \mathbf{C}[x,y]$. Using Lemma 2.4 (except the assertion of irreducibility which we want now to prove), we infer

$$d_k \delta_k(\phi') = d_k n_k = \deg_y(U_{k+1}) = \deg_y(\phi) + \deg_y(\phi') \geq \deg_y(\phi') \geq d_k \delta_k(\phi'),$$

hence $\deg_y(\phi) = 0$ and $\deg_y(\phi') = \deg_y(U_{k+1})$. Now U_{k+1} is unitary in y so $\phi \notin \mathfrak{m}$ and is hence a unit in R. □

This completes the proof of Theorem 2.8.

Remark 2.21. Since U_{k+1} is monic in y we may use Weierstrass' division theorem and the argument above to show that U_{k+1} is also irreducible in $R = \mathbf{C}[[x,y]]$.

2.1.4 From SKP's to Valuations II

We now turn to infinite SKP's.

Theorem 2.22. *Let $[(U_j);(\tilde{\beta}_j)]$ be an infinite SKP and let ν_k the valuation associated to $[(U_j)_0^k;(\tilde{\beta}_j)_0^k]$ for $k \geq 1$ by Theorem 2.8.*

(i) If $n_j \geq 2$ for infinitely many j, then for any $\phi \in R$ there exists $k_0 = k_0(\phi)$ such that $\nu_k(\phi) = \nu_{k_0}(\phi)$ for all $k \geq k_0$. In particular, ν_k converges to a valuation ν_∞.

(ii) If $n_j = 1$ for $j \gg 1$, then U_k converges in R to an irreducible formal power series U_∞ and ν_k converges to a valuation ν_∞. More precisely, for $\phi \in R$ prime to U_∞ we have $\nu_k(\phi) = \nu_{k_0}(\phi)$ for $k \geq k_0 = k_0(\phi)$, and if U_∞ divides ϕ, then $\nu_k(\phi) \to \infty$.

Proof (Theorem 2.22). If $n_j \geq 2$ for infinitely many j, then $\deg_y(U_{k+1}) = d_{k+1} = \prod_1^k n_j$ tends to infinity. Pick $\phi \in R$. By Weierstrass' division theorem we may assume $\deg_y(\phi) < \infty$. By Remark 2.11, $\nu_k(\phi)$ is defined using formula (2.5). For $k \gg 1$, $\deg_y(\phi) < \deg_y(U_{k+1})$, so that $\nu(\phi) = \nu(\phi) = \nu_k(\phi) = \nu_{k+l}(\phi) = \nu_{k+l}(\phi)$ for all $l \geq 0$. Thus ν_k converges towards a valuation ν_∞.

If $n_k = 1$ for $k \geq K$, then set $d := \max_j \deg_y(U_j)$. For $k \geq K$,

$$U_{k+1} = U_k - \theta_k \prod_0^{k-1} U_j^{m_{k,j}}. \tag{2.7}$$

As $\deg_y(U_{k+1}) = \deg_y(U_k) = d$ one has $m_{k,j} = 0$ for any $j \geq K$ by (2.2). Then $\{\tilde{\beta}_k\}_{k \geq K}$ is a strictly increasing sequence of real numbers belonging to the discrete lattice $\sum_0^K \mathbf{Z}\tilde{\beta}_j$, so $\tilde{\beta}_k \to \infty$. Write

$$U_k = y^d + a_{d-1}^k(x)y^{d-1} + \cdots + a_0^k(x),$$

the Weierstrass form of U_k. As $m_{k,j} = 0$ for any $j \geq K$ we get

$$a_n^{k+1}(x) = a_n^k(x) - \theta_k \sum_I a_{i_1}^{j_1}(x) \ldots a_{i_l}^{j_l}(x)$$

where $\#\{j = \alpha\ ;\ j \in I\} = m_{k,\alpha}$, and $i_1 + \cdots + i_l = n$. Hence $a_n^{k+1}(x) - a_n^k(x) \in \mathfrak{m}^{\sum_0^K m_{k,i}}$. But the sequence $\tilde{\beta}_j$ is increasing so

$$\sum_0^K m_{k,i} \geq \tilde{\beta}_K^{-1} \sum_0^K \tilde{\beta}_i m_{k,i} = \frac{\tilde{\beta}_k}{\tilde{\beta}_K} \to \infty,$$

and U_k converges towards a polynomial in y that we denote by U_∞. By (2.7) we have $U_{k+1} = U_k$ modulo ν_{k-1} and $U_{k+l} = U_k$ modulo ν_{k-1} for all $l \geq 0$. So $U_\infty = U_k$ in ν_{k-1}. Therefore $\nu_\infty(U_\infty) := \lim \nu_k(U_\infty) = \lim \nu_k(U_k) = \lim \tilde{\beta}_k = \infty$. Let $\phi \in R$ and write $\phi = \phi_0 + U_\infty \phi_1$ with $\deg_y \phi_0 < \deg_y U_\infty = d$.

When $\phi_0 \equiv 0$, then set $\nu_\infty(\phi) := \lim \nu_k(\phi) = \infty$. Otherwise for $k \geq K$ large enough, for all $l \geq 0$, $\nu_{k+l}(\phi) = \nu_{k+l}(\phi_0) = \nu_k(\phi_0)$. Hence the sequence $\nu_k(\phi)$ is stationary for $k \geq K$ and we can set $\nu_\infty(\phi) := \lim \nu_k(\phi)$. One easily checks that ν_∞ is a valuation. As $\nu_\infty(\phi) = \infty$ iff $U_\infty | \phi$ it follows that U_∞ is irreducible in R. □

2.2 Classification

Having constructed a valuation associated to any SKP, we now make a preliminary classification.

Definition 2.23. *Consider a centered valuation ν on R given by an SKP: say $\nu := \mathrm{val}[(U_j)_0^k; (\tilde{\beta}_j)_0^k]$, where $1 \leq k \leq \infty$. Assume that ν is normalized in the sense that $\nu(\mathfrak{m}) = 1$. We then say that ν is*

(i) monomial *(in coordinates (x, y)) if $k = 1$, $\tilde{\beta}_0 < \infty$ and $\tilde{\beta}_1 < \infty$;*
(ii) quasimonomial *if $k < \infty$, $\tilde{\beta}_0 < \infty$ and $\tilde{\beta}_k < \infty$;*
(iii) divisorial *if ν is quasimonomial and $\tilde{\beta}_k \in \mathbf{Q}$;*
(iv) irrational *if ν is quasimonomial but not divisorial;*
(v) infinitely singular *if $k = \infty$ and $d_j \to \infty$, where $d_j = \deg_y(U_j)$;*
(vi) a curve valuation *if $k = \infty$ and $d_j \not\to \infty$, or $k < \infty$ and $\max\{\tilde{\beta}_0, \tilde{\beta}_k\} = \infty$.*

If ν is not normalized, then the type of ν is defined to be the type of $\nu/\nu(\mathfrak{m})$.

Two remarks are in order. First, this classification may seem unnatural as it only applies to valuations associated to an SKP and since the technique of SKP's uses a fixed choice of local coordinates (x, y).

However, as we shall see, every valuation is associated to an SKP (Theorem 2.29). Moreover, the classification can be rephrased in several equivalent ways, all of which are independent on the choice of coordinates.

Second, we should compare the classification with the "road map" given in Section 1.5. The comparison goes as follows.

(i) If ν is monomial in coordinates (x, y) in the sense above, then its SKP is of length 1. By (2.4), this means that ν is monomial in the sense of Section 1.5.2. Notice that by the definition above, any monomial valuation is also quasimonomial.
(ii) It is not obvious that the definition of quasimonomial, which is based purely on the definition of a valuation as a function on R, coincides with the more geometric point of view presented in Section 1.5.4. We shall prove in Chapter 6 that the two definitions are equivalent: see Proposition 6.41 for a precise statement.
(iii) We shall see at Theorem 2.28 that ν is a divisorial valuation in the sense above, iff $\mathrm{tr.deg}\, \nu = 1$. It follows from Proposition 1.12 that the definition of divisorial valuation given above coincides with the one in Section 1.5.3.

(v) As for infinitely singular valuations, we need to develop more machinery to see that the definition above agrees with the characterizations mentioned in Section 1.5. Informally speaking, however, the sequence $(U_k)_1^\infty$ of key polynomials of increasing and unbounded degrees correspond to the sequence of truncated Puiseux expansions $\sum_1^k x^{\hat{\beta}_j}$ where the $\hat{\beta}_j$'s are rational numbers with unbounded denominators.

(vi) Finally, consider a curve valuation ν in the sense above. First assume it is defined by a finite SKP, say $\nu := \mathrm{val}[(U_j)_0^k; (\tilde{\beta}_j)_0^k]$, where $k < \infty$ and $\tilde{\beta}_k = \infty$. By construction, $\nu(U_k) = \infty$. As U_k is irreducible, we conclude that $\nu = \nu_{U_k}$ (see the discussion following Lemma 1.5). When $k = \infty$ and $\lim \tilde{\beta}_k = \infty$, by Theorem 2.22, U_k converges to an irreducible formal power series U_∞, and $\nu(U_\infty) = \infty$. Again $\nu = \nu_{U_\infty}$.

2.3 Graded Rings and Numerical Invariants

The aim of this section is to give the structure of the valuation associated to a finite or infinite SKP. That is, we describe the structure of the graded ring $\mathrm{gr}_\nu \mathbf{C}(x)[y]$, and compute the three invariants $\mathrm{rk}(\nu)$, $\mathrm{rat.rk}(\nu)$ and $\mathrm{tr.deg}(\nu)$. We refer to [Te4] for general results in higher dimensions.

Given a finite or infinite SKP we denote by $\nu := \mathrm{val}[(U_j); (\tilde{\beta}_j)]$ its associated valuation through Theorem 2.8 or 2.22.

2.3.1 Homogeneous Decomposition I

The following theorem gives the structure of the graded ring of a valuation defined by a finite SKP.

Theorem 2.24. *Let* $\nu := \mathrm{val}[(U_j)_0^k; (\tilde{\beta}_j)_0^k]$*, where* $1 \le k < \infty$*, define* $n_k := \min\{n > 0 \; ; \; n\tilde{\beta}_k \in \sum_0^{k-1} \mathbf{Z}\tilde{\beta}_j\}$*, and pick* $0 \neq \phi \in R$*.*

(i) If $n_k = \infty$ *or equivalently* $\tilde{\beta}_k \notin \mathbf{Q}\tilde{\beta}_0$*, then*

$$\phi = \alpha \prod_{j=0}^{k} U_j^{i_j} \text{ in } \mathrm{gr}_\nu R \text{ and } \mathrm{gr}_\nu R_\nu \tag{2.8}$$

with $\alpha \in \mathbf{C}^*$*,* $0 \le i_j < n_j$ *for* $1 \le j < k$ *and* $i_0, i_k \ge 0$*.*

(ii) If $n_k < \infty$*, write* $n_k\tilde{\beta}_k = \sum_0^{k-1} m_{k,j}\tilde{\beta}_j$ *with* $0 \le m_{k,j} < n_j$ *for* $1 \le j < k$ *and* $m_0 \ge 0$ *as in Lemma 2.7. Then*

$$\phi = p(T) \prod_{j=0}^{k} U_j^{i_j} \text{ in } \mathrm{gr}_\nu R_\nu \tag{2.9}$$

with $0 \le i_j < n_j$ *for* $1 \le j \le k$*,* $i_0, i_k \ge 0$ *and where* p *is a polynomial in* $T := U_k^{n_k} \prod_{j=0}^{k-1} U_j^{-m_{k,j}}$*.*

Both decompositions (2.8), (2.9) *are unique.*

Recall that the ring $\mathrm{gr}_\nu \mathbf{C}(x)[y]$ is a Euclidean domain. Let us describe its irreducible elements.

Corollary 2.25. *Let ν be a valuation as above.*

(i) When $n_k = \infty$ the only irreducible element of $\mathrm{gr}_\nu \mathbf{C}(x)[y]$ is U_k.
(ii) When $n_k < \infty$ the irreducible elements of $\mathrm{gr}_\nu \mathbf{C}(x)[y]$ consist of U_k and all elements of the form $U_k^{n_k} - \theta \prod_{j=0}^{k-1} U_j^{m_{k,j}}$ for some $\theta \in \mathbf{C}^*$.

Proof (Corollary 2.25). Assume $\phi \in \mathrm{gr}_\nu \mathbf{C}(x)[y]$ is irreducible. If $n_k = \infty$, then $\phi = \alpha \prod_{j=0}^{k} U_j^{i_j}$ by (2.8). But U_j is a unit for $j < k$ by Lemma 2.16, so U_k is the only irreducible element in $\mathrm{gr}_\nu \mathbf{C}(x)[y]$.

When $n_k < \infty$, then we use (2.9) and factorize $p(T) = \prod(T - \theta_l)$. Modulo unit factors, we hence get

$$\phi = U_k^{i_k - L n_k} \prod_l \left(U_k^{n_k} - \theta_l \prod_j U_j^{m_{k,j}} \right), \tag{2.10}$$

where $L = \deg p$. On the other hand Lemma 2.18 shows that all elements of the form $U_k^{n_k} - \theta_l \prod_j U_j^{m_{k,j}}$ are irreducible in $\mathrm{gr}_\nu \mathbf{C}(x)[y]$. So (2.10) is the decomposition of ϕ into prime factors in $\mathrm{gr}_\nu \mathbf{C}(x)[y]$. This concludes the proof. □

Proof (Theorem 2.24). First assume that $\deg_y \phi < d_k = \deg_y U_k$. We will show that (2.8) holds. Write $\phi = \sum_0^{n_{k-1}-1} \phi_i U_{k-1}^i$ with $\deg_y \phi_i < d_{k-1}$ for all i. Iterating this procedure we get $\phi = \sum \alpha_I U_0^{i_0} \cdots U_{k-1}^{i_{k-1}}$ with $\alpha_I \in \mathbf{C}$, $i_0 \geq 0$ and $0 \leq i_j < n_j$ for $j \geq 1$. If $\nu(U_0^{i_0} \ldots U_{k-1}^{i_{k-1}}) = \nu(U_0^{j_0} \ldots U_{k-1}^{j_{k-1}})$ then $\sum i_l \tilde{\beta}_l = \sum j_l \tilde{\beta}_l$ so that $(i_{k-1} - j_{k-1})\tilde{\beta}_{k-1} \in \sum_0^{k-2} \mathbf{Z}\tilde{\beta}_j$. Thus n_{k-1} divides $i_{k-1} - j_{k-1}$ so $i_{k-1} = j_{k-1}$ as $0 \leq i_{k-1}, j_{k-1} < n_{k-1}$. Iterating this argument we get (2.8).

In the general case, we can assume $\deg_y \phi < \infty$ by Weierstrass preparation theorem. Write $\phi = \sum \phi_i U_k^i$, with $\deg_y \phi_i < d_k$. By what precedes, we have

$$\phi = \sum \alpha_I U_0^{i_0} \ldots U_{k-1}^{i_{k-1}} U_k^{i_k} \text{ in } \mathrm{gr}_\nu R$$

with $\alpha_I \in \mathbf{C}^*$, $0 \leq i_j < n_j$ for $1 \leq j < k$, $i_0, i_k \geq 0$. We may assume $\sum_0^k i_j \tilde{\beta}_j = \nu(\phi)$ for all I with $\alpha_I \neq 0$. If (2.8) is not valid, there exist $I \neq J$ with $\alpha_I, \alpha_J \neq 0$. Thus $(i_k - j_k)\tilde{\beta}_k \in \sum_0^{k-1} \mathbf{Z}\tilde{\beta}_j$ implying $n_k < \infty$. This shows (2.8) when $n_k = \infty$.

Now assume $n_k < \infty$, and write $n_k \tilde{\beta}_k = \sum_0^{k-1} m_{k,j} \tilde{\beta}_j$ as in Lemma 2.7. Let $I = (i_0, \ldots, i_k)$ be any multiindex with $\alpha_I \neq 0$. Make the Euclidean division $i_k = r_k n_k + \hat{\imath}_k$ with $0 \leq \hat{\imath}_k < n_k$, and write

$$\prod_{j=0}^{k} U_j^{i_j} = U_k^{\widehat{\imath}_k} T^{r_k} \prod_{j=0}^{k-1} U_j^{a_j}$$

with $a_j := i_j + r_k m_{k,j}$ and $T := U_k^{n_k} \prod_{j=0}^{k-1} U_j^{-m_{k,j}}$. The key remark is now that $U_j^{n_j} = \theta_j \prod_0^{j-1} U_l^{m_{j,l}}$ in $\mathrm{gr}_\nu R$ for $1 \le j < k$. Making the Euclidean division $a_{k-1} = r_{k-1} n_{k-1} + \widehat{\imath}_{k-1}$ with $0 \le \widehat{\imath}_{k-1} < n_{k-1}$, we get

$$\prod_{j=0}^{k-1} U_j^{a_j} = \theta_{k-1} U_k^{\widehat{\imath}_{k-1}} \prod_{j=0}^{k-2} U_j^{a'_j}$$

for some $a'_j \in \mathbf{N}$. We finally get by induction that

$$\prod_{j=0}^{k} U_j^{i_j} = \theta_I T^{r_I} \prod_{0}^{k} U_j^{\widehat{\imath}_j}$$

in $\mathrm{gr}_\nu R_\nu$, with $\theta_I \in \mathbf{C}^*$, $r_I (= r_k) \ge 0$, $0 \le \widehat{\imath}_j < n_j$ for $1 \le j \le k$ and $\widehat{\imath}_0 \ge 0$.

Now $\sum_0^k i_j \tilde{\beta}_j = \nu(\phi)$ and $\nu(T) = 0$, hence $\sum_0^k \widehat{\imath}_j \tilde{\beta}_j = \nu(\phi)$. Suppose $\sum_0^k \widehat{\imath}_j \tilde{\beta}_j = \sum_0^k \tilde{\imath}_j \tilde{\beta}_j$ with $0 \le \tilde{\imath}_j < n_j$ for $1 \le j \le k$. Since $|\widehat{\imath}_k - \tilde{\imath}_k| < n_k$, the definition of n_k gives $\widehat{\imath}_k = \tilde{\imath}_k$. By induction we get $\widehat{\imath}_j = \tilde{\imath}_j$ for all $j \ge 0$. We have proved that ϕ can be written in the form (2.9).

Uniqueness of both decompositions (2.8), (2.9) comes from unique factorization in $\mathrm{gr}_\nu \mathbf{C}(x)[y]$. Indeed, when $n_k = \infty$, $i_k = \delta_k(\phi)$. From $\sum i_j \tilde{\beta}_j = \nu(\phi)$ and $i_j < n_j$ for $1 \le j < k$ we deduce uniqueness of the decomposition (2.8). When $n_k < \infty$, $\nu(\phi)$ determines i_j for all $j \ge 0$, and the polynomial $p(T)$ is determined as ϕ admits a unique decomposition into prime factors (see Corollary 2.25). This concludes the proof of the theorem. □

2.3.2 Homogeneous Decomposition II

We now turn to infinite SKP's.

Theorem 2.26. *Let* $\nu := \mathrm{val}[(U_j)_0^\infty; (\tilde{\beta}_j)_0^\infty]$. *Pick* $0 \ne \phi \in R$.

(i) If $n_j \ge 2$ *for infinitely many* j, *then there exists* $k = k(\phi)$ *such that*

$$\phi = \alpha \prod_{j=0}^{k} U_j^{i_j} \quad \textit{in } \mathrm{gr}_\nu R \tag{2.11}$$

with $\alpha \in \mathbf{C}^*$, $0 \le i_j < n_j$ *for* $1 \le j \le k$, *and* $i_0 \ge 0$.

(ii) If $n_j = 1$ *for* $j \gg 1$, *then there exists* $k = k(\phi)$ *such that*

$$\phi = \alpha \cdot U_\infty^n \cdot \prod_{j=0}^{k} U_j^{i_j} \quad \textit{in } \mathrm{gr}_\nu R \tag{2.12}$$

with $\alpha \in \mathbf{C}$, $0 \le i_j < n_j$ *for* $1 \le j \le k$, $i_0 \ge 0$ *and* $n \ge 0$.

Both decompositions (2.11), (2.12) *are unique.*

The analogue of Corollary 2.25 is

Corollary 2.27. *Assume ν is associated to an infinite SKP.*

(i) When $n_j \geq 2$ for infinitely many j, the ring $\mathrm{gr}_\nu \mathbf{C}(x)[y]$ *is a field.*
(ii) When $n_j = 1$ for $j \gg 1$, the only irreducible element of $\mathrm{gr}_\nu \mathbf{C}(x)[y]$ *is* U_∞.

Proof (Theorem 2.26). Fix $\phi \in R$, and suppose it is in Weierstrass form, polynomial in y. First assume that $n_j \geq 2$ for infinitely many j. For $j \gg 1$ $\deg_y \phi < \deg_y U_j$, hence $\phi = \alpha \prod_{l=0}^{j-1} U_l^{i_l}$ in $\mathrm{gr}_{\nu_j} R$ for some $\alpha \in \mathbf{C}^*$, and $i_l < n_l$ for $l \geq 1$ (see the beginning of the proof of Theorem 2.24). As $\nu \geq \nu_j$, we get $\phi = \alpha \prod_{l=0}^{j-1} U_l^{i_l}$ also in $\mathrm{gr}_\nu R$.

When $n_j = 1$ for $j \gg 1$, write $\phi = U_\infty^n \phi'$ with ϕ', U_∞ prime. For $j \gg 1$ $\nu_{j+1}(\phi') = \nu_j(\phi') = \nu(\phi')$. Hence $\phi' = \alpha \prod_{l=0}^{j-1} U_l^{i_l}$ in $\mathrm{gr}_{\nu_j} R$ and $\mathrm{gr}_\nu R$ as above. □

Proof (Corollary 2.27). The first assertion is immediate by Lemma 2.16. Assume $n_j = 1$ for $j \gg 1$ and pick $\phi \in R$. If U_∞ does not divide ϕ, then (2.11) holds. For $k \gg 1$ we get $\delta_k(\phi) = 0$. Thus ϕ is invertible in $\mathrm{gr}_{\nu_k} \mathbf{C}(x)[y]$, hence in $\mathrm{gr}_\nu \mathbf{C}(x)[y]$.

On the other hand $U_\infty = \phi\psi$ in $\mathrm{gr}_\nu \mathbf{C}(x)[y]$ implies either $\nu(\phi) = \infty$ or $\nu(\psi) = \infty$. Hence U_∞ divides either ϕ or ψ as formal power series hence in $\mathrm{gr}_\nu \mathbf{C}(x)[y]$. We conclude noting that U_∞ cannot be a unit in $\mathrm{gr}_\nu \mathbf{C}(x)[y]$. Indeed $\nu(U_\infty \phi) = \nu(1)$ implies that $U_\infty \phi$ is prime with U_∞. □

2.3.3 Value Semigroups and Numerical Invariants

Knowing the structure of the graded rings allows us to compute the value semigroups and numerical invariants introduced in Section 1.3. The first assumption in the next theorem is in fact redundant as we shall see below in Theorem 2.29.

Theorem 2.28. *Let ν be a valuation associated to a finite or infinite SKP. Then the value semi-group*[1] $\nu(R)$ *is equal to* $\sum \tilde{\beta}_j \mathbf{N}$, *Moreover, we have the following intrinsic characterization of the type of a valuation as introduced in Definition 2.23:*

(i) ν is a curve valuation, iff $\mathrm{rk}(\nu) = \mathrm{rat.rk}(\nu) = 2$, $\mathrm{tr.deg}(\nu) = 0$.
(ii) ν is divisorial, iff $\mathrm{rk}(\nu) = \mathrm{rat.rk}(\nu) = 1$ *and* $\mathrm{tr.deg}(\nu) = 1$.
(iii) ν is irrational, iff $\mathrm{rk}(\nu) = 1$, $\mathrm{rat.rk}(\nu) = 2$ *and* $\mathrm{tr.deg}(\nu) = 0$.
(iv) ν is infinitely singular, iff $\mathrm{rk}(\nu) = \mathrm{rat.rk}(\nu) = 1$ *and* $\mathrm{tr.deg}(\nu) = 0$.

[1] Note, however, that a curve valuation ν may take the value ∞ on nonzero elements.

Proof (Theorem 2.28). The computation of the value semigroup is a simple exercise that is left to the reader. As for the numerical invariants we treat the cases of finite and infinite SKP's separately.

First assume that ν is associated to a finite SKP $[(U_j)_0^k; (\tilde{\beta}_j)_0^k]$, $1 \leq k < \infty$. The two invariants $\mathrm{rk}(\nu)$ and $\mathrm{rat.rk}(\nu)$ can be read directly from the value semigroup $\nu(R)$ so their computation is straightforward thanks to the first part of the proof. For the computation of $\mathrm{tr.deg}(\nu)$ we rely on Theorem 2.24 and proceed as follows.

When $\tilde{\beta}_0 = \infty$, $\tilde{\beta}_k = \infty$ or $\tilde{\beta}_k \notin \mathbf{Q}^*$ (so $\tilde{\beta}_k \notin \sum_0^{k-1} \mathbf{Z}\tilde{\beta}_j$), then (2.8) applies. Pick $\phi, \psi \in R$ with $\nu(\phi) = \nu(\psi)$. Then $\phi = \alpha \prod_{l=0}^k U_l^{i_l}$, $\psi = \gamma \prod_{l=0}^k U_l^{j_l}$ modulo ν, with $\sum_0^k i_l \tilde{\beta}_l = \sum_0^k j_l \tilde{\beta}_l$. Hence $i_k = j_k$ since $n_k = \infty$. Now $\sum_0^{k-1} i_l \tilde{\beta}_l = \sum_0^{k-1} j_l \tilde{\beta}_l$ and $0 \leq i_l, j_l < n_l$ for all $1 \leq j < k$ so $i_l = j_l$ for all l by (2.2). Hence $\phi/\psi = \alpha/\gamma \in \mathbf{C}^*$ in $\mathrm{gr}_\nu R_\nu$. We have shown that $k_\nu := R_\nu/\mathfrak{m}_\nu \cong \mathbf{C}$ so $\mathrm{tr.deg}(\nu) = 0$.

If $\tilde{\beta}_0 < \infty$ and $\tilde{\beta}_k \in \mathbf{Q}^*$, then $n_k < \infty$. For $\phi, \psi \in R$ with $\nu(\phi) = \nu(\psi)$ write

$$\phi = p(T) \prod_{l=0}^{k} U_l^{i_l} \quad \text{and} \quad \psi = q(T) \prod_{l=0}^{k} U_l^{j_l} \quad \text{in } \mathrm{gr}_\nu R_\nu$$

as in (2.9), with $0 \leq i_l, j_l < n_l$ for $1 \leq l \leq k$, $i_0, j_0 \geq 0$ and $p, q \in \mathbf{C}[T]$. As $\sum_0^k i_l \tilde{\beta}_l = \sum_0^k j_l \tilde{\beta}_l$, one has $i_l = j_l$ for all l. Hence $\phi/\psi = p(T)/q(T)$ in $\mathrm{gr}_\nu R_\nu$. This shows $k_\nu \cong \mathbf{C}(T)$ so $\mathrm{tr.deg}(\nu) = 1$.

Now assume ν is associated to an infinite SKP, $\nu := \mathrm{val}[(U_j); (\tilde{\beta}_j)]$. First assume $n_j \geq 2$ for infinitely many j. Set $\nu_k := \mathrm{val}[(U_j)_0^k; (\tilde{\beta}_j)_0^k]$. Then the sequence $\nu_k(\phi)$ is eventually stationary for any $\phi \in R$. One easily checks that $\mathrm{rk}(\nu) = \mathrm{rat.rk}(\nu) = 1$. Let us compute $\mathrm{tr.deg}(\nu)$. Fix $\phi, \psi \in R$ with $\nu(\phi) = \nu(\psi)$. For $k \gg 1$ we have $\nu_k(\phi) = \nu(\phi) = \nu(\psi) = \nu_k(\psi)$. Write $\phi = \alpha \prod_{l=0}^k U_l^{i_l}$ and $\psi = \gamma \prod_{l=0}^k U_l^{j_l}$ where $\alpha, \gamma \in \mathbf{C}^*$, $0 \leq i_l, j_l < n_l$ for $l \geq 1$ and apply the preceding arguments. We get $\phi/\psi = \alpha/\gamma$ in $\mathrm{gr}_{\nu_k} R_{\nu_k}$ hence in $\mathrm{gr}_\nu R_\nu$. This shows $k_\nu \cong \mathbf{C}$.

The last case is when $n_j = 1$ for $j \gg 1$. We leave it to the reader to verify that $\mathrm{rk}(\nu) = \mathrm{rat.rk}(\nu) = 2$ and $\mathrm{tr.deg}(\nu) = 0$ in this case. □

2.4 From Valuations to SKP's

We now show that every centered valuation on R is represented by an SKP.

Theorem 2.29. *For any centered valuation ν on R, there exists a unique SKP $[(U_j)_0^n; (\tilde{\beta}_j)_0^n]$, $1 \leq n \leq \infty$, such that $\nu = \mathrm{val}[(U_j); (\tilde{\beta}_j)]$. We have $\nu(U_j) = \tilde{\beta}_j$ for all j. Further, if $k < n$ and $\nu_k := \mathrm{val}[(U_j)_0^k; (\tilde{\beta}_j)_0^k]$, then $\nu(\phi) \geq \nu_k(\phi)$ for all $\phi \in R$ and $\nu(\phi) > \nu_k(\phi)$ iff U_{k+1} divides ϕ in $\mathrm{gr}_{\nu_k} \mathbf{C}(x)[y]$.*

Remark 2.30. If ν is a Krull valuation, the same result holds for a Γ-SKP.

Remark 2.31. In Spivakovsky's terminology [Sp], the subset of (U_j) for which $n_j > 1$ forms a minimal generating sequence for the valuation ν. The associated divisorial valuations ν_j will play an important role in the sequel. They form what we call the approximating sequence of ν, see Section 3.5.

Proof (Theorem 2.29). We construct by induction on k a valuation ν_k so that $\nu_k(U_j) = \tilde{\beta}_j$ for $j \leq k$. Let $U_0 = x$, $U_1 = y$ and $\tilde{\beta}_0 = \nu(x)$, $\tilde{\beta}_1 = \nu(y)$. Assume $\nu_k := \mathrm{val}[(U_j)_0^k; (\tilde{\beta}_j)_0^k]$ has been defined. By Theorem 2.8, $\nu(\phi) \geq \nu_k(\phi)$ for $\phi \in R$. As $\nu(x) = \nu_k(x)$ this also holds for $\phi \in \mathbf{C}(x)[y]$. If $\nu = \nu_k$, then we are done. If not, set $\mathcal{D}_k := \{\phi \in \mathbf{C}(x)[y] \ ; \ \nu(\phi) > \nu_k(\phi)\}$.

Lemma 2.32. *$\mathcal{D}_k$ defines a prime ideal in* $\mathrm{gr}_{\nu_k} \mathbf{C}(x)[y]$.

Proof. Let us check that $\mathcal{D}_k$ is well defined in $\mathrm{gr}_{\nu_k} \mathbf{C}(x)[y]$. Pick $\phi, \phi' \in \mathbf{C}(x)[y]$ with $\phi = \phi'$ modulo ν_k. If $\phi \in \mathcal{D}_k$ but $\phi' \notin \mathcal{D}_k$, then $\nu(\phi') = \nu_k(\phi') = \nu_k(\phi) < \nu(\phi)$. So $\nu(\phi - \phi') = \nu(\phi') = \nu_k(\phi') < \nu_k(\phi - \phi')$, a contradiction. To show that $\mathcal{D}_k$ is a prime ideal is easy and left to the reader. □

We continue the proof of the theorem. Recall that $\mathrm{gr}_{\nu_k} \mathbf{C}(x)[y]$ is a Euclidean domain. Hence Corollary 2.25 and Lemma 2.7 show that $\mathcal{D}_k$ is generated by a unique irreducible element $U_{k+1} = U_k^{n_k} - \theta_k \prod_{j=0}^{k-1} U_j^{m_{k,j}}$ with $\theta_k \in \mathbf{C}^*$, $0 \leq m_{k,j} < n_j$ for $j \geq 1$ and $m_{k,0} \geq 0$. Define $\tilde{\beta}_{k+1} := \nu(U_{k+1})$. It is easy to check that $[(U_j)_0^{k+1}; (\tilde{\beta}_j)_0^{k+1}]$ is an SKP. This completes the induction step.

Either the induction terminates at some finite k, or we get an infinite SKP $[(U_j); (\beta_j)]$. In the latter case we claim that $\nu = \mathrm{val}[(U_j); (\beta_j)]$. This amounts to showing that if $\phi \in R$, then the increasing sequence $\nu_k(\phi)$ converges to $\nu(\phi)$.

If $\nu_k(\phi) = \nu(\phi)$ for $k \gg 1$ then we are done, so assume $\nu_k(\phi) < \nu(\phi)$ for all k. Then $\phi \in \mathcal{D}_k$ so U_{k+1} divides ϕ in $\mathrm{gr}_{\nu_k} \mathbf{C}(x)[y]$. In particular, $\nu_{k+1}(\phi) > \nu_k(\phi)$, and $\deg_y(\phi) \geq \deg_y(U_{k+1}) = d_{k+1}$. Hence d_k is bounded, and $n_k = 1$ for $k \gg 1$. By Theorem 2.22, $U_k \to U_\infty$ in R. Moreover U_∞ divides ϕ in R or else $\nu_k(\phi)$ would be stationary for k large. But then $\nu(\phi) \geq \lim \nu_k(\phi) = \infty$ so $\nu_k(\phi) \to \nu(\phi)$. This completes the proof of the theorem. □

2.5 A Computation

We now compute $\nu(\phi)$ for a normalized valuation ν and $\phi \in \mathfrak{m}$ irreducible in terms of SKP's. This computation will be crucial to describe the parameterization of valuation space; see Lemma 3.32.

Let ν_ϕ be the curve valuation associated to ϕ. Write $\nu = \mathrm{val}[(U_j); (\tilde{\beta}_j)]$ and $\nu_\phi = \mathrm{val}[(U_j^\phi); (\tilde{\beta}_j^\phi)]$. Assume $\nu \neq \nu_\phi$ and define the contact order of ν and ν_ϕ by

$$\mathrm{con}(\nu, \nu_\phi) = \max\{j \ ; \ U_j = U_j^\phi\}. \tag{2.13}$$

Let n_j^ϕ be the integers defined by (2.2) for ν_ϕ and set $\gamma_k^\phi = \prod_{j \geq k} n_j^\phi$ for $k \geq 1$. These products are in fact finite as $n_j^\phi = 1$ for $j \gg 1$.

Proposition 2.33. *If $\phi = x$ (up to a unit), then $\nu(\phi) = \tilde{\beta}_0$. Otherwise*

$$\nu(\phi) = \gamma_k^\phi \min\{\tilde{\beta}_k, \tilde{\beta}_k^\phi\} \min\{1, \tilde{\beta}_0/\tilde{\beta}_0^\phi\}, \tag{2.14}$$

where $k = \mathrm{con}(\nu, \nu_\phi)$.

Later on we will be interested in the quotient $\nu(\phi)/m(\phi)$. Applying Proposition 2.33 to $\nu = \nu_{\mathfrak{m}}$ we obtain $m(\phi) = \gamma_1^\phi/\tilde{\beta}_0^\phi$ if $\phi \neq x$; this leads to

Proposition 2.34. *If $\nu \in \mathcal{V}$ and $\phi \in \mathfrak{m}$ is irreducible, then*

$$\frac{\nu(\phi)}{m(\phi)} = d_k^{-1} \min\{\tilde{\beta}_k, \tilde{\beta}_k^\phi\} \min\{\tilde{\beta}_0, \tilde{\beta}_0^\phi\},$$

where $k = \mathrm{con}(\nu, \nu_\phi)$ and $d_k = \deg_y U_k = \prod_1^{k-1} n_j$.

Proof (Proposition 2.33). The case $\phi = x$ is trivial, so assume $\phi \neq x$, i.e. $\tilde{\beta}_0^\phi < \infty$. We may then assume $\phi = U_l^\phi$, where $l \in [k, \infty]$ is the length of the SKP attached to ν_ϕ. Note that $\min\{\tilde{\beta}_0, \tilde{\beta}_1\} = \min\{\tilde{\beta}_0^\phi, \tilde{\beta}_1^\phi\} = 1$ since ν and ν_ϕ are normalized.

Let us first consider the case when $\tilde{\beta}_0 = \tilde{\beta}_0^\phi$ and $\tilde{\beta}_1 = \tilde{\beta}_1^\phi$; this holds e.g. if $k \geq 2$. Then $\tilde{\beta}_j = \tilde{\beta}_j^\phi$ for $0 \leq j < k$. If $l = k$, then $\phi = U_k$ so $\nu(\phi) = \tilde{\beta}_k$ and $\tilde{\beta}_k^\phi = \infty$, which implies (2.14). Hence assume $l > k$. Define $\xi_j = \tilde{\beta}_{k+j}^\phi$ and $\eta_j = \min\{\tilde{\beta}_k, \tilde{\beta}_k^\phi\}\gamma_k^\phi/\gamma_{k+j}^\phi$ for $0 \leq j \leq l-k$. Then $\nu_\phi(U_{k+j}^\phi) = \xi_j$ for $j \geq 0$ and $\nu(U_j^\phi) = \tilde{\beta}_j$ for $0 \leq j \leq k$. We will prove inductively that $\nu(U_{k+j}^\phi) = \eta_j$ for $j \geq 1$; when $j = l-k$ this gives (2.14). The induction is based on (2.3), which reads

$$U_{k+j+1}^\phi = (U_{k+j}^\phi)^{n_{k+j}^\phi} - \theta_{k+j}^\phi \prod_{i=0}^{k+j-1} (U_i^\phi)^{m_{k+j,i}^\phi} =: A_j - B_j. \tag{2.15}$$

Let us first show that $\nu(U_{k+1}^\phi) = \eta_1$, using (2.15) for $j = 0$. There are three cases, depending on $\tilde{\beta}_k$ and $\tilde{\beta}_k^\phi$. The first case is when $\tilde{\beta}_k > \tilde{\beta}_k^\phi$. Then $n_k^\phi \tilde{\beta}_k > n_k^\phi \tilde{\beta}_k^\phi = \sum_0^{k-1} m_{k,i}^\phi \tilde{\beta}_i$, so using (2.15) we obtain $\nu(U_{k+1}^\phi) = \tilde{\beta}_k^\phi n_k^\phi = \eta_1$. Similarly, in the second case, $\tilde{\beta}_k < \tilde{\beta}_k^\phi$, then $\sum_0^{k-1} m_{k,i}^\phi \tilde{\beta}_i = n_k^\phi \tilde{\beta}_k^\phi > n_k^\phi \tilde{\beta}_k$, so that $\nu(U_{k+1}^\phi) = n_k^\phi \tilde{\beta}_k = \eta_1$. The third case is when $\tilde{\beta}_k = \tilde{\beta}_k^\phi$. Set $\nu_k := \mathrm{val}[(U_j)_0^k; (\tilde{\beta}_j)_0^k]$. Then U_{k+1}^ϕ is irreducible in $\mathrm{gr}_{\nu_k} \mathbf{C}(x)[y]$ and $\nu_k(U_{k+1}^\phi) = \tilde{\beta}_k n_k^\phi = \eta_1$. If the length of the SKP associated to ν is equal to k then $\nu = \nu_k$ and $\nu(U_{k+1}^\phi) = \eta_1$ so assume this length is strictly greater than k. By assumption $U_{k+1}^\phi \neq U_{k+1}$, hence U_{k+1}^ϕ does not belong to the ideal generated by U_{k+1} in $\mathrm{gr}_{\nu_k} \mathbf{C}(x)[y]$. But this ideal coincides with $\{\nu > \nu_k\}$ by Theorem 2.29, so $\nu(U_{k+1}^\phi) = \nu_k(U_{k+1}^\phi) = \eta_1$.

Now fix $1 \leq j < l-k$ and assume that $\nu(U^{\phi}_{k+i}) = \eta_i$ for $1 \leq i \leq j$. We will prove that $\nu(U^{\phi}_{k+j+1}) = \eta_{j+1}$ using (2.15). Write $a_i = m^{\phi}_{k+j,k+i}$ for $0 \leq i < j$ and $c = n^{\phi}_{k+j}$. Then $\nu(A_j) = c\eta_j$ and

$$\begin{aligned}
\nu(B_j) &= \sum_{i=1}^{k} m^{\phi}_{k+j,i}\tilde{\beta}_i + \sum_{i=1}^{j-1} a_i\eta_i \\
&= \sum_{i=1}^{k+j-1} m^{\phi}_{k+j,i}\tilde{\beta}^{\phi}_i + a_0(\tilde{\beta}_k - \xi_0) + \sum_{i=1}^{j-1} a_i(\eta_i - \xi_i) \\
&\geq \sum_{i=1}^{k+j-1} m^{\phi}_{k+j,i}\tilde{\beta}^{\phi}_i + \sum_{i=0}^{j-1} a_i(\eta_i - \xi_i) = c\xi_j + \sum_{i=0}^{j-1} a_i(\eta_i - \xi_i).
\end{aligned}$$

To show that $\nu(U^{\phi}_{k+j+1}) = \eta_{j+1}$ we only need to show $\nu(B_j) > c\eta_j$ or simply $\sum_0^{j-1} a_i(\xi_i - \eta_i) > c(\xi_j - \eta_j)$. Note that $\sum_0^{j-1} a_i\xi_i \leq c\xi_j$ and that the sequence $(\eta_i/\xi_i)_0^j$ is strictly decreasing. Set $p_i = a_i\xi_i / \sum_0^{j-1} a_i\xi_i$. Then $\sum_0^{j-1} p_i = 1$ so

$$\sum_{i=0}^{j-1} a_i(\xi_i - \eta_i) = \left(\sum_{i=0}^{j-1} a_i\xi_i\right) \sum_{i=0}^{j-1} p_i \left(1 - \frac{\eta_i}{\xi_i}\right) < c\xi_j \left(1 - \frac{\eta_j}{\xi_j}\right) = c(\xi_j - \eta_j).$$

This completes the proof when $\tilde{\beta}_0 = \tilde{\beta}^{\phi}_0$ and $\tilde{\beta}_1 = \tilde{\beta}^{\phi}_1$. The remaining cases all have $k = 1$ and are as follows: $\tilde{\beta}_0 \geq \tilde{\beta}_1 = 1$ and $1 = \tilde{\beta}^{\phi}_0 \leq \tilde{\beta}^{\phi}_1$; $\tilde{\beta}_0 \geq \tilde{\beta}_1 = 1$ and $\tilde{\beta}^{\phi}_0 \geq \tilde{\beta}^{\phi}_1 = 1$; $1 = \tilde{\beta}_0 \leq \tilde{\beta}_1$ and $1 = \tilde{\beta}^{\phi}_0 < \tilde{\beta}^{\phi}_1$; $1 = \tilde{\beta}_0 \leq \tilde{\beta}_1$ and $\tilde{\beta}^{\phi}_0 \geq \tilde{\beta}^{\phi}_1 = 1$. These are handled in the same way as above. The details are left to the reader.
□

3

Tree Structures

In this third chapter we show that valuation space $\mathcal{V}$ has the structure of a tree; we shall subsequently refer to it as *the valuative tree.*

Roughly speaking, a tree[1] is a union of real intervals welded together in such a way that no cycles appear. We will more precisely distinguish between three types of tree structures, see below. The main result of the chapter is then that valuation space can be naturally equipped with all three of these structures and even a bit more. What we obtain is effectively a coordinate free visualization of the encoding of valuations by SKP's (which are defined using a fixed choice of local coordinates). Indeed, SKP's play an instrumental role in most proofs in this chapter.

It is good to keep in mind that the quite intricate structure of the valuative tree that we are about to develop all derives from the definition of an element of $\mathcal{V}$ as a function on R satisfying certain axioms. In Chapter 6 we shall arrive at the same tree structure using a purely geometric construction.

The organization of this chapter is as follows. In Section 3.1, we discuss three different types of trees. First we have *nonmetric trees*, defined as partially ordered sets satisfying certain axioms. In particular, every "full", totally ordered subset can be parameterized (in a noncanonical way) by a real interval. This notion seems to be new, although related to the approach of Berkovich [Be]. Then we have *parameterized trees.* These are nonmetric trees that come with a fixed parameterization. Finally we have *metric trees* (typically called **R**-trees in the literature). Metric trees are closely related to parameterized trees, and are defined as metric spaces satisfying certain conditions.

As we show in Section 3.2, the natural partial ordering on $\mathcal{V}$ induces a nonmetric tree structure. We analyze this structure in detail. In particular, we show that the set $\mathcal{V}_{\mathrm{qm}}$ of quasimonomial valuations is exactly the tree $\mathcal{V}$ with all ends removed, hence in itself is a nonmetric tree.

---

[1]We use the term "tree" instead of "**R**-tree" when the tree is modeled on the real line.

In Sections 3.3, 3.4, 3.6, we introduce two natural parameterizations of $\mathcal{V}$ and $\mathcal{V}_{\mathrm{qm}}$ and show they can be used to exhibit these spaces as metric trees. These parameterizations will play a fundamental role in applications [FJ1] [FJ2], [FJ3].

We first define in Section 3.3 a numerical invariant of a valuation, its *skewness*. This gives the first parameterization of both $\mathcal{V}$ and $\mathcal{V}_{\mathrm{qm}}$.

We then define in Section 3.4 a discrete invariant, the *multiplicity* of a valuation. This invariant naturally extends the notion of multiplicity of a curve. To a divisorial valuation is also associated a *generic multiplicity*.

Multiplicity is an increasing function on $\mathcal{V}$ with values in $\overline{\mathbf{N}}$. By analyzing the points where it jumps, we can define a canonical *approximating sequence* of a given valuation (Section 3.5).

Using multiplicity and skewness, we define in Section 3.6 a third important invariant of a valuation: its *thinness*. It has a particular geometric interpretation which makes it extremely useful for applications. Thinness is obtained by integrating multiplicity with respect to skewness and hence gives the second parameterization of $\mathcal{V}$ and $\mathcal{V}_{\mathrm{qm}}$ as mentioned above.

In Section 3.7 we use the approximating sequences to compute the value group of a valuation, as well as the generic multiplicity of a divisorial valuation.

After that, we present a different but intriguing approach to the valuative tree. Namely, we show in Section 3.8 that a quasimonomial valuation can be identified with a ball of irreducible curves in a particular (ultra-)metric. The partial ordering, skewness and multiplicity on the valuative tree then have natural interpretations as statements about balls of curves.

We conclude the chapter by a study of the *relative valuative tree*. By definition, this is the (closure of) the set of centered valuations on R normalized by the condition $\nu(x) = 1$, where $\{x = 0\}$ is a smooth formal curve; such a normalization is natural in several situations. As we show, the relative valuative tree $\mathcal{V}_x$ has a structure which is very similar to that of $\mathcal{V}$.

3.1 Trees

In this section we discuss different types of trees.

3.1.1 Rooted Nonmetric Trees

Consider a partially ordered set, or *poset* $(\mathcal{T}, \leq)$. Let us say that a totally ordered subset $\mathcal{S} \subset \mathcal{T}$ is *full* if $\sigma, \sigma' \in \mathcal{S}$, $\tau \in \mathcal{T}$ and $\sigma \leq \tau \leq \sigma'$ imply $\tau \in \mathcal{S}$.

Definition 3.1. *A* rooted nonmetric tree *is a poset* $(\mathcal{T}, \leq)$ *such that*

(T1) $\mathcal{T}$ *has a unique minimal element* τ_0, *called the* root *of* $\mathcal{T}$;
(T2) if $\tau \in \mathcal{T}$, *then the set* $\{\sigma \in \mathcal{T}\ ;\ \sigma \leq \tau\}$ *is isomorphic to a real interval;*
(T3) every full, totally ordered subset of $\mathcal{T}$ *is isomorphic to a real interval.*

Statements (T2) and (T3) assert that there exists an order preserving bijection from a real interval onto the corresponding set. However, there may not be a canonical choice of bijection.

Remark 3.2. Condition (T3) may seem superfluous in view of (T2) but is necessary to avoid a "long half-line", i.e. a totally ordered set $(\mathcal{T}, \leq)$ with a (unique) minimal element τ_0 for which every set $\{\sigma \in \mathcal{T} \ ; \ \tau_0 \leq \sigma \leq \tau\}$ is isomorphic to a real interval but the full set $\mathcal{T}$ is not.

Remark 3.3. In fact, it is useful—and not hard to see—that if (T1)-(T2) hold, then (T3) is equivalent to

(T3') if $\mathcal{S}$ is a totally ordered subset of $\mathcal{T}$ without upper bound in $\mathcal{T}$, then there exists a countable increasing sequence in $\mathcal{S}$ without upper bound in $\mathcal{T}$.

Remark 3.4. More generally, if Λ is a totally ordered set, then a *rooted non-metric Λ-tree* is a partially ordered set such that (T1)-(T3) hold, with the intervals in (T2) and (T3) being intervals in Λ. Besides $\Lambda = \mathbf{R}$, interesting examples include $\Lambda = \mathbf{N}$, $\Lambda = \overline{\mathbf{N}}$ and $\Lambda = \mathbf{Q}$.

It follows from the completeness of $\mathbf{R}$ that every subset $S \subset \mathcal{T}$ admits an *infimum*, denoted by $\wedge_{\tau \in S} \tau$. Indeed, the set $\{\sigma \in \mathcal{T} \ ; \ \sigma \leq \tau \ \forall \tau \in S\}$ is isomorphic to the intersection of closed real intervals with common left endpoint.

If $\mathcal{T}$ is a rooted, nonmetric tree and τ_1, τ_2 are two points in $\mathcal{T}$, then we set

$$[\tau_1, \tau_2] := \{\tau \in \mathcal{T} \ ; \ \tau_1 \wedge \tau_2 \leq \tau \leq \tau_1 \quad \text{or} \quad \tau_1 \wedge \tau_2 \leq \tau \leq \tau_2\}.$$

We call $[\tau_1, \tau_2]$ a *segment*, and define $[\tau_1, \tau_2[:= [\tau_1, \tau_2] \setminus \{\tau_2\}$. The segments $]\tau_1, \tau_2]$ and $]\tau_1, \tau_2[$ are defined similarly.

If $\mathcal{S}$, $\mathcal{T}$ are rooted nonmetric trees with roots σ_0, τ_0, then a mapping $\Phi : \mathcal{S} \to \mathcal{T}$ is a *morphism of rooted nonmetric trees* if for any $\sigma \in \mathcal{S}$, Φ gives an order preserving bijection of $[\sigma_0, \sigma]$ onto $[\tau_0, \Phi(\sigma)]$. If Φ is also bijective, then it is an *isomorphism of rooted nonmetric trees.* and we say that $\mathcal{S}$ and $\mathcal{T}$ are *isomorphic.*

A *subtree* of a rooted nonmetric tree $(\mathcal{T}, \leq)$ is a subset $\mathcal{S}$ such that $\sigma \in \mathcal{S}$, $\tau \in \mathcal{T}$ and $\tau \leq \sigma$ implies $\tau \in \mathcal{S}$. Clearly $\mathcal{S}$ is then a rooted nonmetric tree with root τ_0. We say that $\mathcal{S}$ is a *finite* (*countable*) subtree if $\mathcal{S}$ is a union of finitely (countably) many segments.

A rooted nonmetric tree $\mathcal{T}$ is *complete* if every increasing sequence $(\tau_i)_{i \geq 1}$ in $\mathcal{T}$ has a majorant, i.e. an element $\tau_\infty \in \mathcal{T}$ with $\tau_i \leq \tau_\infty$ for every i. Thanks to (T3), any rooted nonmetric tree $\mathcal{T}$ has a *completion* $\bar{\mathcal{T}}$ obtained by adding points corresponding to unbounded increasing sequences τ_i in $\mathcal{T}$. A maximal point in $\bar{\mathcal{T}}$ (under $\leq$) is called an *end* of $\mathcal{T}$. Hence all points in $\bar{\mathcal{T}} \setminus \mathcal{T}$ are ends. We sometimes use the notation $\mathcal{T}^o$ for the sets of elements of $\mathcal{T}$ that are not ends, i.e. the set of nonmaximal elements of $\mathcal{T}$.

3.1.2 Nonmetric Trees

In some situations, the choice of root in a rooted nonmetric tree is not so important. By "forgetting" the root, we obtain an object called a (nonrooted) *nonmetric tree*.

Let us be more precise. Consider a rooted, nonmetric tree $(\mathcal{T}, \leq)$ with root τ_0. Pick any point $\tau_0' \in \mathcal{T}$ and define a new partial ordering $\leq'$ on $\mathcal{T}$ by declaring $\tau_1 \leq' \tau_2$ iff $[\tau_0', \tau_1] \subset [\tau_0', \tau_2]$. Then $(\mathcal{T}, \leq')$ is a nonmetric tree rooted at τ_0'. One easily verifies that segments in $(\mathcal{T}, \leq')$ are the same as segments in $(\mathcal{T}, \leq)$.

More generally, for any set $\mathcal{T}$, let $P = P(\mathcal{T})$ be the (possibly empty) set of partial orderings $\leq$ on $\mathcal{T}$ such that $(\mathcal{T}, \leq)$ is a rooted, nonmetric tree. If $\leq_1$ and $\leq_2$ are partial orderings in P, then we say that $\leq_1$ and $\leq_2$ are *equivalent* if the segments in $(\mathcal{T}, \leq_1)$ and $(\mathcal{T}, \leq_2)$ are the same. It is straightforward to verify that two partial orderings are equivalent iff one is obtained from the other by changing the root as above.

This leads to the following definition.

Definition 3.5. *A* nonmetric tree *is a set $\mathcal{T}$ with $P(\mathcal{T}) \neq \emptyset$, together with a nonempty equivalence class in $P(\mathcal{T})$.*

If $\mathcal{T}$ is a nonmetric tree, we obtain a canonical rooted, nonmetric tree $(\mathcal{T}, \leq)$ by fixing a point (the root) in $\mathcal{T}$.

Concepts on rooted, nonmetric trees that can be formulated purely in terms of segments carry over to the nonrooted setting. For instance, a nonmetric tree $\mathcal{T}$ is *complete* if the rooted nonmetric tree $(\mathcal{T}, \leq)$ is complete for some choice of root. This makes sense since $\mathcal{T}$ is complete iff every open segment in $\mathcal{T}$ is contained in a closed segment.

Given a nonmetric tree and a point $\tau \in \mathcal{T}$ we define an equivalence relation on $\mathcal{T} \setminus \{\tau\}$ by declaring σ, σ' to be equivalent if the segments $]\tau, \sigma]$ and $]\tau, \sigma']$ intersect. An equivalence class is called a *tangent vector* at τ and the set of tangent vectors is called the *tangent space* at τ, denoted $T\tau$. We say that the point σ *represents* the tangent vector. The tangent spaces should be thought of as projectivized. This is natural since there is no canonical parameterization of segments by real intervals.

A point in $\mathcal{T}$ is an *end* if its tangent space has only one element. It is a *regular point* if the tangent space has two elements, and a *branch point* otherwise.

Notice that the concept of end differs slightly between rooted and nonrooted nonmetric trees: the root of a rooted, nonmetric tree $(\mathcal{T}, \leq)$ may be an end in the associated nonrooted tree but is never an end in $(\mathcal{T}, \leq)$ itself.

Two nonmetric trees $\mathcal{T}_1$ and $\mathcal{T}_2$ are *isomorphic* if there exists a bijection $\Phi : \mathcal{T}_1 \to \mathcal{T}_2$ and partial orderings $\leq_1$ and $\leq_2$ such that Φ is an isomorphism of the rooted nonmetric trees $(\mathcal{T}_1, \leq_1)$ and $(\mathcal{T}_2, \leq_2)$, i.e. Φ is order-preserving. Then Φ is called an *isomorphism of nonmetric trees.*

3.1.3 Parameterized Trees

A *parameterization* of a rooted, nonmetric tree $(\mathcal{T}, \leq)$ is an increasing (or decreasing) mapping $\alpha : \mathcal{T} \to [-\infty, +\infty]$ whose restriction to any full, totally ordered subset of $\mathcal{T}$ gives a bijection onto a real interval. A rooted, nonmetric tree is *parameterizable* if it admits a parameterization.

Note that by postcomposing with a suitable monotone function, we may require the parameterizations to be increasing with values in $[0, \infty]$ or even in $[0, 1]$. In the rest of this chapter, we shall always work with such parameterizations.

Let us point out that in some situations (notably for the much of the analysis in Chapter 7 and its applications in Chapter 8) the choice of parameterization is important. At any rate, on the valuative tree, all the parameterizations that we shall be concerned with are increasing, with values in $[0, \infty]$.

Consider a rooted nonmetric tree $(\mathcal{T}, \leq)$ with root τ_0 and a parameterization $\alpha : \mathcal{T} \to [0, 1]$. Pick any τ_0' and consider the equivalent partial ordering $\leq'$ rooted in τ_0'. Then the function $\alpha' : \mathcal{T}' \to [0, 2]$ defined by $\alpha'(\tau) = \alpha(\tau) + \alpha(\tau_0') - 2\alpha(\tau \wedge \tau_0')$, where $\wedge$ defines the minimum with respect to $\leq$, gives a parameterization of the rooted, nonmetric tree $(\mathcal{T}, \leq')$. We may therefore define a (nonrooted) nonmetric tree $\mathcal{T}$ to be *parameterizable* if $(\mathcal{T}, \leq)$ is parameterizable for any choice of root.

It follows easily that a nonmetric tree $\mathcal{T}$ is parameterizable iff its completion $\bar{\mathcal{T}}$ is. We do not know if there exists a non-parameterizable nonmetric tree. All nonmetric trees we consider in this monograph will be parameterizable.

A *morphism of parameterized trees* between $(\mathcal{T}_1, \alpha_1)$ and $(\mathcal{T}_2, \alpha_2)$ is a morphism $\Phi : \mathcal{T}_1 \to \mathcal{T}_2$ of the underlying rooted, nonmetric trees, such that $\alpha_2 \circ \Phi = \alpha_1$. If Φ is moreover a bijection then it is an *isomorphism of parameterized trees* and we say that $(\mathcal{T}_1, \alpha_1)$ and $(\mathcal{T}_2, \alpha_2)$ are *isomorphic*. Of course Φ is then also an isomorphism of the underlying (rooted) nonmetric trees $\mathcal{T}_1$ and $\mathcal{T}_2$.

Example 3.6. Fix a set X and set $\mathcal{T} = X \times [0, \infty)/\sim$, where $(x, 0) \sim (y, 0)$ for any $x, y \in X$. Then $\mathcal{T}$ is a rooted, nonmetric tree under the partial ordering $\leq$ defined by $(x, s) \leq (y, t)$ iff either $s = t = 0$, or $x = y$ and $s \leq t$.

Given $g : X \to (0, \infty)$ define a parameterization $\alpha_g : \mathcal{T} \to [0, \infty)$ on $\mathcal{T}$ by $\alpha_g(x, s) = g(x)s$. The identity map $(\mathcal{T}, \alpha_g) \to (\mathcal{T}, \alpha_h)$ is then an isomorphism of parameterized trees iff $g \equiv h$.

3.1.4 The Weak Topology

A nonmetric tree carries a natural *weak topology* defined as follows. If $\vec{v} \in T\tau$ is a tangent vector at a point $\tau \in \mathcal{T}$, set

$$U(\vec{v}) := \{\sigma \in \mathcal{T} \setminus \{\tau\} \; ; \; \sigma \text{ represents } \vec{v}\}.$$

See Figure 3.1. Then the weak topology is generated by the sets $U(\vec{v})$ in the

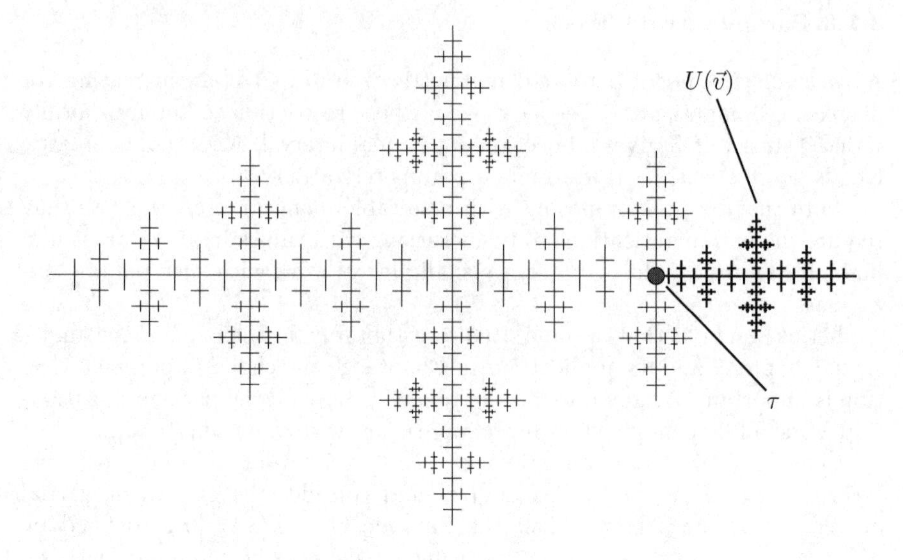

Fig. 3.1. The open set $U(\vec{v})$ associated to a tangent vector $\vec{v}$ at a point τ in a tree.

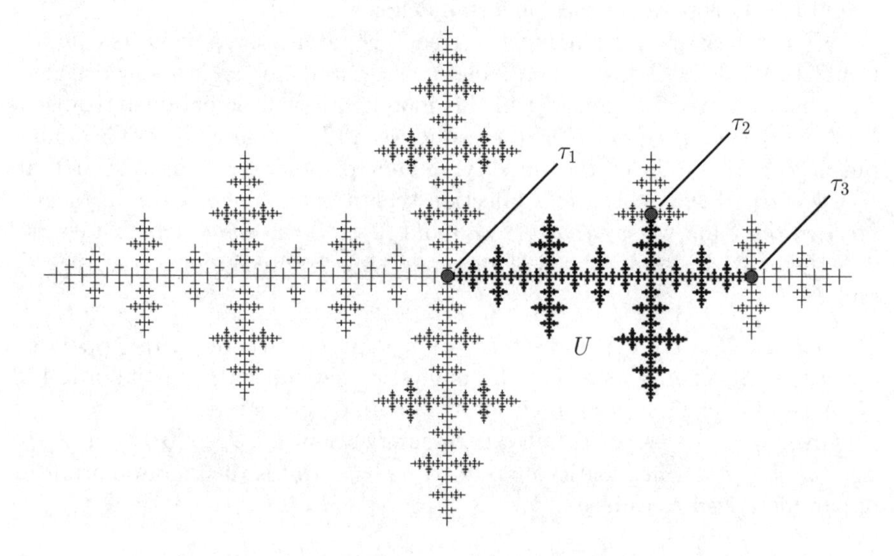

Fig. 3.2. An open set U which is the intersection of three open sets $U(\vec{v}_i)$, where $\vec{v}_i$ is a tangent vector at τ_i, $i = 1, 2, 3$. Any weak open set is the union of sets of the type above.

sense that the open sets are unions of finite intersections of such sets.

We shall study the weak topology in more detail in Section 7.2, but let us summarize the main features. The weak topology is Hausdorff. Any complete subtree $\mathcal{S}$ of $\mathcal{T}$ is weakly closed in $\mathcal{T}$ and the injection $\mathcal{S} \to \mathcal{T}$ is an embedding. In particular, any segment $\gamma = [\tau, \tau']$ in $\mathcal{T}$ is closed, and the induced topology on γ coincides with the standard topology on $[0, 1]$ under the identification of γ with the latter set. However, the branching of a nonmetric tree does play an important role for the weak topology:

Example 3.7. Let $\mathcal{T} = X \times [0.\infty[/ \sim$ be as in Example 3.6. Assume that X is infinite and pick a sequence $(x_n)_1^\infty$ of distinct elements of X. Then $(x_n, 1)$ converges weakly to the root $(x, 0)$ as $n \to \infty$.

See Proposition 7.5 for a precise criterion of weak sequential convergence. In Section 7.2 we shall prove that *any parameterizable, complete, nonmetric tree is weakly compact.*[2] Note that a nonmetric tree $\mathcal{T}$ which is not complete cannot be weakly compact: a sequence in $\mathcal{T}$ that increases (with respect to some choice of partial ordering) to an element in $\bar{\mathcal{T}} \setminus \mathcal{T}$ cannot have a convergent subsequence.

Proposition 3.8. *If $\alpha : \mathcal{T} \to [0, \infty)$ is a parameterization of a rooted, nonmetric tree, then the function α is weakly lower semicontinuous.*

Proof. We have to show that the superlevel set $\mathcal{T}_t = \{\tau \; ; \; \alpha(\tau) > t\}$ is open for every t. Consider t such that $\mathcal{T}_t$ is nonempty and pick $\tau \in \mathcal{T}_t$. Pick $\tau' < \tau$ with $\alpha(\tau') \geq t$ and let $\vec{v}$ be the tree tangent vector at σ' represented by τ. Then $U(\vec{v})$ is an open neighborhood of τ on which $\alpha > t$. This completes the proof. ☐

Remark 3.9. The function α is *not* weakly continuous in general as can be seen from Example 3.6 with $g \equiv 1$ and X an infinite set. See also Propositions 3.31 and 3.47.

3.1.5 Metric Trees

Closely connected to parameterized trees are *metric trees.*[3] These are metric spaces in which every two points is joined by a unique arc, or segment, and this segment is isometric to a real interval. It is known [MO] that a metrizable topological space is a metric tree (i.e. admits a compatible metric under which it becomes a metric tree) iff it is uniquely pathwise connected and locally pathwise connected.

---

[2]In fact, a parameterizable, complete, nonmetric tree is exactly the same as a compact, simply connected, special one-dimensional quasipolyhedron in the sense of Berkovich [Be, Section 4.1].

[3]Metric trees are normally called **R**-trees in the literature but this terminology would not be precise enough for our purposes.

Roughly speaking, metric trees are to parameterized trees what nonmetric trees are to rooted, nonmetric trees.[4] Let us make this precise. First, a metric tree $(\mathcal{T}, d)$ gives rise to a nonmetric tree. Indeed, fix $\tau_0 \in \mathcal{T}$ (the root) and define a partial ordering $\leq$ on $\mathcal{T}$: $\tau \leq \tau'$ iff τ belongs to the segment between τ_0 and τ'. Clearly (T1)-(T3) hold. Moreover, different choices of τ_0 give rise, by the very definition, to equivalent partial orderings on $\mathcal{T}$. Thus $\mathcal{T}$ is naturally a (nonrooted) nonmetric tree. We say that the metric d on $\mathcal{T}$ is *compatible* with the nonmetric tree structure.

Second, if $(\mathcal{T}, d)$ is a metric tree, then given any choice of root $\tau_0 \in \mathcal{T}$, the function $\alpha : \mathcal{T} \to [0, \infty)$ defined by $\alpha(\tau) = d(\tau, \tau_0)$ gives a parameterization of the rooted, nonmetric tree $(\mathcal{T}, \leq)$.

Conversely, consider a rooted, nonmetric tree $(\mathcal{T}, \leq)$ with a parameterization $\alpha : \mathcal{T} \to [0, 1]$. Define a function $d : \mathcal{T} \times \mathcal{T} \to [0, 2]$ by

$$d(\sigma, \tau) = (\alpha(\sigma) - \alpha(\sigma \wedge \tau)) + (\alpha(\tau) - \alpha(\sigma \wedge \tau)).$$

Proposition 3.10. *$(\mathcal{T}, d)$ is a metric tree.*

Proof. It is straightforward to verify that $(\mathcal{T}, d)$ is a metric space in which every two points is joined by an arc isometric to a real interval. What remains to be seen is that there is a *unique* arc between any two points in $\mathcal{T}$. For this, consider a continuous injection $\imath : I \to (\mathcal{T}, d)$ of a real interval $I = [t_1, t_2]$ into $\mathcal{T}$, and let $\tau_i = \imath(t_i)$. By declaring τ_1 to be the root of $\mathcal{T}$, we may suppose $\tau_1 \leq \tau_2$. Define $\pi(t) = \imath(t) \wedge \tau_2$, and suppose $\pi^{-1}\{t\}$ has an interior point for some $t \in I$. If $[a, b]$, $a < b$ denotes a non trivial connected component of $\pi^{-1}\{t\}$, the tree structure of $\mathcal{T}$ implies $\imath(a) = \imath(b)$. This contradicts the injectivity of $\imath$. From the fact that the preimage by π of any point in $[\tau_1, \tau_2]$ has empty interior, we infer that $\imath$ maps I into the segment $[\tau_1, \tau_2]$. The injectivity of $\imath$ then gives that $\imath$ is an increasing homeomorphism of I onto $[\tau_1, \tau_2]$. This completes the proof. □

If a metric tree $\mathcal{T}$ is complete as a nonmetric tree, then $(\mathcal{T}, d)$ is a complete metric space for every compatible tree metric d on $\mathcal{T}$. As a partial converse, if a nonmetric tree $\mathcal{T}$ admits a compatible, complete tree metric d of finite diameter, then $\mathcal{T}$ is complete as a tree. Notice, however, that $\mathbf{R}$ with its standard metric is complete as a metric space but not as a nonmetric tree.

If a metric tree $\mathcal{T}$ has finite diameter, then the completion of $\mathcal{T}$ as a metric space agrees with the completion of $\mathcal{T}$ as a nonmetric tree. Moreover, it is always possible to find an equivalent metric in which $\mathcal{T}$ has finite diameter.

---

[4]This analogy may be strengthened by allowing metric trees to be metric spaces in which some points are at infinite distance to each other, but we shall not pursue this.

3.1.6 Trees from Ultrametric Spaces

There is a natural way to construct trees from ultrametric spaces. This procedure will be used to illustrate the connection between quasimonomial valuations and curves in Section 3.8.

Recall that a metric d on a space X is an *ultrametric* if it satisfies the stronger triangle inequality $d(x,y) \leq \max\{d(x,z), d(y,z)\}$ for any $x, y, z \in X$.

Let (X, d_X) be an ultrametric space of diameter 1. Define an equivalence relation $\sim$ on $X \times (0,1)$ by declaring $(x,s) \sim (y,t)$ iff $d(x,y) \leq s = t$. Note that $(x,1) \sim (y,1)$ for any x, y. The set $\mathcal{T}_X$ of equivalence classes is a nonmetric tree rooted at $(x,1)$ under the partial order $(x,s) \leq (y,t)$ iff $d(x,y) \leq s \geq t$. It is a metric tree under the metric defined by $d((x,s),(x,t)) = |s-t|$ and $d(\sigma,\tau) = d(\sigma, \sigma \wedge \tau) + d(\tau, \sigma \wedge \tau)$ for general $\sigma, \tau \in \mathcal{T}_X$. It can also be parameterized by declaring $\alpha(x,t) = t^{-1}$.

If the diameter of any ball of radius r equals r (!), then we may think of $(x,t) \in \mathcal{T}_X$ as the closed ball of radius t in X centered at x.

3.1.7 Trees from Simplicial Trees

Next we show how the classical notion of simplicial tree fits into our framework.

Recall that a *simplicial tree* is a set V (vertices) together with a collection E (edges) of subsets of V of cardinality 2, such that the following holds: for any two distinct vertices $\sigma, \tau \in V$ there exist a unique sequence $\sigma = \sigma_0, \sigma_1, \dots, \sigma_n = \tau$ of distinct vertices such that $\{\sigma_{i-1}, \sigma_i\} \in E$ for all $i = 1, \dots, n$. We write $[\sigma, \tau] = \{\sigma_i\}_0^n$. A *rooted simplicial tree* is a triple (V, E, τ_0) consisting of a simplicial tree (V, E) together with a marked vertex $\tau_0 \in E$ (the root).

If (V, E, τ_0) is a rooted, simplicial tree, then we can define a partial ordering on V by declaring $\tau \leq \tau'$ iff $[\tau_0, \tau] \subset [\tau_0, \tau']$. It is straightforward to verify that $(V, \leq)$ is a rooted, nonmetric $\mathbf{N}$-tree in the sense of Section 3.1.1. Conversely, to any rooted, nonmetric $\mathbf{N}$-tree $(\mathcal{T}, \leq)$ we can associate a rooted, simplicial tree (V, E, τ_0) as follows: $V = \mathcal{T}$, τ_0 is the root of $\mathcal{T}$, and $\{\tau, \tau'\} \in E$ iff the segment $[\tau, \tau']$ in $\mathcal{T}$ contains exactly two elements. One easily verifies that (V, E, τ_0) is a rooted simplicial tree, and that the two operations just defined are inverse to each other. Moreover, changing the root of (V, E, τ_0) leads to an equivalent partial ordering on $\mathcal{T} = V$, and vice versa. We thus conclude that *(rooted) simplicial trees can be identified with (rooted) nonmetric* $\mathbf{N}$*-trees.*

Notice that any rooted, nonmetric $\mathbf{N}$-tree $(\mathcal{T}, \leq)$ has a unique parameterization $\alpha : \mathcal{T} \to \mathbf{N}$ satisfying $\alpha(\tau_0) = 1$: this is given by $\alpha(\tau) = \#\{\sigma \leq \tau\}$.

To any rooted (nonmetric) $\mathbf{N}$-tree $(\mathcal{T}_\mathbf{N}, \leq)$ with root τ_0 we can associate a rooted, nonmetric $\mathbf{R}$-tree $(\mathcal{T}_\mathbf{R}, \leq)$ by "adding the edges" as follows: we set

$$\mathcal{T}_\mathbf{R} = \{(\tau_0, 0)\} \cup ((\mathcal{T}_\mathbf{N} \setminus \{\tau_0\}) \times]-1, 0])$$

and define the partial ordering on $\mathcal{T}_{\mathbf{R}}$ by lexicographic ordering: $(\tau, s) \leq (\tau', s')$ iff $\tau < \tau'$ or $\tau = \tau'$ and $s \leq s'$. It is straightforward to verify that $\mathcal{T}_{\mathbf{R}}$ is a nonmetric $\mathbf{R}$-tree rooted in $(\tau_0, 0)$. Notice that $\mathcal{T}_{\mathbf{R}}$ naturally contains $\mathcal{T}_{\mathbf{N}} \simeq \mathcal{T}_{\mathbf{N}} \times \{0\}$ as a subset. In fact, $\mathcal{T}_{\mathbf{R}}$ is the minimal rooted, nonmetric $\mathbf{R}$-tree with this property. We leave it to the reader to make the last statement precise, as well as to verify that equivalent partial orderings on $\mathcal{T}_{\mathbf{N}}$ give rise to equivalent partial orderings on $\mathcal{T}_{\mathbf{R}}$. Finally notice that the canonical parameterization $\alpha : \mathcal{T}_{\mathbf{N}} \to \mathbf{N}$ extends to a parameterization $\alpha : \mathcal{T}_{\mathbf{R}} \to \mathbf{R}_+$ by setting $\alpha(\tau, s) = \alpha(\tau) + s$.

3.1.8 Trees from Q-trees

As we show in this section, there is a natural way of passing from a $\mathbf{Q}$-tree to an $\mathbf{R}$-tree by "adding irrational points". This construction will be used in Section 6.1.

First, notice that definition of a $\mathbf{Q}$-tree does not involve any arithmetic properties of the set $\mathbf{Q}$ but only its structure as a totally ordered set. As such, it is characterized by

Lemma 3.11. *A totally ordered set Λ is isomorphic to an interval in $\mathbf{Q}$ iff Λ is countable and has no gaps in the sense that if $\lambda, \lambda' \in \Lambda$ and $\lambda < \lambda'$ then there exists $\lambda'' \in \Lambda$ with $\lambda < \lambda'' < \lambda'$.*

Proof. Clearly any interval in $\mathbf{Q}$ has the stated property. For the converse, fix a countable, totally ordered set Λ with no gaps. We wish to show it is isomorphic to an interval in $\mathbf{Q}$. There is no loss of generality in assuming that Λ contains its infimum and supremum. The proposition will follow if we can produce an isomorphism χ from Λ onto the set D of dyadic rational numbers in $[0, 1]$, i.e. 0,1 and all numbers of the form $i/2^j$, $j \geq 1$, $0 < i < 2^j$. Indeed, we can then post-compose with an isomorphism from D onto $[0, 1] \cap \mathbf{Q}$.

Let $(\lambda_n)_0^\infty$ be an enumeration of Λ. Assume that $\lambda_0 = \min \Lambda$ and $\lambda_1 = \max \Lambda$. Set $\chi(\lambda_0) = 0$ and $\chi(\lambda_1) = 1$. Inductively, suppose $n \geq 2$ and that we have defined $\chi(\lambda_m)$ for $m < n$ in an order-preserving way. Define $\lambda' = \max\{\lambda_m \; ; \; m < n, \lambda_m < \lambda_n\}$ and $\lambda'' = \min\{\lambda_m \; ; \; m < n, \lambda_m > \lambda_n\}$. Then pick $\chi(\lambda_n)$ to be a dyadic rational in $]0, 1[$ with minimal denominator such that $\chi(\lambda') < \chi(\lambda) < \chi(\lambda'')$.

It is clear that this gives an order-preserving mapping of Λ into the set D. The fact that it is onto follows from the assumption that Λ has no gaps. □

We now formulate the way in which a $\mathbf{Q}$-tree is canonically embedded in an $\mathbf{R}$-tree.

Proposition 3.12. *Given a rooted, nonmetric $\mathbf{Q}$-tree $\mathcal{T}_{\mathbf{Q}}$ there exists a rooted, nonmetric $\mathbf{R}$-tree $\mathcal{T}_{\mathbf{R}}$ and an order-preserving injection $\imath : \mathcal{T}_{\mathbf{Q}} \to \mathcal{T}_{\mathbf{R}}$ such that:*

(i) $\imath(\mathcal{T}_{\mathbf{Q}})$ is weakly dense in $\mathcal{T}_{\mathbf{R}}$;
(ii) every point in $\mathcal{T}_{\mathbf{R}} \setminus \imath(\mathcal{T}_{\mathbf{Q}})$ is a regular point of $\mathcal{T}_{\mathbf{R}}$;

(iii) if $\imath' : \mathcal{T}_{\mathbf{Q}} \to \mathcal{T}'_{\mathbf{R}}$ is an order-preserving injection into another rooted, nonmetric $\mathbf{R}$-tree with weakly dense image, then there exists an injective morphism $\Phi : \mathcal{T}_{\mathbf{R}} \to \mathcal{T}'_{\mathbf{R}}$ of rooted, nonmetric trees such that $\Phi \circ \imath = \imath'$ and such that Φ extends to an isomorphism of rooted, nonmetric trees between the completions of $\mathcal{T}_{\mathbf{R}}$ and $\mathcal{T}'_{\mathbf{R}}$.

Moreover, given a parameterization $\alpha_{\mathbf{Q}} : \mathcal{T}_{\mathbf{Q}} \to \mathbf{Q}_+$ of $\mathcal{T}_{\mathbf{Q}}$ there exists a unique parameterization $\alpha_{\mathbf{R}} : \mathcal{T}_{\mathbf{R}} \to \mathbf{R}_+$ such that $\alpha_{\mathbf{R}} \circ \imath = \alpha_{\mathbf{Q}}$.

Proof. We first show how to define $\mathcal{T}_{\mathbf{R}}$ and $\imath$. Define $\mathcal{T}_{\mathbf{R}}$ as the set of equivalence classes $[I_\bullet]$ of decreasing sequences $I_\bullet = (I_n)_0^\infty$ of closed segments in $\mathcal{T}_{\mathbf{Q}}$ so that $\cap_{n\geq 0} I_n$ contains at most one point. Here $I_\bullet$ and $J_\bullet$ are equivalent iff for every n there exists m such that $I_m \subset J_n$ and $J_m \subset I_n$. Further, if $\sigma \in \mathcal{T}_{\mathbf{Q}}$, then $\imath(\sigma) \in \mathcal{T}_{\mathbf{R}}$ is defined to be the equivalence class containing $I_\bullet = (I_n)_0^\infty$, where $I_n = [\sigma, \sigma]$ for all n.

Let us show that $\mathcal{T}_{\mathbf{R}}$ is naturally a rooted nonmetric $\mathbf{R}$-tree, i.e. that it admits a natural partial ordering satisfying (T1)-(T3). This partial ordering is defined as follows: $[I_\bullet] \leq [J_\bullet]$ iff there exist increasing sequences $(n_k)_1^\infty$, $(m_k)_1^\infty$ and, for every k, elements $\sigma_k \in I_{n_k}$, $\tau_k \in J_{m_k}$ with $\sigma_k \leq \tau_k$. We leave it to the reader to verify that this is well-defined and that $\imath : \mathcal{T}_{\mathbf{Q}} \to \mathcal{T}_{\mathbf{R}}$ is an order-preserving injection. If σ_0 is the root of $\mathcal{T}_{\mathbf{Q}}$, then clearly $\imath(\sigma_0)$ is the unique minimal element of $\mathcal{T}_{\mathbf{R}}$, so $\mathcal{T}_{\mathbf{R}}$ satisfies (T1). We now consider (T2). We have to show that if $[I_\bullet] \in \mathcal{T}_{\mathbf{R}}$, then $\{[J_\bullet] \in \mathcal{T}_{\mathbf{R}}\ ;\ [J_\bullet] \leq [I_\bullet]\}$ is a totally ordered set isomorphic to a real interval. First suppose $[I_\bullet] = \imath(\sigma)$ for some $\sigma \in \mathcal{T}_{\mathbf{Q}}$. The segment $[\sigma_0, \sigma]$ is a totally ordered set isomorphic to $[0,1] \cap \mathbf{Q}$. But it is well-known that if we perform the construction above on the nonmetric $\mathbf{Q}$-tree $[0,1] \cap \mathbf{Q}$, then we end up with $[0,1] \cap \mathbf{R}$. Hence we are done in this case. If $[I_\bullet]$ is not of the form $\imath(\sigma)$, then there still exists a decreasing sequence σ_n in $\mathcal{T}_{\mathbf{Q}}$ such that $\imath(\sigma_n)$ decreases to $[I_\bullet]$. Then $\{[J_\bullet] \leq [I_\bullet]\}$ is isomorphic to the intersection of real intervals $[0, s_n] \subset \mathbf{R}$, where $(s_n)_1^\infty$ is a decreasing sequence of real numbers with $s_\infty = \lim s_n > 0$. As $\{[J_\bullet] \leq [I_\bullet]\}$ has a maximal element, it must be isomorphic to the real interval $[0, s_\infty]$.

Thus (T2) holds. Instead of proving (T3) we prove the equivalent statement (T3') in Remark 3.3. Consider a totally ordered subset $\mathcal{S}_{\mathbf{R}} \subset \mathcal{T}_{\mathbf{R}}$ without upper bound in $\mathcal{T}_{\mathbf{R}}$. We may assume that $\mathcal{S}_{\mathbf{R}}$ is full. Then $\mathcal{S}_{\mathbf{Q}} := \imath^{-1}(\mathcal{S}_{\mathbf{R}})$ is a nonempty, totally ordered subset of $\mathcal{T}_{\mathbf{Q}}$ without upper bound. Since $\mathcal{T}_{\mathbf{Q}}$ is a $\mathbf{Q}$-tree, there exists an increasing sequence (σ_n) in $\mathcal{S}_{\mathbf{Q}}$ without upper bound in $\mathcal{T}_{\mathbf{Q}}$. Then it is easy to see that $(\imath(\sigma_n))$ is an increasing sequence in $\mathcal{S}_{\mathbf{R}}$ without upper bound in $\mathcal{T}_{\mathbf{R}}$.

We conclude that $\mathcal{T}_{\mathbf{R}}$ is an $\mathbf{R}$-tree. It follows from the construction that $\gamma \cap \imath(\mathcal{T}_{\mathbf{Q}})$ is weakly dense in γ for every segment $\gamma \subset \mathcal{T}_{\mathbf{R}}$, where γ is equipped with the topology induced from $\mathbf{R}$. Hence $\imath(\mathcal{T}_{\mathbf{Q}})$ is weakly dense in $\mathcal{T}_{\mathbf{R}}$. That all points of $\mathcal{T}_{\mathbf{R}} \setminus \imath(\mathcal{T}_{\mathbf{Q}})$ are regular points of $\mathcal{T}_{\mathbf{R}}$ is clear from the construction. Thus we have proved (i) and (ii).

Finally, suppose $\imath'$ is an order-preserving injection of $\mathcal{T}_{\mathbf{Q}}$ into an $\mathbf{R}$-tree $\mathcal{T}'_{\mathbf{R}}$ with weakly dense image. Let us construct the mapping $\Phi : \mathcal{T}_{\mathbf{R}} \to \mathcal{T}'_{\mathbf{R}}$ as in (iii). The construction is based on the following result, the proof of which is left to the reader:

Lemma 3.13. *If X and Y are countable dense subsets of the real interval* $[0,1]$ *and* $\chi : X \to Y$ *is an order-preserving bijection, then* χ *extends uniquely to an order-preserving bijection (i.e. an increasing homeomorphism) of* $[0,1]$ *onto itself.*

We continue the proof of the proposition. The fact that $\imath'(\mathcal{T}_{\mathbf{Q}})$ is weakly dense in $\mathcal{T}'_{\mathbf{R}}$ implies that the root of $\mathcal{T}'_{\mathbf{R}}$ is $\imath'(\sigma_0)$, and that $\imath'([\sigma_0,\sigma])$ is dense in the segment $[\imath'(\sigma_0),\imath(\sigma)]$ in $\mathcal{T}'_{\mathbf{R}}$ for any $\sigma \in \mathcal{T}_{\mathbf{Q}}$ (this uses the fact that $\mathcal{T}_{\mathbf{Q}}$ is a **Q**-tree and does not follow from the fact that $\imath'(\mathcal{T}_{\mathbf{Q}})$ is a weakly dense subset of $\mathcal{T}'_{\mathbf{R}}$). Lemma 3.13 then implies that for any $\sigma \in \mathcal{T}_{\mathbf{Q}}$ there is a unique order-preserving bijection $\Phi_\sigma : [\imath(\sigma_0),\imath(\sigma)] \to [\imath'(\sigma_0),\imath'(\sigma)]$ such that $\Phi_\sigma \circ \imath = \imath'$ on $[\sigma_0,\sigma]$. These bijections patch together to form an order-preserving injection $\Phi : \mathcal{T}_{\mathbf{R}} \to \mathcal{T}'_{\mathbf{R}}$ with $\Phi \circ \imath = \imath'$ on $\mathcal{T}_{\mathbf{Q}}$. Since $\imath'(\mathcal{T}_{\mathbf{Q}})$ is dense in $\mathcal{T}'_{\mathbf{R}}$, Φ extends to an isomorphism of rooted trees between the completions of $\mathcal{T}_{\mathbf{R}}$ and $\mathcal{T}'_{\mathbf{R}}$, completing the proof of (iii).

Finally, the fact that any parameterization of $\mathcal{T}_{\mathbf{Q}}$ extends uniquely to a parameterization of $\mathcal{T}_{\mathbf{R}}$ is again a consequence of Lemma 3.13. □

3.2 Nonmetric Tree Structure on $\mathcal{V}$

We now show that the natural partial ordering on valuation space $\mathcal{V}$ turns it into a rooted, nonmetric tree, the *valuative tree*, and describe its structure.

3.2.1 Partial Ordering

Recall that we have normalized the valuations in $\mathcal{V}$ by $\nu(\mathfrak{m}) = 1$ and that the partial ordering $\leq$ on $\mathcal{V}$ is given by $\nu \leq \mu$ iff $\nu(\phi) \leq \mu(\phi)$ for all $\phi \in R$. The multiplicity valuation is given by $\nu_{\mathfrak{m}}(\phi) = m(\phi)$.

Theorem 3.14. *Valuation space* $\mathcal{V}$ *is a complete nonmetric tree rooted at* $\nu_{\mathfrak{m}}$.

The general properties of nonmetric trees then give:

Corollary 3.15. *Any subset of* $\mathcal{V}$ *admits an infimum.*

Remark 3.16. If (ν_i) are valuations in $\mathcal{V}$, then the infimum $\nu := \wedge \nu_i \in \mathcal{V}$ can be constructed using the following properties: $\nu(\phi) = \inf_i \nu_i(\phi)$ for all irreducible $\phi \in \mathfrak{m}$ and $\nu(\phi\psi) = \nu(\phi) + \nu(\psi)$ for all $\phi, \psi \in \mathfrak{m}$.

This construction does not work on more general rings. For instance, consider monomial valuations ν_i, $i = 1,2,3$ on $\mathbf{C}[[x_1,x_2,x_3]]$ defined by $\nu_i(x_j) = 3$ if $i \neq j$ and $\nu_i(x_i) = 1$. Define $\nu = \min_i \nu_i$ by the construction above and consider $\phi = x_1x_2 - x_2x_3 + x_3x_1$, $\psi = x_1x_2 + x_2x_3 - x_3x_1$. Then $\nu(\phi) = \nu(\psi) = 4$ but $\nu(\phi + \psi) = 2$, so ν is not a valuation.

A similar calculation shows that the natural partial ordering on the set of normalized valuations on $\mathbf{C}[[x_1,x_2,x_3]]$ does *not* define a tree structure.

Any invertible formal mapping $f : (\mathbf{C}^2, 0) \to (\mathbf{C}^2, 0)$ induces a ring automorphism $f^* : R \to R$. Since $f^*\mathfrak{m} = \mathfrak{m}$, we have an induced mapping $f_* : \mathcal{V} \to \mathcal{V}$ given by $f_*(\nu)(\phi) = \nu(f^*\phi)$. If $\mu \le \nu$, then clearly $f_*\mu \le f_*\nu$. Hence we get:

Proposition 3.17. *Any invertible formal mapping $f : (\mathbf{C}^2, 0) \to (\mathbf{C}^2, 0)$ induces an isomorphism $f_* : \mathcal{V} \to \mathcal{V}$ of rooted, nonmetric trees.*

We now turn to the proof of Theorem 3.14. It is proved by describing the partial ordering on $\mathcal{V}$ in terms of SKP's:

Proposition 3.18. *Let ν and ν' be valuations in $\mathcal{V}$, $\nu \ne \nu'$. Pick local coordinates (x, y) such that $1 = \nu(x) = \nu'(x) \le \min\{\nu(y), \nu'(y)\}$. Write $\nu = \mathrm{val}[(U_j); (\tilde\beta_j)]$ and $\nu' = \mathrm{val}[(U'_j); (\tilde\beta'_j)]$. Then $\nu < \nu'$ iff*

$$\mathrm{length}(\nu') \ge \mathrm{length}(\nu) =: k < \infty, \quad U_j = U'_j \text{ for } 0 \le j \le k \quad \text{and} \quad \tilde\beta'_k \ge \tilde\beta_k.$$

As a direct consequence, the infimum of any family of normalized valuations exists and is computable in terms of SKP's, at least as long as we can choose the coordinates (x, y) conveniently.

Corollary 3.19. *Let $(\nu^i)_{i \in I}$ be a family of valuations in $\mathcal{V}$ and suppose we can find local coordinates (x, y) such that $1 = \nu^i(x) \le \nu^i(y)$ for all $i \in I$. Write $\nu^i = \mathrm{val}[(U^i_j); (\tilde\beta^i_j)]$. Then the infimum of the ν^i exists and is given in terms of SKP's by*

$$\bigwedge_{i \in I} \nu^i = \mathrm{val}[(U_j)_0^k; (\tilde\beta_j)_0^{k-1}, \inf_{i \in I} \tilde\beta^i_k],$$

where $1 \le k < \infty$ is maximal such that $U^i_j = U^{i'}_j =: U_j$ for $0 \le j \le k$ and $i, i' \in I$, and $\tilde\beta^i_j = \tilde\beta^{i'}_j =: \tilde\beta_j$ for $0 \le j < k$ and $i, i' \in I$.

Proof (Proposition 3.18). First suppose the three displayed conditions hold. Then property (Q2) of Theorem 2.8 implies $\nu' \ge \nu_0 = \mathrm{val}[(U_j)_0^k; (\tilde\beta_j)_0^{k-1}, \tilde\beta'_k]$. But $\nu_0 \ge \nu$, so $\nu' \ge \nu$.

Conversely assume $\mathrm{val}[(U'_j); (\tilde\beta'_j)] = \nu' > \nu = \mathrm{val}[(U_j)_0^k; (\tilde\beta_j)_0^k]$ with $1 \le k \le \infty$. Let us show inductively that $U'_j = U_j$ for $j \le k$. This is true by definition for $j = 0, 1$. Assume we proved it for $j < k$. Define $\nu_{j-1} = \mathrm{val}[(U_l)_0^{j-1}; (\tilde\beta_l)_0^{j-1}]$. Then $\nu'(U_j) \ge \nu(U_j) > \nu_{j-1}(U_j)$. Hence Theorem 2.29 implies $U'_j = U_j$. Finally, $k < \infty$ (or else $\nu = \nu'$) and $\tilde\beta'_k = \nu'(U_k) \ge \nu(U_k) = \tilde\beta_k$. □

Proof (Theorem 3.14). It is clear that $(\mathcal{V}, \le)$ is a partially ordered set with unique minimal element $\nu_{\mathfrak{m}}$. Thus (T1) holds.

To prove (T2), fix $\nu \in \mathcal{V}$ with $\nu > \nu_{\mathfrak{m}}$. We will show that the set $I = \{\mu\ ;\ \nu_{\mathfrak{m}} \le \mu \le \nu\}$ is a totally ordered set isomorphic to an interval in $\overline{\mathbf{R}}_+$.

We may pick local coordinates (x, y) such that $1 = \nu(x) \le \nu(y)$. Write $\nu = \mathrm{val}[(U_j)_0^k; (\tilde\beta_j)_0^k]$. First assume $k < \infty$. Set $d_j = \deg_y(U_j)$ for $1 \le j \le k$ and

$d_0 = \infty$ by convention. Recall that the sequence $(\tilde{\beta}_j/d_j)$ is strictly increasing (see Lemma 2.4). We claim that I is isomorphic to the interval $J = [1, \tilde{\beta}_k/d_k]$. To see this, pick $t \in J$. There exists a unique integer $l \in [1, k]$ such that $\tilde{\beta}_{l-1}/d_{l-1} < t \leq \tilde{\beta}_l/d_l$. Set $\nu_t = \mathrm{val}[(U_j)_0^l; (\tilde{\beta}_j)_{j=0}^{l-1}, td_l]$. Proposition 3.18 then shows that this gives an isomorphism from J onto I. The case $k = \infty$ is treated in a similar way.

As for (T3), it is easier to prove the equivalent statement (T3') given in Remark 3.3. Moreover (T3') clearly follows if we can prove that every totally ordered subset of $\mathcal{V}$ has a majorant in $\mathcal{V}$. This will in fact also prove that $\mathcal{V}$ is a complete tree. Thus consider such a totally ordered subset $\mathcal{S} \subset \mathcal{V}$. We may pick local coordinates (x, y) such that $1 = \nu(x) \leq \nu(y)$ for every $\nu \in \mathcal{S}$. By Proposition 3.18 the SKP defining $\nu \in \mathcal{S}$ has a length that is a nondecreasing function of ν. When $\mathrm{length}\,\nu \to \infty$, Theorem 2.22 shows that ν tends to a curve valuation or an infinitely singular valuation dominating all the ν's. Otherwise $\mathrm{length}\,\nu$ is constant for large ν, $\nu \in \mathcal{S}$, and we can write $\nu = \mathrm{val}[(U_j)_0^n; (\tilde{\beta}_j)_0^{n-1}, \tilde{\beta}_n^{(\nu)}]$. By Proposition 3.18, $\tilde{\beta}_n^{(\nu)}$ is an increasing function of ν, hence is dominated by some $\tilde{\beta}_n \in \overline{\mathbf{R}}_+$. The valuation $\mu = \mathrm{val}[(U_j)_0^n; (\tilde{\beta}_j)_0^n]$ dominates all the ν's. Thus $\mathcal{V}$ is a complete tree. ☐

3.2.2 Dendrology

We now undertake a more detailed study of the nonmetric tree structure on valuation space. Recall from Section 2.2 that we have classified the valuations in $\mathcal{V}$ into four categories: divisorial, irrational, infinitely singular and curve valuations. Moreover, we called a valuation quasimonomial if it is either divisorial or irrational; see Definition 2.23.

This classification was defined in terms of SKP's and a priori depended on a choice of local coordinates. We saw in Theorem 2.28 that the classification could also be formulated in terms of numerical invariants and in particular did not depend on coordinates. Here we show how the nonmetric tree structure interacts with the classification.

Proposition 3.20. *The rooted, nonmetric tree structure on valuation space $\mathcal{V}$ has the following properties:*

(i) the root of $\mathcal{V}$ is the multiplicity valuation $\nu_{\mathfrak{m}}$;
(ii) the ends of $\mathcal{V}$ are the infinitely singular and curve valuations;
(iii) any tangent vector in $\mathcal{V}$ is represented by a curve valuation as well as by an infinitely singular valuation;
(iv) the regular points of $\mathcal{V}$ are the irrational valuations;
(v) the branch points of $\mathcal{V}$ are the divisorial valuations. Further, the tangent space at a divisorial valuation is in bijection with $\mathbf{P}^1$.

Remark 3.21. The bijection in (v) will be made much more precise in Appendix B. In particular we will show that each tangent vector has an interpretation both as a point on a rational curve and as a Krull valuation.

Remark 3.22. The nonmetric tree structure on $\mathcal{V}$ does not allow us to distinguish between a curve valuation and an infinitely singular valuation. In Section 3.4 we will define the *multiplicity* of a valuation. This multiplicity is an increasing function on $\mathcal{V}$ with values in $\overline{\mathbf{N}}$ and the infinitely singular valuations are the ones with infinite multiplicity.

Proof (Proposition 3.20). We have already proved (i). For (ii), let $\nu \in \mathcal{V}$ be quasimonomial and pick local coordinates (x, y). Then $\nu = \mathrm{val}[(U_j)_0^k; (\tilde{\beta}_j)_0^k]$, with $k < \infty$ and $\tilde{\beta}_k < \infty$. By Proposition 3.18, ν is dominated by any valuation of the form $\mathrm{val}[(U_j)_0^k; (\tilde{\beta}_j)_0^{k-1}, \tilde{\beta}'_k]$, where $\tilde{\beta}'_k > \tilde{\beta}_k$. Thus no quasimonomial valuation is an end in $\mathcal{V}$. The same proposition also shows that no valuation with an SKP of infinite length or with length $k < \infty$ and $\tilde{\beta}_k = \infty$ can be dominated by another valuation. Thus curve valuations and infinitely singular valuations are ends in $\mathcal{V}$, proving (ii).

The proof of (iii) is similar. Indeed, consider $\nu \in \mathcal{V}$ and a tangent vector $\vec{v}$ at ν. First assume $\vec{v}$ is not represented by $\nu_{\mathfrak{m}}$. By (ii), ν is then quasimonomial and $\vec{v}$ is represented by another quasimonomial valuation μ. Pick local coordinates (x, y) and write $\mu = \mathrm{val}[(U_j)_0^k; (\tilde{\beta}_j)_0^k]$, where $k, \tilde{\beta}_k < \infty$. The curve valuation $\mathrm{val}[(U_j)_0^k; (\tilde{\beta}_j)_0^{k-1}, \infty]$ then dominates μ and hence represents $\vec{v}$. We can also construct an infinitely singular valuation dominating μ: pick $\tilde{\beta}'_k \in \mathbf{Q}$ with $\tilde{\beta}'_k \geq \tilde{\beta}_k$ and inductively extend the SKP $[(U_j)_0^k; (\tilde{\beta}_j)_0^{k-1}, \tilde{\beta}'_k]$ to an infinite SKP $[(U_j)_0^\infty; (\tilde{\beta}'_j)_0^\infty]$ where $\tilde{\beta}'_j = \tilde{\beta}_j$ for $0 \leq j < k$ and such that $d_j := \deg_y(U_j) \to \infty$. The valuation $\mathrm{val}[(U_j)_0^\infty; (\tilde{\beta}'_j)_0^\infty]$ is infinitely singular and dominates μ; hence it represents $\vec{v}$.

If $\vec{v}$ if represented by $\nu_{\mathfrak{m}}$, then $\nu \neq \nu_{\mathfrak{m}}$. Assume the local coordinates have been picked so that $\nu(y) \geq \nu(x) = 1$. By Corollary 3.19 the curve valuation ν_x with SKP $[(x, y); (\infty, 1)]$ represents a tangent vector $\vec{w}$ at $\nu_{\mathfrak{m}}$ different from the one represented by ν. From the preceding argument, $\vec{w}$ is represented by both a curve valuation (e.g. ν_y) and by an infinitely singular valuation. These valuations then also represent the tangent vector $\vec{v}$. Thus we have proved (iii).

Next consider a quasimonomial valuation ν and a tangent vector $\vec{v} \in T\nu$ represented by a curve valuation ν_C. First assume that $\nu_C \wedge \nu < \nu$. Then the segment $]\nu, \nu_C]$ intersects $]\nu, \nu_{\mathfrak{m}}]$ at $\nu_C \wedge \nu$ so $\vec{v}$ is the tangent vector represented by $\nu_{\mathfrak{m}}$.

Otherwise, $\nu_C > \nu$. First suppose ν is irrational. We claim that all curve valuations ν_C with $\nu_C > \nu$ represent the same tangent vector. Indeed, in general, if ν_C and ν_D are distinct curve valuations, it follows from Corollary 3.19 that $\nu_C \wedge \nu_D$ is divisorial. Thus if $\nu_C, \nu_D > \nu$ and ν is irrational, then $\nu_C \wedge \nu_D > \nu$ so ν_C and ν_D represent the same tangent vector at ν. This proves (iv).

When ν is divisorial, write $\nu = \mathrm{val}[(U_j)_0^k, (\tilde{\beta}_j)_0^k]$ with $k < \infty$ and $\tilde{\beta}_k \in \mathbf{Q}$. For $\mu > \nu$, either $\mu(U_k) > \nu(U_k) = \tilde{\beta}_k$, or $\mu(U_k) = \tilde{\beta}_k$. In the former case, μ represents the same vector as $\nu_\infty = \mathrm{val}[(U_j)_0^k, (\tilde{\beta}_j)_0^{k-1}, \infty]$. In the latter case, Theorem 2.8 (Q2) shows that there exists a polynomial $U_{k+1} = U_k^{n_k} - \theta \prod_0^{k-1} U_j^{m_{kj}}$ with $\theta = \theta(\mu) \in \mathbf{C}^*$, such that if $\phi \in R$ then $\mu(\phi) > \nu(\phi)$ iff

U_{k+1} divides ϕ in $\mathrm{gr}_\mu\, \mathbf{C}(x)[y]$. Define $\nu_\theta = \nu_{U_{k+1}}$ and $\tilde{\beta}_{k+1} = \mu(U_{k+1})$. Then $\mu \wedge \nu_\theta \geq \mathrm{val}[(U_j)_0^{k+1}, (\tilde{\beta}_j)_0^{k+1}] > \nu$ so that μ and ν_θ define the same tangent vector. Conversely $\nu_\theta \wedge \nu_{\theta'} = \nu$ as soon as $\theta \neq \theta'$. The set of tangent vectors at ν is hence in bijection with $\{\nu_{\mathfrak{m}}, \nu_\infty, \nu_\theta\}_{\theta \in \mathbf{C}^*}$ which is in bijection with $\mathbf{P}^1$, proving (v). □

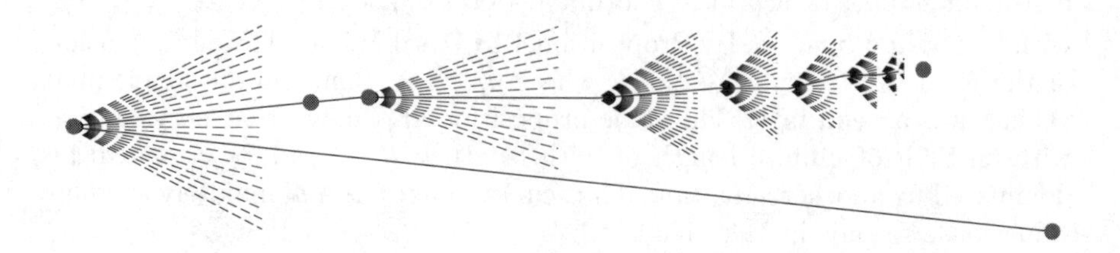

Fig. 3.3. The nonmetric tree structure on the valuative tree $\mathcal{V}$. The valuations marked by dots are, from left to right: the multiplicity valuation $\nu_{\mathfrak{m}}$, an irrational valuation, a divisorial valuation, an infinitely singular valuation and a curve valuation.

3.2.3 A Model Tree for $\mathcal{V}$

The identification of valuations with SKP's gives an explicit model for the tree structure on $\mathcal{V}$. Namely, let $\mathcal{T}$ be the set consisting of (0) and of pairs $(\bar{s}, \bar{\theta})$ with $\bar{s} = (s_1, \ldots, s_{m+1})$, $\bar{\theta} = (\theta_1, \ldots, \theta_m)$, $0 \leq m \leq \infty$, $s_j \in \mathbf{Q}_+^*$ for $1 \leq j < m+1$, $s_{m+1} \in (0, \infty]$ if $m < \infty$, $\theta_j \in \mathbf{C}^*$. Define a partial ordering on $\mathcal{T}$ by $(\bar{s}, \bar{\theta}) \leq (\bar{s}', \bar{\theta}')$ iff $m \leq m'$, $s_j = s_j'$ and $\theta_j = \theta_j'$ for $j < m$ and $s_m \leq s_m'$, and by declaring (0) to be the unique minimal element. Then $(\mathcal{T}, \leq)$ is a complete nonmetric tree rooted in (0).

Define $\imath : \mathcal{T} \to \mathcal{V}$ as follows. Fix local coordinates (x, y). First set $\imath(0) = \nu_x$. If $(\bar{s}, \bar{\theta}) \in \mathcal{T}$, then set $\tilde{\beta}_0 = 1/\min\{1, s_1\}$, $\tilde{\beta}_1 = s_1/\min\{1, s_1\}$, and define, inductively, $n_k = \min\{n\ ;\ n\tilde{\beta}_k \in \sum_0^{k-1} \tilde{\beta}_j \mathbf{Z}\}$ and $\tilde{\beta}_{k+1} := n_k \tilde{\beta}_k + s_{k+1}$. Define U_k by (P2) and set $\imath(\bar{s}, \bar{\theta}) := \mathrm{val}[(U_k); (\tilde{\beta}_k)]$. Proposition 3.18 shows that $\imath$ is an isomorphism of nonmetric trees.

The map $\imath$ is in fact an isomorphism of *rooted* nonmetric trees if we take ν_x to be the root of $\mathcal{V}$. We shall study this situation more closely in Section 3.9.

3.3 Parameterization of $\mathcal{V}$ by Skewness

While the nonmetric tree structure on $\mathcal{V}$ induced by the partial ordering is quite appealing, it only reflects some of the features of the valuative tree. For applications it is crucial to *parameterize* $\mathcal{V}$. As we will see, there are two canonical parameterizations. Here we will discuss the parameterization by skewness. The second one—thinness—will be introduced in Section 3.6 after we have defined the concept of the multiplicity of a valuation.

3.3.1 Skewness

We first introduce a new invariant of a valuation. It measures in an intrinsic way how far the valuation is from the multiplicity valuation $\nu_{\mathfrak{m}}$.

Definition 3.23. *For $\nu \in \mathcal{V}$, define the* skewness $\alpha(\nu) \in [1, \infty]$ *by*

$$\alpha(\nu) := \sup \left\{ \frac{\nu(\phi)}{m(\phi)} \; ; \; \phi \in \mathfrak{m} \right\}. \tag{3.1}$$

This quantity is an invariant of a valuation, in the following sense:

Proposition 3.24. *If $f : (\mathbf{C}^2, 0) \to (\mathbf{C}^2, 0)$ is an invertible formal map, and $f_* : \mathcal{V} \to \mathcal{V}$ the induced map, then $\alpha(f_*\nu) = \alpha(\nu)$ for any $\nu \in \mathcal{V}$.*

Proof. This is immediate since

$$\alpha(f_*\nu) = \sup_{\phi \in \mathfrak{m}} \frac{\nu(f^*\phi)}{m(f^*\phi)} = \sup_{\psi \in \mathfrak{m}} \frac{\nu(\psi)}{m(\psi)} = \alpha(\nu).$$

□

A main reason why skewness is useful is the following result that will be used repeatedly in the sequel.

Proposition 3.25. *For any valuation $\nu \in \mathcal{V}$ and any irreducible $\phi \in \mathfrak{m}$ we have*

$$\nu(\phi) = \alpha(\nu \wedge \nu_\phi) m(\phi).$$

In particular $\nu(\phi) \leq \alpha(\nu) m(\phi)$ with equality iff $\nu_\phi \geq \nu$.

3.3.2 Parameterization

As the following result asserts, skewness provides a good parameterization of the valuative tree.

Theorem 3.26. *Skewness defines a parameterization $\alpha : \mathcal{V} \to [1, \infty]$ of the valuative tree $\mathcal{V}$ rooted in the multiplicity valuation $\nu_{\mathfrak{m}}$. Moreover:*

(i) if ν is a divisorial valuation, then $\alpha(\nu)$ is rational;
(ii) if ν is an irrational valuation, then $\alpha(\nu)$ is irrational;
(iii) if ν is a curve valuation, then $\alpha(\nu) = \infty$;
(iv) if ν is infinitely singular, then $\alpha(\nu) \in (1, \infty]$.

It follows that skewness also defines a parameterization $\alpha : \mathcal{V}_{\mathrm{qm}} \to [1, \infty)$. We refer to Appendix A for a construction of an infinitely singular valuation with prescribed skewness $t \in (1, \infty]$.

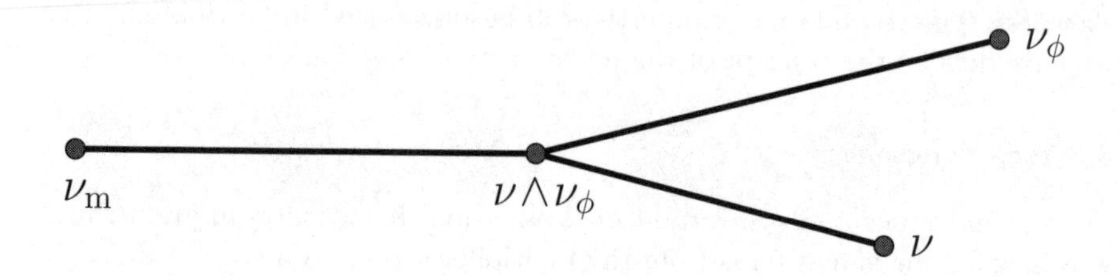

Fig. 3.4. The value $\nu(\phi)$ for $\phi \in \mathfrak{m}$ irreducible depends on the multiplicity $m(\phi)$ and the relative position of ν and the curve valuation ν_ϕ in the valuative tree $\mathcal{V}$. See Proposition 3.25.

Definition 3.27. *Fix an irreducible curve C and pick $t \in [1, \infty]$. We denote by $\nu_{C,t}$ the unique valuation in the segment $[\nu_{\mathfrak{m}}, \nu_C]$ having skewness $\alpha(\nu) = t$. If $C = \{\phi = 0\}$ for $\phi \in \mathfrak{m}$ irreducible, then we also write $\nu_{\phi,t} = \nu_{C,t}$. See Figure 3.5.*

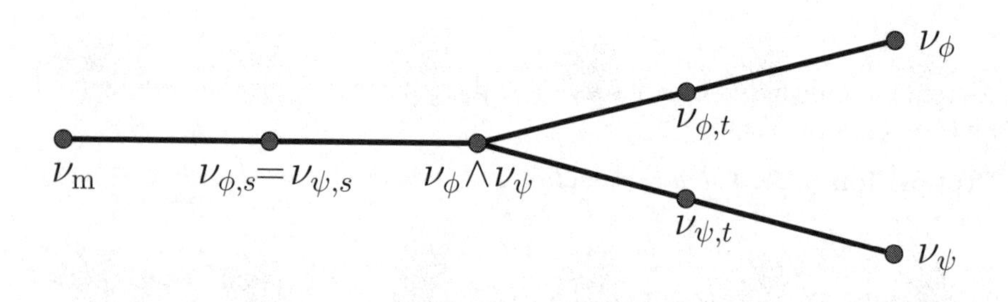

Fig. 3.5. Skewness is used to parameterize the segment $[\nu_{\mathfrak{m}}, \nu_\phi]$ by $\nu_{\phi,t}$, $1 \le t \le \infty$. We have $\nu_{\phi,t} = \nu_{\psi,s}$ iff $s = t \le \alpha(\nu_\phi \wedge \nu_\psi)$. See Theorem 3.26 and Definition 3.27.

Remark 3.28. The parameterization of $\mathcal{V}$ by skewness does not allow us to distinguish between curve valuations associated to smooth and singular curves since both have infinite skewness, as do some infinitely singular valuations. The distinction will be made in Section 3.4 by the multiplicity function on $\mathcal{V}$.

Before proving the theorem, let us note two additional results.

Corollary 3.29. *If $\nu \in \mathcal{V}$ and $\phi \in \mathfrak{m}$ is irreducible, then $\nu \ge \nu_{\phi,t}$, where $t = \nu(\phi)/m(\phi)$.*

Proof. By Proposition 3.25 we have $t = \alpha(\nu \wedge \nu_\phi)$. Since $\nu \wedge \nu_\phi \in [\nu_{\mathfrak{m}}, \nu_\phi]$ we must have $\nu \wedge \nu_\phi = \nu_{\phi,t}$. Thus $\nu \ge \nu_{\phi,t}$. □

Corollary 3.30. *If $\nu \in \mathcal{V}$, $\phi \in \mathfrak{m}$ is irreducible and $\nu(\phi)$ is irrational, then $\nu = \nu_{\phi,t}$ where $t = \nu(\phi)/m(\phi)$.*

Proof. Define $t = \nu(\phi)/m(\phi)$. By Corollary 3.29 we have $\nu \geq \nu_{\phi,t}$. Since $\nu_{\phi,t}$ is irrational, either equality holds, or else $\nu \geq \nu_{\phi,s}$ for some $s > t$, in which case $\nu(\phi) \geq sm(\phi) > tm(\phi)$, a contradiction. □

Proposition 3.31. *Skewness $\alpha : \mathcal{V} \to [1, \infty]$ is lower semicontinuous, but not continuous, in the weak tree topology on $\mathcal{V}$.*

Proof. Lower semicontinuity follows from Proposition 3.8 and the fact that skewness defines a parameterization of $\mathcal{V}$. Consider the sequence $\nu_n = \nu_{y-nx,2}$ in $\mathcal{V}$. We have $\alpha(\nu_n) = 2$ for all n, but $\nu_n \to \nu_{\mathfrak{m}}$ and $\alpha(\nu_{\mathfrak{m}}) = 1$. Thus skewness is not weakly continuous. □

3.3.3 Proofs

The proofs of all the statements about skewness go by translating them into properties of SKP's. Hence we first prove

Lemma 3.32. *Let $\nu \in \mathcal{V}$ and write $\nu = \mathrm{val}[(U_j)_0^k; (\tilde{\beta}_j)_0^k]$ in coordinates (x, y). Then the following hold:*

(i) if ν is quasimonomial, then $\alpha(\nu) = d_k^{-1}\tilde{\beta}_k\tilde{\beta}_0$; where $d_k = \deg_y U_k$; in particular $\alpha(\nu)$ is rational iff ν is divisorial;
(ii) if ν is a curve valuation, then $\alpha(\nu) = \infty$;
(iii) if ν is infinitely singular, then $\alpha(\nu) = \lim_{j\to\infty} d_j^{-1}\tilde{\beta}_j\tilde{\beta}_0 \in (1, \infty]$.

Proof. If $\nu = \nu_\phi$ is a curve valuation, then $\alpha(\nu) \geq \nu(\phi)/m(\phi) = \infty$, which proves (ii). As for (i) and (iii) note that it suffices to use ϕ irreducible in (3.1). Indeed, if $\nu(\phi) \leq \alpha\ m(\phi)$ for $\phi \in \mathfrak{m}$ irreducible and $\phi = \prod \phi_i$ is reducible, then

$$\nu(\phi) = \sum_i \nu(\phi_i) \leq \alpha \sum_i m(\phi_i) = \alpha\, m(\phi).$$

So assume ϕ irreducible and write $\nu_\phi = \mathrm{val}[(U_j^\phi); (\tilde{\beta}_j^\phi)]$. Let $l = \mathrm{con}(\nu, \nu_\phi)$ as defined in (2.13). Then $l \leq k$. Recall that the sequence $(d_j^{-1}\tilde{\beta}_j)_1^k$ is increasing. We apply Proposition 2.34:

$$\frac{\nu(\phi)}{m(\phi)} = d_l^{-1} \min\{\tilde{\beta}_l, \tilde{\beta}_l^\phi\} \min\{\tilde{\beta}_0, \tilde{\beta}_0^\phi\} \leq \sup_{j\leq k} d_j^{-1}\tilde{\beta}_j\tilde{\beta}_0. \tag{3.2}$$

If ν is quasimonomial, i.e. $k < \infty$, then equality holds in (3.2) if $l = k$ and $\tilde{\beta}_k^\phi = \tilde{\beta}_k$, proving (i).

If ν is infinitely singular, so that $k < \infty$, then for any j we may pick ϕ with $l = j$ and $\tilde{\beta}_l^\phi = \tilde{\beta}_l$. Then $\nu(\phi)/m(\phi) \geq d_j^{-1}\tilde{\beta}_j\tilde{\beta}_0$. Letting $j \to \infty$ yields (iii). □

Proof (Theorem 3.26). Assertions (i)-(iv) follow from Lemma 3.32. To show that α defines a parameterization it suffices to show that if ν is an end in $\mathcal{V}$, i.e. a curve or infinitely singular valuation, then α restricts to an increasing mapping of $[\nu_{\mathrm{m}}, \nu[$ onto $[1, \alpha(\nu)[$.

Pick local coordinates (x, y) such that $\nu(x) = 1 \leq \nu(y)$. In terms of SKP's, write $\nu = \mathrm{val}[(U_j)_0^k; (\tilde{\beta}_j)_0^k]$. Here either $k = \infty$ or $\tilde{\beta}_k = \infty$. Let $d_l = \deg_y U_l$. Recall that the sequence $(\tilde{\beta}_l/d_l)_1^k$ is strictly increasing. By Proposition 3.18 any valuation $\mu \in]\nu_{\mathrm{m}}, \nu[$ can be written (uniquely) in the form

$$\mu = \mathrm{val}[(U_j)_0^l; (\tilde{\beta}_j)_0^{l-1}, \tilde{\beta}]$$

where $1 \leq l \leq \min\{k, \infty\}$, $\tilde{\beta} < \infty$ if $l = k < \infty$, and where $d_{l-1}^{-1} d_l \tilde{\beta}_{l-1} < \tilde{\beta} \leq \tilde{\beta}_l$. By Lemma 3.32 we have $\alpha(\mu) = \tilde{\beta}/d_l$. This easily implies that α is an increasing bijection of $[\nu_{\mathrm{m}}, \nu[$ onto $[1, \alpha(\nu)[$. □

Proof (Proposition 3.25). Keep the notation of Lemma 3.32 and its proof. If $\nu = \nu_\phi$, then $\nu \wedge \nu_\phi = \nu_\phi$ and the result follows from (ii). Otherwise $l := \mathrm{con}(\nu_\phi, \nu) < \infty$ and $\nu \wedge \nu_\phi = \mathrm{val}[(U_j)_0^l; (\tilde{\beta}_j)_0^{l-1}, \tilde{\beta}]$, where $\tilde{\beta} = \min(\tilde{\beta}_l, \tilde{\beta}_l^\phi)$. Then

$$\alpha(\nu \wedge \nu_\phi) = d_l^{-1} \tilde{\beta} \tilde{\beta}_0 = \nu(\phi)/m(\phi),$$

which completes the proof. □

Remark 3.33. Skewness is closely related to *volume* as defined in [ELS]. Indeed, if we define $\mathrm{Vol}(\nu) = \limsup_{c \to \infty} 2c^{-2} \dim_{\mathbf{C}} R/\{\nu \geq c\}$, then we have

$$\mathrm{Vol}(\nu) = \alpha(\nu)^{-1}. \tag{3.3}$$

We sketch the proof, referring to [ELS, Example 3.15] for details. Write $\nu = \mathrm{val}[(U_j)_0^k; (\tilde{\beta}_j)_0^k]$ and suppose for simplicity that k is finite. For any $c > 0$, a basis for the vector space $R/\{\nu \geq c\}$ is given by the monomials $U_0^{r_0} \ldots U_k^{r_k}$ where $0 \leq r_j < n_j$ for $1 \leq j \leq k-1$, and $\sum_0^k r_j \tilde{\beta}_j < c$. The number of such monomials is given (up to a bounded function) by $\prod_1^{k-1} n_j \cdot \mathrm{Area}\{r_0 + \tilde{\beta}_k r_k \leq c\ ;\ r_0, r_k \geq 0\} \simeq d_k c^2 / 2\tilde{\beta}_k$. This gives (3.3).

3.3.4 Tree Metrics

Skewness defines a parameterization of the valuative tree. As we have seen, there is a close connection between parameterized trees and metric trees. Here we use this relationship to define natural tree metrics both on the valuative tree $\mathcal{V}$ and on the subtree $\mathcal{V}_{\mathrm{qm}}$ consisting of quasimonomial valuations.

We start with the quasimonomial case. For $\nu, \mu \in \mathcal{V}_{\mathrm{qm}}$ set

$$d_{\mathrm{qm}}(\mu, \nu) = (\alpha(\mu) - \alpha(\mu \wedge \nu)) + (\alpha(\nu) - \alpha(\mu \wedge \nu)). \tag{3.4}$$

This makes sense since any quasimonomial valuation has finite skewness. As an immediate consequence of Proposition 3.10 we have

Theorem 3.34. *The metric d_{qm} defines a metric tree structure on the set $\mathcal{V}_{\mathrm{qm}}$ of quasimonomial valuations.*

As a general valuation can have infinite skewness, (3.4) does not define a metric on $\mathcal{V}$, at least not in a standard sense. This problem can be resolved by postcomposing skewness by a positive, monotone, bounded function on $[1, \infty]$. In view of a later application (see Theorem 3.57) we use the function $\alpha \mapsto \alpha^{-1}$. Hence we define

$$d(\mu, \nu) = \left(\frac{1}{\alpha(\mu \wedge \nu)} - \frac{1}{\alpha(\mu)}\right) + \left(\frac{1}{\alpha(\mu \wedge \nu)} - \frac{1}{\alpha(\nu)}\right). \tag{3.5}$$

for any valuations ν, μ in $\mathcal{V}$.

Theorem 3.35. *The metric d gives valuation space $\mathcal{V}$ the structure of a metric tree. Further, $(\mathcal{V}, d)$ is complete.*

Proof. A simple adaptation of the proof of Proposition 3.10 shows that $(\mathcal{V}, d)$ is a metric tree. Completeness of $(\mathcal{V}, d)$ follows from completeness of $\mathcal{V}$ as a nonmetric tree. □

Remark 3.36. Proposition 3.24 shows that the metrics d and d_{qm} are invariant metrics in the sense that if $f : (\mathbf{C}^2, 0) \to (\mathbf{C}^2, 0)$ is an invertible formal map, then the induced maps $f_* : \mathcal{V} \to \mathcal{V}$ and $f_* : \mathcal{V}_{\mathrm{qm}} \to \mathcal{V}_{\mathrm{qm}}$ are isometries for $(\mathcal{V}, d)$ and $(\mathcal{V}_{\mathrm{qm}}, d_{\mathrm{qm}})$, respectively.

3.4 Multiplicities

Consider the maximal, increasing function $m : \mathcal{V} \to \overline{\mathbf{N}}$ such that $m(\nu_C) = m(C)$ for every irreducible formal curve C. Concretely, this is given as follows: the *multiplicity* $m(\nu)$ of a quasimonomial valuation $\nu \in \mathcal{V}_{\mathrm{qm}}$ is defined by

$$m(\nu) = \min\{m(C) \ ; \ C \text{ irreducible}, \ \nu_C \geq \nu\}.$$

This multiplicity can be extended to arbitrary valuations in $\mathcal{V}$ by observing that $\nu \mapsto m(\nu)$ is increasing. We then have to allow for the possibility that $m = \infty$.

Proposition 3.37. *If $\mu \leq \nu$ then $m(\mu)$ divides $m(\nu)$. Further:*

(i) $m(\nu) = \infty$ iff ν is infinitely singular;
(ii) $m(\nu) = 1$ iff ν is monomial in some local coordinates (x, y).

Proposition 3.38. *The multiplicity function $m : \mathcal{V} \to \overline{\mathbf{N}}$ is lower semicontinuous, but not continuous, in the weak topology on $\mathcal{V}$.*

We postpone the proofs to the end of this section.

We also define $m(\vec{v})$ for a tangent vector $\vec{v} \in T\nu$ as follows: if $\vec{v}$ is represented by $\nu_{\mathfrak{m}}$, then $m(\vec{v}) := m(\nu)$. Otherwise, $m(\vec{v})$ is the minimum of $m(C)$ over all ν_C representing $\vec{v}$. As there are uncountably many tangent vectors at a divisorial valuation, it is useful to know that their multiplicities are well behaved:

Proposition 3.39. *Let ν be a divisorial valuation. Then there exists a positive integer $b(\nu)$ divisible by $m(\nu)$ such that exactly one of the following holds:*

(i) all tangent vectors at ν share the same multiplicity $m(\nu)$; in this case we set $b(\nu) = m(\nu)$;
(ii) among the tangent vectors at ν, there are exactly two with multiplicity $m(\nu)$ and one of them is determined by $\nu_{\mathfrak{m}}$; all other tangent vectors at ν share a common multiplicity $b(\nu) > m(\nu)$.

Definition 3.40. *We call $b(\nu)$ the* generic multiplicity *of ν.*

Remark 3.41. As we shall show in Chapter 6, the generic multiplicity $b = b(\nu)$ has the following geometric interpretation. There exists a composition of blowups $\pi : X \to (\mathbf{C}^2, 0)$ and an irreducible component E of the exceptional divisor $\pi^{-1}(0)$ such that $\pi_* \operatorname{div}_E = b\,\nu$. Moreover, if C is an irreducible curve whose strict transform intersects E transversely at a point of E which is smooth on $\pi^{-1}(0)$, then C has multiplicity $m(C) = b$. The situation $b(\nu) = m(\nu)$ happens exactly when $\pi = \pi' \circ \tilde{\pi}$ and E is the exceptional divisor of the blowup $\tilde{\pi}$ of a *smooth* point on $(\pi')^{-1}(0)$.

Let us now turn to the proofs of Propositions 3.37, 3.38 and 3.39. They all rely on a translation to statements about SKP's.

Lemma 3.42. *Consider a valuation $\nu \in \mathcal{V}$ and pick local coordinates (x, y) such that $1 = \nu(x) \le \nu(y)$. Write $\nu = \operatorname{val}[(U_j)_0^k; (\tilde{\beta}_j)_0^k]$, $1 \le k \le \infty$. Then the multiplicity $m(\nu)$ is the maximum of the degree in y of the polynomials U_j.*

Proof. By Lemma 2.5 we know that $m(U_j) = d_j = \deg_y U_j$, which is an increasing function in j. The result then follows easily from the definition of $m(\nu)$ and from Proposition 3.18. The details are left to the reader. □

Proof (Proposition 3.37). First pick $\mu \le \nu$. Write $\nu = \operatorname{val}[(U_j)_0^k; (\tilde{\beta}_j)_0^k]$, $1 \le k \le \infty$. By Proposition 3.18, we have $\nu = \operatorname{val}[(U_j)_0^{k'}; (\tilde{\beta}_j)_0^{k'-1}, \tilde{\beta}'_{k'}]$, with $k' \le k$ and $\tilde{\beta}'_{k'} \le \tilde{\beta}_{k'}$. By Lemma 3.42 and Lemma 2.4, we get $m(\mu) = \prod_0^{k'-1} n_j$ and $m(\nu) = \prod_0^{k-1} n_j$. Whence $m(\mu)$ divides $m(\nu)$.

Definition 2.23 and Lemma 3.42 immediately imply that $m(\nu) = \infty$ iff ν is infinitely singular. Thus (i) holds.

As for (ii) it is clear that if ν is monomial in coordinates (x, y), then $\nu \le \nu_x$ or $\nu \le \nu_y$, which implies $m(\nu) = 1$. Conversely, suppose that $m(\nu) = 1$. Pick an irreducible formal curve C such that $\nu_C \ge \nu$ and $m(C) = m(\nu) = 1$. We may then pick local coordinates (x, y) with $C = \{y = 0\}$. Thus $\nu_C = \nu_y$. For each $t \in [1, \infty]$, let ν_t be the monomial valuation in coordinates (x, y) with $\nu_t(x) = 1$, $\nu_t(y) = t$ in the sense of (2.4). It is then clear that $\nu_t \leq \nu_y$ and that $\alpha(\nu_t)$ has skewness t. As $\mathcal{V}$ is a tree, $\nu_t = \nu_{y,t}$. This shows that all valuations in the segment $[\nu_{\mathfrak{m}}, \nu_y]$ are monomial. In particular ν is monomial. □

Proof (Proposition 3.38). We first show that m is lower semicontinuous on each segment of $\mathcal{V}$. Pick an increasing sequence of valuations $\nu_n \to \nu$. We want to prove $m(\nu) = \lim m(\nu_n)$. To do so write $\nu = \operatorname{val}[(U_j)_0^k; (\tilde{\beta}_j)_0^k]$. Then Proposition 3.18 imply that for n large enough, $\nu_n = \operatorname{val}[(U_j)_0^k; (\tilde{\beta}_j)_0^{k-1}, \tilde{\beta}_k^n]$ with $\tilde{\beta}_k^n$ increasing to $\tilde{\beta}_k$. We conclude by Lemma 3.42 above.

Lower semicontinuity on all of $\mathcal{V}$ is then proved in much the same way as Proposition 3.8. We have to show that the superlevel set $\{\nu \,;\, m(\nu) > t\}$ is weakly open for any real t. Pick any $\nu_0 \in \mathcal{V}$ with $m(\nu_0) > t$. If $\nu_0 = \nu_{\mathfrak{m}}$, then $t < 1$ and $m(\nu) > t$ for all $\nu \in \mathcal{V}$. Hence suppose $\nu_0 \neq \nu_{\mathfrak{m}}$. Since m is lower semicontinuous and increasing on the segment $[\nu_{\mathfrak{m}}, \nu_0]$, we may find $\mu < \nu$ such that $m(\mu) = m(\nu_0) > t$. Let $\vec{v}$ be the tangent vector at μ represented by ν_0. Then $m > t$ on the open neighborhood $U(\vec{v})$ of ν_0.

Thus m is lower semicontinuous. The example $\nu_n = \nu_{(y+nx)^2+x^3}$ in coordinates (x, y) shows that it is not continuous: $m(\nu_n) = 2$, $\nu_n \to \nu_{\mathfrak{m}}$ but $m(\nu_{\mathfrak{m}}) = 1$. □

Proof (Proposition 3.39). We use the description of the tangent space at a divisorial valuation in terms of SKP's given in the proof of Proposition 3.18. Fix a divisorial valuation $\nu = \operatorname{val}[(U_j)_0^k; (\tilde{\beta}_j)_0^k]$. Let $n_k = \min\{l \,;\, l\tilde{\beta}_k \in \sum_0^{k-1} \mathbf{Z}\tilde{\beta}_j\}$, and write $n_k\tilde{\beta}_k = \sum m_{kj}\tilde{\beta}_j$ with $0 \leq m_{kj} \leq n_j$ as in (2.2). For any $\theta \in \mathbf{C}^*$ we define $U_{k+1}(\theta) = U_k^{n_k} - \theta \cdot \prod U_j^{m_{kj}}$.

Then any tangent vector at ν which is not determined by $\nu_{\mathfrak{m}}$ is represented by one and only one of the following curve valuations: ν_ϕ with $\phi = U_k$, or $\phi = U_{k+1}(\theta)$ for some $\theta \in \mathbf{C}^*$. This representation gives a bijection between $T\nu$ and $\mathbf{P}^1$.

Let us now prove the proposition. Pick coordinates (x, y) such that $\nu(y) \geq \nu(x)$. Write $\nu = \operatorname{val}[(U_j)_0^k; (\tilde{\beta}_j)_0^k]$ and pick $\vec{v} \in T\nu$. When $\vec{v}$ is represented by $\nu_{\mathfrak{m}}$, its multiplicity is $m(\nu)$ by definition. When $\vec{v}$ is represented by the curve valuation associated to U_k, the multiplicity of $\vec{v}$ is bounded from above by $m(U_k) = m(\nu)$ and hence equals $m(\nu)$.

Otherwise, $\vec{v}$ is represented by a curve valuation associated to $U_{k+1} = U_k^{n_k} - \theta \prod_0^{k-1} U_j^{m_{kj}}$, for some $\theta \in \mathbf{C}^*$. The multiplicity of $\vec{v}$ is hence at most $m(U_{k+1}) = n_k m(U_k) = n_k m(\nu)$ (see Lemma 2.5). On the other hand, pick any curve valuation ν_ϕ representing $\vec{v}$. As $\nu_\phi \geq \nu$ the SKP defining ν_ϕ starts with $(U_j)_0^k$. When $\nu_\phi(U_k) > \tilde{\beta}_k$, ν_ϕ determines the same tangent vector as U_k. Therefore $\nu_\phi(U_k) = \tilde{\beta}_k$ and the next polynomial in the SKP of ν_ϕ is of the form $U_k^{n_k} - \theta' \prod_0^{k-1} U_j^{m_{kj}}$. As ν_ϕ and U_{k+1} determine the same tangent vector, $\theta' = \theta$. The multiplicity of ϕ is the supremum of the multiplicity of the polynomials appearing in the SKP of ν_ϕ, hence $m(\phi) \geq m(U_{k+1}) = n_k m(\nu)$. We conclude that $m(\vec{v}) = n_k m(\nu)$.

When $n_k = 1$, we have proved that all tangent vectors have the same multiplicity. When $n_k \geq 2$ all tangent vectors but two have the same multiplicity, and this multiplicity is a multiple of $m(\nu)$. The remaining two tangent vectors have multiplicity $m(\nu)$. This completes the proof. □

Note that the proof gives

Lemma 3.43. *Consider a divisorial valuation $\nu \in \mathcal{V}$ and pick local coordinates (x, y) such that $1 = \nu(x) \leq \nu(y)$. Write $\nu = \mathrm{val}[(U_j)_0^k; (\tilde{\beta}_j)_0^k]$, $1 \leq k \leq \infty$. Then the generic multiplicity $b(\nu)$ is equal to the product $n_k d_k$, where $d_k = \deg_y(U_k)$ and n_k is defined by condition (P2) in Definition 2.1.*

3.5 Approximating Sequences

As the multiplicity function m is increasing, it is piecewise constant on any segment and the set of jump points is discrete. By Propositions 3.37, 3.38 and 3.39 this observation immediately gives

Proposition 3.44. *For any valuation $\nu \in \mathcal{V}$ of finite multiplicity, there exists a finite sequence of divisorial valuations ν_i and a strictly increasing sequence of integers m_i such that:*

$$\nu_{\mathfrak{m}} = \nu_0 < \nu_1 < \cdots < \nu_g < \nu_{g+1} = \nu \tag{3.6}$$

and such that $m(\mu) = m_i$ for $\mu \in]\nu_{i-1}, \nu_i]$, $1 \leq i \leq g+1$. Moreover $m_1 = 1$, m_i divides m_{i+1} and ν_i has generic multiplicity $b(\nu_i) = m_{i+1}$. See Figure 3.6.

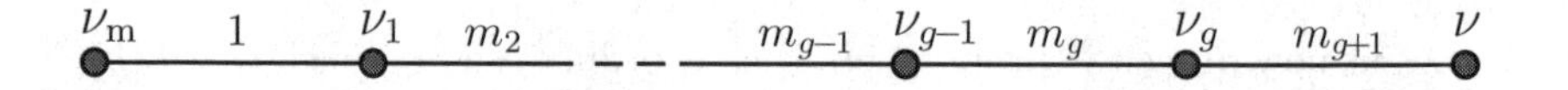

Fig. 3.6. The approximating sequence of a valuation of finite multiplicity.

Definition 3.45. *We call the sequence $(\nu_i)_{i=0}^g$ the* approximating sequence *associated to ν. The concept of approximating sequences extends naturally to infinitely singular valuations: in this case they are infinite, of the form $(\nu_i)_0^\infty$.*

In Chapter 6 we shall give a geometric interpretation of the approximating sequence: see Remark 6.44.

As we show in Appendix D.1, the approximating sequence of a curve valuation can be used to extract the classical invariants of the associated irreducible curve.

3.6 Thinness

As we have indicated above, skewness is the first out of (at least) two natural parameterizations of the valuative tree. Having defined multiplicities we are now in position to introduce the second. Namely, we define the *thinness*[5] of a valuation $\nu \in \mathcal{V}$ by

$$A(\nu) = 2 + \int_{\nu_{\mathrm{m}}}^{\nu} m(\mu)\, d\alpha(\mu). \tag{3.7}$$

Here the integral in (3.7) is defined by $\int_{\nu_{\mathrm{m}}}^{\nu} m(\nu)\, d\alpha(\mu) = \int_1^t m(\nu_{C,s})\, ds$, if $\nu = \nu_{C,t}$ is quasimonomial or a curve valuation, and as a suitable increasing limit if ν is infinitely singular.

Concretely, if ν has finite multiplicity and $(\nu_i)_{i=0}^g$ is its approximating sequence, then

$$A(\nu) = 2 + \sum_{i=1}^{g+1} m_i(\alpha_i - \alpha_{i-1}), \tag{3.8}$$

where $\alpha_i = \alpha(\nu_i)$. This formula extends naturally to infinitely singular valuations, in which case $g = \infty$.

Proposition 3.46. *Thinness defines a parameterization $A : \mathcal{V} \to [2, \infty]$ of the valuative tree. Moreover:*

(i) if ν is divisorial, then $A(\nu)$ is rational;
(ii) if ν is irrational, then $A(\nu)$ is irrational;
(iii) if ν is a curve valuation, then $A(\nu) = \infty$;
(iv) if ν is infinitely singular, then $A(\nu) \in (2, \infty]$.

It follows that thinness also defines a parameterization $A : \mathcal{V}_{\mathrm{qm}} \to [2, \infty)$ of the tree $\mathcal{V}_{\mathrm{qm}}$ of quasimonomial valuations.

We refer to Appendix A for a construction of an infinitely singular valuation with prescribed thinness $t \in (2, \infty]$.

Proof. All of this follows from (3.8) and Theorem 3.26. □

Proposition 3.47. *Thinness $A : \mathcal{V} \to [2, \infty]$ is lower semicontinuous, but not continuous, in the weak tree topology on $\mathcal{V}$.*

Proof. Lower semicontinuity follows from Proposition 3.8 and the fact that thinness defines a parameterization of $\mathcal{V}$. The same example $\nu_n = \nu_{y-nx,2}$ that was used for skewness (see Proposition 3.31) serves to show that thinness is not continuous in the weak topology. □

[5] We normalize A such that $A(\nu_{\mathrm{m}}) = 2$. This is natural in view of Remarks 3.49 and 3.50.

Let us further analyze the relationship between skewness, thinness and multiplicity. When viewed as functions on $\mathcal{V}$, any two of them determines the third. Indeed, (3.7) gives thinness in terms of skewness and multiplicity. We can then recover skewness from thinness and multiplicity:

$$\alpha(\nu) = 1 + \int_{\nu_{\mathfrak{m}}}^{\nu} \frac{1}{m(\mu)}\, dA(\mu). \tag{3.9}$$

Finally, multiplicity can easily be recovered from thinness and skewness by differentiating either (3.7) or (3.9).

Alternatively, we can try to relate skewness, thinness and multiplicity for a fixed valuation. There is no general formula, but the following elementary bounds are useful for applications.

Proposition 3.48. *For any valuation $\nu \in \mathcal{V}$ we have*

(i) $A(\nu) \geq 1 + \alpha(\nu)$ with equality iff $m(\nu) = 1$, i.e. if ν is monomial in some local coordinates;
(ii) if $m(\nu) > 1$, then $A(\nu) < m(\nu)\alpha(\nu)$.

Proof. The first statement is immediate. As for the second we use the approximating sequence $(\nu_i)_0^g$. Then $m(\nu) = m_{g+1}$, $\alpha(\nu) = \alpha_{g+1}$ and

$$\begin{aligned} A(\nu) - m(\nu)\alpha(\nu) &= 1 - (m_2 - 1)\alpha_1 - (m_3 - m_2)\alpha_2 + \cdots + (m_{g+1} - m_g)\alpha_g \\ &\leq 1 - \alpha_1 < 0. \end{aligned}$$

This concludes the proof. □

Remark 3.49. The name "thinness" was chosen with the following picture in mind. Suppose ν is quasimonomial and write $\nu = \nu_{\phi,t}$ with $m(\phi) = m(\nu)$. Assume we have picked local coordinates (x, y) with $\nu_x \wedge \nu_\phi = \nu_{\mathfrak{m}}$. Then, for $r > 0$ small, the region

$$\Omega(r) = \left\{(x,y) \in \mathbf{C}^2 \ ; \ |x| < r, \ |\phi(x,y)| < |x|^{tm(\phi)}\right\} \subset \mathbf{C}^2$$

is a small neighborhood of the curve $\phi = 0$ (with the origin removed). See Figure 3.7. A large value of t (and hence $A(\nu)$) corresponds to a "thin" neighborhood. In fact the volume of $\Omega(r)$ is roughly $r^{2A(\nu)}$. Regions of this type, called *characteristic regions*, are important in [FJ2], where they are used to define the values of ν on plurisubharmonic functions. See also Section 1.5.4.

Remark 3.50. In case ν is a divisorial valuation, thinness has the following algebraic characterization. We shall freely use results from Chapter 6. As the value group of ν is isomorphic to $\mathbf{Z}$, the maximal ideal $\mathfrak{m}_\nu$ of the valuation ring R_ν is principal, and any other ideal is a power of $\mathfrak{m}_\nu$. We may define an ideal $J_{R_\nu/R}$ inside the valuation ring R_ν called its Jacobian ideal. By definition, this is the 0-th Fitting ideal of the sheaf of differentials of R_ν over R.

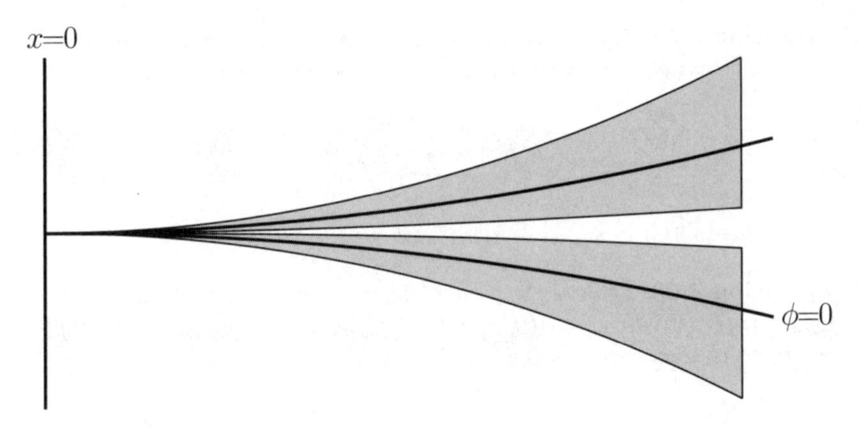

Fig. 3.7. A characteristic region

Geometrically the situation is as follows. Let π be a composition of blowups above the origin, and $E \subset \pi^{-1}(0)$ an irreducible component of the exceptional divisor $\pi^{-1}(0)$ such that $\pi_* \operatorname{div}_E = b\nu$ for some $b \in \mathbf{N}^*$. Then an element $\phi \in R_\nu \subset K$ belongs to $\mathfrak{m}_\nu$ iff $\pi^*\phi$ defines a regular function at a generic point on E and vanishes to order at least one. Similarly, $\phi \in J_{R_\nu/R}$ iff $\pi^*\phi$ defines a regular function at a generic point on E and vanishes to at least the same order as the Jacobian determinant $J\pi$. Write $a - 1 = \operatorname{div}_E(J\pi)$. Then, we have $\nu(\mathfrak{m}_\nu) = b^{-1}$, and $\nu(J_{R_\nu/R}) = (a-1)/b$. Thanks to Theorem 6.22, the Farey parameter coincides with the thinness, i.e. we have $A(\nu) = a/b$, whence $A(\nu) = \nu(J_{R_\nu/R}\,\mathfrak{m}_\nu)$.

Remark 3.51. Associated to the parameterization by thinness are tree metrics D_{qm} and D on $\mathcal{V}_{\mathrm{qm}}$ and $\mathcal{V}$, respectively. They are defined similarly to the tree metrics d_{qm} and d induced by skewness: see (3.4) and (3.5). More precisely, set

$$D_{\mathrm{qm}}(\mu, \nu) = (A(\mu) - A(\mu \wedge \nu)) + (A(\nu) - A(\mu \wedge \nu)), \tag{3.10}$$

for $\nu, \mu \in \mathcal{V}_{\mathrm{qm}}$ and

$$D(\mu, \nu) = (A^{-1}(\mu \wedge \nu) - A^{-1}(\mu)) + (A^{-1}(\mu \wedge \nu) - A^{-1}(\nu)) \tag{3.11}$$

for $\nu, \mu \in \mathcal{V}$. We shall compare these metrics with d_{qm} and d in Chapter 5.

3.7 Value Semigroups and Approximating Sequences

The approximating sequence encapsulates most of the structure of a valuation. We use it here to compute the value semigroup of a general valuation, and the generic multiplicity of a divisorial valuation. Write $R^* = R \setminus \{0\}$.

Proposition 3.52. *Let $\nu \in \mathcal{V}$ be a valuation and let $(\nu_i)_0^g$ be its approximating sequence (g possibly infinite). Then the value semigroup $\nu(R^*)$ is given by*

$$\nu(R^*) = \sum_{i=0}^{g+1} m_i \alpha_i \mathbf{N}, \tag{3.12}$$

where $\nu_{g+1} = \nu$ if $g < \infty$, $\alpha_i = \alpha(\nu_i)$ and $m_i = m(\nu_i)$.

Proposition 3.53. *Let $\nu \in \mathcal{V}$ be a divisorial valuation with approximating sequence $(\nu_i)_0^g$. Write $\alpha_i = \alpha(\nu_i)$ for $0 \le i \le g$. Then the generic multiplicity of ν can be characterized by the equations:*

$$b(\nu) = \inf\{b \in m\mathbf{N} \; ; \; b\alpha \in \sum_{i=0}^{g} m_i \alpha_i \mathbf{N}\} = \inf\{b \in m\mathbf{N} \; ; \; b\alpha \in \sum_{i=0}^{g} m_i \alpha_i \mathbf{Z}\}, \tag{3.13}$$

where $m = m(\nu)$, $\alpha = \alpha(\nu)$ and where we set $m_0 = 1$ by convention.

Notice that this implies that for $1 \le j \le g$, m_{j+1} is the smallest integer divisible by m_j such that $m_{j+1}\alpha_j \in \sum_{i=0}^{j-1} m_i \alpha_i \mathbf{N}$.

Proposition 3.54. *Consider a valuation $\nu \in \mathcal{V}$. Then the following hold:*

(i) if ν is a curve valuation with multiplicity $m = m(\nu)$, then $\nu(R^*) \subset m^{-1}\overline{\mathbf{N}}$; moreover, if $a \in \mathbf{N}$, then $a\,\nu(R^*) \subset \overline{\mathbf{N}}$ iff m divides a;
(ii) if ν is divisorial with generic multiplicity $b = b(\nu)$, then $\nu(R^*) \subset b^{-1}\mathbf{N}$; moreover, if $a \in \mathbf{N}$, then $a\,\nu(R^*) \subset \mathbf{N}$ iff b divides a;
(iii) if ν is irrational with multiplicity $m = m(\nu)$ and skewness $\alpha = \alpha(\nu)$, then $\nu(R^*) \subset \mathbf{Q} + m\alpha\mathbf{N}$ but $\nu(R^*) \not\subset \mathbf{Q}$; moreover, if $a \in \mathbf{N}$, then $a\,\nu(R^*) \subset \mathbf{N} + \alpha\mathbf{N}$ iff m divides a;
(iv) if ν is infinitely singular, then $\nu(R^*) \subset \mathbf{Q}$ but $\nu(R^*)$ is not finitely generated.

The proofs of these three propositions are based on the representation of a valuation by an SKP. Before embarking on the individual proofs, we note some common facts regarding the translation of approximating sequences into SKP's.

We are working with a fixed valuation $\nu \in \mathcal{V}$. Fix local coordinates (x, y) such that $\nu(y) \ge \nu(x) = 1$. Write $\nu = \mathrm{val}[(U_j)_0^k; (\tilde{\beta}_j)_0^k]$ with $\tilde{\beta}_1 \ge \tilde{\beta}_0 = 1$. Also write $d_j = \deg_y U_j$. Define n_j as in (2.2), i.e. $n_j := \min\{n \in \mathbf{N} \; ; \; n\tilde{\beta}_j \in \sum_{j'=0}^{j-1} \tilde{\beta}_{j'}\mathbf{Z}\}$. Then $d_k = \prod_0^{k-1} n_j$. Define a sequence $(k_i)_0^{g+1}$ by $k_0 = 0$ and by the conditions $n_{k_i} \ge 2$ and $n_j = 1$ if $k_i < j < k_{i+1}$.

By the proof of Proposition 3.44 we have $\nu_i = \mathrm{val}[(U_j)_0^{k_i}; (\tilde{\beta}_j)_0^{k_i}]$ and $m_i := m(\nu_i) = d_{k_i}$. Moreover, by Lemma 3.32, $\alpha_i := \alpha(\nu_i) = \tilde{\beta}_{k_i}/d_{k_i}$. Also recall that m_{i+1} is the generic multiplicity of ν_i for $1 \le i < g+1$. In particular, if ν is divisorial, then its generic multiplicity is given by $b(\nu) = d_k n_k = \prod_1^k n_j$.

Proof (Proposition 3.52). By Theorem 2.28, $\nu(R^*) = \sum_0^k \tilde{\beta}_j \mathbf{N}$. But if $k_i < j < k_{i+1}$, then $n_j = 1$, i.e. $\tilde{\beta}_j \in \sum_{j'<j} \tilde{\beta}_{j'} \mathbf{N}$. This gives $\nu(R^*) = \sum_0^{g+1} \tilde{\beta}_{k_i} \mathbf{N}$ which amounts to $\nu(R^*) = \sum_0^{g+1} m_i \alpha_i \mathbf{N}$ using the translation above. □

Proof (Proposition 3.53). We have $b(\nu) = d_k n_k$. Here $n_k = \min\{a \in \mathbf{N} \; ; \; a\tilde{\beta}_k \in \sum_{j=0}^{k-1} \tilde{\beta}_j \mathbf{Z}\}$. As in the previous proof, we have $\sum_0^{k-1} \tilde{\beta}_j \mathbf{Z} = \sum_0^g m_i \alpha_i \mathbf{Z}$. Moreover, $\tilde{\beta}_k = m\alpha$, so multiplying with $d_k = m$ we get

$$b(\nu) = \min\{b \in m\mathbf{N} \; ; \; b\alpha \in \sum_0^g m_i \alpha_i \mathbf{Z}\}.$$

Finally we can replace $\mathbf{Z}$ by $\mathbf{N}$ by Lemma 2.7. □

Proof (Proposition 3.54). We first consider (ii), so assume ν is divisorial. Then $k < \infty$. Define $a = \min\{a \in \mathbf{N} \; ; \; a\nu(R^*) \subset \mathbf{N}\}$. Then (ii) will follow if we can show that $a = b(\nu)$. By Theorem 2.28 we have $\nu(R^*) = \sum_0^k \tilde{\beta}_j \mathbf{N}$. Write $\tilde{\beta}_j = r_j / s_j$, where r_j and s_j are integers without common factor and set $S_j = \mathrm{lcm}\{s_0, s_1, \ldots, s_j\}$ for $0 \le j \le k$. Then $S_0 = 1$ and $a = S_k^{-1}$. Moreover, it is an elementary arithmetic fact that $\sum_0^j \tilde{\beta}_{j'} \mathbf{Z} = S_j^{-1} \mathbf{Z}$, although $\sum_0^j \tilde{\beta}_{j'} \mathbf{N} \subsetneq S_j^{-1} \mathbf{N}$ in general. Since $b(\nu) = \prod_1^k n_j$ we are done if we can show that $n_j = S_j / S_{j-1}$ for $1 \le j \le k$. But

$$\begin{aligned} n_j = \min\{n \in \mathbf{N} \; ; \; n\tilde{\beta}_j \in \sum_0^{j-1} \tilde{\beta}_j \mathbf{Z}\} &= \min\{n \in \mathbf{N} \; ; \; n\tilde{\beta}_j \in S_{j-1}^{-1} \mathbf{Z}\} \\ = \min\{n \in \mathbf{N} \; ; \; n r_j S_{j-1} \in s_j \mathbf{Z}\} &= \min\{n \in \mathbf{N} \; ; \; n S_{j-1} \in s_j \mathbf{Z}\} \\ &= \mathrm{lcm}\{s_j, S_{j-1}\}/S_{j-1} = S_j / S_{j-1}. \end{aligned}$$

This completes the proof of (ii).

Next suppose that ν is a curve valuation. Let $(\nu_i)_0^g$ be its approximating sequence. Here $0 \le g < \infty$. It follows from Proposition 3.52 that $\nu(R^*) = \nu_g(R^*) \cup \{\infty\}$. Moreover, ν_g is a divisorial valuation with generic multiplicity $b(\nu_g) = m(\nu)$. Hence (i) follows from(ii).

As for (iii), suppose ν is irrational, and let $(\nu_i)_0^g$ be its (finite) approximating sequence. Then (3.12) gives $\nu(R^*) = \nu_g(R^*) + m\alpha\mathbf{N}$. As ν_g is divisorial, this immediately yields $\nu(R^*) \subset \mathbf{Q} + m\alpha\mathbf{N}$ but $\nu(R^*) \not\subset \mathbf{Q}$. Further, if $a \in \mathbf{N}$, then $a\,\nu(R^*) \subset \mathbf{N} + \alpha\mathbf{N}$ iff $a\,\nu_g(R^*) \subset \mathbf{N}$, which by (ii) is the case iff $b(\nu_g) = m(\nu)$ divides a.

Finally consider an infinitely singular valuation ν, with approximating sequence $(\nu_i)_0^\infty$. Then $\alpha(\nu_i) \in \mathbf{Q}$ so (3.12) yields $\nu(R^*) \subset \mathbf{Q}$. Suppose that $\nu(R^*)$ is a finitely generated semigroup. Then there exists $b \in \mathbf{N}$ such that $b\nu(R^*) \subset \mathbf{N}$. But by (3.12) this implies $b\nu_i(R^*) \subset \mathbf{N}$, which by (i) gives that b_i divides b. But this is impossible as $b_i \to \infty$ as $i \to \infty$.

The proof is complete. □

3.8 Balls of Curves

Above we have shown that the valuative tree $\mathcal{V}$ carries a very rich structure, induced by the partial ordering, skewness, multiplicity and thinness. The key to this structure was the detailed analysis of valuations as functions on the ring R.

However, it is an intriguing fact that there are many paths leading to the valuative tree. Here we show how to identify quasimonomial valuations with balls of irreducible curves in a particular (ultra-)metric. There are then natural interpretations of the partial ordering, skewness and multiplicity on $\mathcal{V}_{\mathrm{qm}}$ as statements about balls of curves.

The description of a quasimonomial valuations as a ball of curves adds to the observation by Spivakovsky [Sp, p.109] that the classification of valuations (in dimension 2) is essentially equivalent to the classification of plane curve singularities.

3.8.1 Valuations Through Intersections

The starting point is the fact that a curve valuation acts by intersection (see Section 1.5.5). Given formal curves C, D, let $C \cdot D$ denote the intersection product between C and D (see Section 1.4). If C is irreducible, then for any $\psi \in \mathfrak{m}$:

$$\nu_C(\psi) = \frac{C \cdot \{\psi = 0\}}{m(C)}. \tag{3.14}$$

Spivakovsky [Sp, Theorem 7.2] showed that a divisorial valuation ν acts (up to normalization) by intersection with a ν-generic curve (or ν-curvette). See Section 6.6.1 for a precise statement.

Here we give a different way of realizing a valuation as the intersection with curves. In fact this works for general quasimonomial valuations.

Proposition 3.55. *If $\nu \in \mathcal{V}$ is quasimonomial and $\psi \in \mathfrak{m}$ then*

$$\nu(\psi) = \min\{\nu_C(\psi);\ C \text{ irr.},\ \nu_C \geq \nu\} = \min\left\{\frac{C \cdot \{\psi = 0\}}{m(C)}\ ;\ C \text{ irr.},\ \nu_C \geq \nu\right\}.$$

Proof. By (3.14) we only have to show the first equality. Pick $\psi \in \mathfrak{m}$. The inequality $\nu(\psi) \leq \nu_C(\psi)$ for $\nu_C \geq \nu$ is trivial. For the other inequality write $\psi = \psi_1 \cdots \psi_n$, with $\psi_i \in \mathfrak{m}$ irreducible. First assume that ν is divisorial. Pick a tangent vector $\vec{v}$ at ν which is not represented by any ν_{ψ_i} nor $\nu_{\mathfrak{m}}$, and choose C irreducible such that ν_C represents $\vec{v}$. Then $\nu_C \geq \nu$ and $\nu_C \wedge \nu_{\psi_i} = \nu$ for all i. so by Proposition 3.25 we get $\nu(\psi_i) = \nu_C(\psi_i)$ so that $\nu(\psi) = \nu_C(\psi)$.

Thus Proposition 3.55 holds for ν divisorial. But if $\nu \in \mathcal{V}$ is irrational, then we may pick ν_n divisorial with $\nu_n > \nu$ and ν_n decreasing to ν. Fix $\psi \in \mathfrak{m}$. For each n there exists an irreducible curve C_n with $\nu_{C_n} > \nu_n > \nu$ such that $\nu_{C_n}(\psi) = \nu_n(\psi)$. Since $\nu_n(\psi) \to \nu(\psi)$ we obtain the desired equality. □

3.8.2 Balls of Curves

Proposition 3.55 shows that a quasimonomial valuation ν is determined by the irreducible curves C satisfying $\nu_C \geq \nu$. We proceed to show that this gives an isometry between the subtree of quasimonomial valuations, and the set of balls of irreducible curves in a particular (ultra)-metric. Namely, let $\mathcal{C}$ be the set of local irreducible curves. Let us define a metric on $\mathcal{C}$ by

$$d_{\mathcal{C}}(C_1, C_2) = \frac{m(C_1)m(C_2)}{C_1 \cdot C_2}, \tag{3.15}$$

where $C_1 \cdot C_2$ denotes the intersection multiplicity between the curves C_1 and C_2.

Lemma 3.56. *[Ga, Corollary 1.2.3]. Equation* (3.15) *defines an ultrametric on $\mathcal{C}$. For $C_1, C_2 \in \mathcal{C}$, we have*

$$d_{\mathcal{C}}(C_1, C_2) = \frac{1}{2} d(\nu_{C_1}, \nu_{C_2}) = \frac{1}{\alpha(\nu_{C_1} \wedge \nu_{C_2})}. \tag{3.16}$$

Further, the ultrametric space $(\mathcal{C}, d_{\mathcal{C}})$ has diameter 1.

The proof is given at the end of this section.

In Section 3.1.6 we constructed a parameterized tree as well as a metric tree associated to any ultrametric space (of diameter 1). Write $\mathcal{T}_{\mathcal{C}}$ for the parameterized tree associated to $(\mathcal{C}, d_{\mathcal{C}})$. Let us recall its definition. A point in $\mathcal{T}_{\mathcal{C}}$ is a closed ball in $\mathcal{C}$ of positive radius. The partial order on $\mathcal{T}_{\mathcal{C}}$ is given by reverse inclusion. The parameterization on $\mathcal{T}_{\mathcal{C}}$ sends a ball of radius r to the real number r^{-1}, and the distance between two balls B_1, B_2 in the metric tree is given by $(r_1 - r) + (r_2 - r)$, where r_1, r_2 and r are the radii of B_1, B_2 and $B_1 \cap B_2$, respectively.

There is a natural multiplicity function on $\mathcal{T}_{\mathcal{C}}$. Its value on a ball is the minimum multiplicity of any curve in the ball.

The completion of $\mathcal{T}_{\mathcal{C}}$ can be taken in the sense of nonmetric trees or, equivalently, in the sense of metric spaces. An element in the completion is represented by a decreasing sequence of balls. The multiplicity function and the parameterization on $\mathcal{T}_{\mathcal{C}}$ both extend naturally to the completion. An element in the completion, represented by a decreasing sequence of balls, has finite multiplicity iff the sequence has nonempty intersection in $\mathcal{C}$.

Let us define a map from $\mathcal{V}_{\mathrm{qm}}$ to $\mathcal{T}_{\mathcal{C}}$ by sending a quasimonomial valuation ν to the ball $B_\nu := \{C \,;\, \nu_C > \nu\}$ (we will see shortly that this is indeed a ball). Similarly, $B \mapsto \nu_B := \wedge_{C \in B} \nu_C$ gives a map from $\mathcal{T}_{\mathcal{C}}$ to $\mathcal{V}_{\mathrm{qm}}$.

Theorem 3.57. *The mappings $\nu \mapsto B_\nu$ and $B \mapsto \nu_B$ preserve multiplicity and give inverse isomorphisms between the parameterized trees $(\mathcal{V}_{\mathrm{qm}}, \alpha)$ and $(\mathcal{T}_{\mathcal{C}}, r^{-1})$.*

Further, these mappings extend uniquely to isomorphisms between $\mathcal{V}$ and the completion of $\mathcal{T}_{\mathcal{C}}$ and induce isometries between the corresponding metric trees.

Proof. We claim that if B is centered at C and of radius $r \in (0,1]$, then $\nu_B = \nu_{C,r^{-1}}$. Indeed, if $D \in \mathcal{C}$, then $\nu_{C,r^{-1}}$ and $\nu_C \wedge \nu_D$ both belong to the segment $[\nu_{\mathfrak{m}}, \nu_C]$. Hence (3.15) implies

$$d_{\mathcal{C}}(C,D) \leq r \Leftrightarrow \alpha(\nu_C \wedge \nu_D) \geq r^{-1} \Leftrightarrow \nu_C \wedge \nu_D \geq \nu_{C,r^{-1}} \Leftrightarrow \nu_D \geq \nu_{C,r^{-1}}.$$

Since every quasimonomial valuation is of the form $\nu_{C,t}$, this shows that the mappings $B \mapsto \nu_B$ and $\nu \mapsto B_\nu$ are well defined and each others inverse. They are clearly increasing, hence define isomorphisms between the rooted, nonmetric trees $\mathcal{V}_{\mathrm{qm}}$ and $\mathcal{T}_{\mathcal{C}}$. As $\alpha(\nu_{C,r^{-1}}) = r^{-1}$, they are in fact isomorphism of parameterized trees. Multiplicity is preserved by its very definition.

The remaining facts, namely that we get an isomorphism between $\mathcal{V}$ and the completion of $\mathcal{T}_{\mathcal{C}}$ as well as isometries between the corresponding metric trees, are easy consequences and left to the reader. □

Remark 3.58. Similarly one may interpret a tree tangent vector $\vec{v}$ at a divisorial valuations $\nu \in \mathcal{V}$ in terms of irreducible curves. If $\vec{v}$ is represented by $\nu_{\mathfrak{m}}$, then $\mathcal{C} \cap U(\vec{v})$ is the complement of the ball B_ν in $\mathcal{C}$. If $\vec{v}$ is not represented by $\nu_{\mathfrak{m}}$, then $\mathcal{C} \cap U(\vec{v})$ is the *open* ball of radius $\alpha(\nu)^{-1}$ centered at any $\phi \in \mathcal{C}$ such that ν_ϕ represents $\vec{v}$.

Proof (Lemma 3.56). Write $C_i = \{\phi_i = 0\}$ for $i = 1, 2$. Let us first show (3.16). The second equality follows from (3.5) as ν_{C_1} and ν_{C_2} have infinite skewness. For the first equality we use Proposition 3.25. Assume $\nu_{C_1} \neq \nu_{C_2}$ and set $\nu = \nu_{C_1} \wedge \nu_{C_2}$. This is divisorial and $\alpha(\nu) = \nu(\phi_1)/m(C_1)$ as $\nu_{C_1} > \nu$. But $\nu(\phi_1) = \nu_{C_2}(\phi_1) = C_1 \cdot C_2/m(C_2)$. Thus (3.16) holds.

The fact that $d_{\mathcal{C}}$ is an ultrametric for which $\mathcal{C}$ has diameter 1 follows easily from the formula $d_{\mathcal{C}}(C_1, C_2) = \alpha^{-1}(\nu_{C_1} \wedge \nu_{C_2})$ and the tree structure of $\mathcal{V}$. The details are left to the reader. □

3.9 The Relative Tree Structure

We have normalized the valuations in $\mathcal{V}$ by $\nu(\mathfrak{m}) = 1$. This condition is natural when we study properties invariant by the group of (formal) automorphisms of $(\mathbf{C}^2, 0)$. However, in some situations, other normalization may be more natural. We shall refrain from undertaking a general study of normalizations, but rather focus on one particular, but important, case.

More precisely, we will describe the set of valuations $\mathcal{V}_x$ normalized by the condition $\nu(x) = 1$, where $x \in \mathfrak{m}$ and $m(x) = 1$, i.e. $\{x = 0\}$ defines a smooth formal curve at the origin in $\mathbf{C}^2$. By going back and forth between the two normalizations $\nu(x) = 1$ and $\nu(\mathfrak{m}) = 1$, we shall see that $\mathcal{V}_x$ can be endowed with a natural partial ordering $\leq_x$ which induces a nonmetric tree structure. We hence refer to $\mathcal{V}_x$ as *the relative valuative tree.* It also has two natural parameterizations, the *relative skewness* α_x, and the *relative thinness* A_x, as well as a *relative multiplicity* m_x. We shall see that the theory of the relative valuative tree is much the same as that of its nonrelative counterpart.

The relative valuative tree will appear in two different contexts in this monograph. The first place is in Chapter 4, where we describe valuations in terms of Puiseux series. The relative valuative tree appears as the quotient of the set of valuations on $\hat{k}[[y]]$ where $\hat{k}$ denotes the set of Puiseux series in x (Theorem 4.17); and also as a natural (metric) subtree of the Bruhat-Tits building of $\mathrm{PGL}_2(\mathbf{C}((x)))$ (see Section 4.6).

In Chapter 6, the relative valuative tree appears in the following situation. Let $\pi : X \to (\mathbf{C}^2, 0)$ be a composition of blowups, and p be a smooth point on the exceptional divisor $\pi^{-1}(0)$. Then π induces a map between the space of centered valuations on the local ring $\mathcal{O}_p$ at p, into the space of centered valuations on R. If the exceptional divisor is given by $\{z = 0\}$ at p, then π induces a natural isomorphism of parameterized trees between the relative valuative tree $\mathcal{V}_z$ and its image in the valuative tree $\mathcal{V}$: this image is the closure of a set of the form $U(\vec{v})$ as in Section 3.1.4. See Theorem 6.51 and Figure 6.12.

3.9.1 The Relative Valuative Tree

Fix a smooth formal curve at the origin in $\mathbf{C}^2$, say $\{x = 0\}$, where $x \in \mathfrak{m}$ and $m(x) = 1$. We let $\mathcal{V}_x$ be the set of centered valuations $\nu : R \to \overline{\mathbf{R}}_+$ vanishing on $\mathbf{C}^*$ and such that $\nu(x) = 1$, together with the valuation div_x, defined by $\mathrm{div}_x(\phi) = \max\{n \,;\, x^n \mid \phi\}$. (The center of div_x is the curve $\{x = 0\}$.) Note that $\mathcal{V}_x$ is compact for the weak topology.

While it is perfectly feasible to study $\mathcal{V}_x$ using SKP's normalized by $\tilde{\beta}_0 = 1$, we shall instead take advantage of the theory that we have already developed for $\mathcal{V}$. Indeed, when ν is a valuation normalized by $\nu(\mathfrak{m}) = 1$ and different from ν_x, then $\nu/\nu(x)$ belongs to $\mathcal{V}_x$. Conversely, $\nu \in \mathcal{V}_x \setminus \{\mathrm{div}_x\}$ implies $\nu/\nu(\mathfrak{m}) \in \mathcal{V}$. Whence

Lemma 3.59. *The map $N : \mathcal{V} \to \mathcal{V}_x$ sending ν to $\nu/\nu(x)$, and ν_x to div_x is a homeomorphism for the weak topologies.*

An immediate consequence is that, except for ν_x and div_x, the valuations ν and $N\nu$ define equivalent Krull valuations on R, hence have the same invariants $\mathrm{rk}, \mathrm{rat.rk}, \mathrm{tr.deg}$, and also the same type (see Theorem 2.28). We shall thus say a valuation in $\mathcal{V}_x$ is quasimonomial, divisorial, curve or infinitely singular, when its image in $\mathcal{V}$ has the required property. In the sequel, we denote by $\mathcal{V}_{\mathrm{qm},x}$ the set of all quasimonomial valuations in $\mathcal{V}_x$. By convention we consider $\mathcal{V}_{\mathrm{qm},x}$ to contain the valuation div_x.

For $\nu_1, \nu_2 \in \mathcal{V}_x$, we write $\nu_1 \leq_x \nu_2$ when $\nu_1(\phi) \leq \nu_2(\phi)$ for all $\phi \in R$. It is clear that div_x is the unique minimal element of $\mathcal{V}_x$.

Lemma 3.60. *Let $\nu_1, \nu_2 \in \mathcal{V}_x \setminus \{\mathrm{div}_x\}$. Then $\nu_1 \leq_x \nu_2$ if and only if we have $[\nu_x, N^{-1}\nu_1] \subset [\nu_x, N^{-1}\nu_2]$ in the valuative tree $\mathcal{V}$.*

As an immediate consequence (see Section 3.1.2), we get

Proposition 3.61. *The poset $(\mathcal{V}_x, \leq_x)$ is a complete, nonmetric tree rooted at* div_x*. The map $N : \mathcal{V} \to \mathcal{V}_x$ is an isomorphism of (nonrooted) nonmetric trees.*

We shall write $\nu \wedge_x \mu$ for the infimum of ν and μ with respect to the tree structure on $(\mathcal{V}_x, \leq_x)$.

Remark 3.62. The ends (i.e. the maximal elements) of $\mathcal{V}_x$ under $\leq_x$ are exactly of the form $N\nu$, where $\nu \in \mathcal{V}$ is not quasimonomial and $\nu \neq \nu_x$. Hence $(\mathcal{V}_{\mathrm{qm},x}, \leq_x)$ is also a nonmetric tree rooted in div_x. Recall our convention that $\mathrm{div}_x \in \mathcal{V}_{\mathrm{qm},x}$.

Remark 3.63. Both $\mathcal{V}$ and $\mathcal{V}_x$ can be viewed as subsets of the set $\tilde{\mathcal{V}}$ of centered valuations on R. It follows from Lemma 3.60 that

$$\mathcal{V} \cap \mathcal{V}_x = \{\nu \in \mathcal{V}\ ;\ N(\nu) = \nu\} = \{\nu \in \mathcal{V}\ ;\ \nu \wedge \nu_x = \nu_{\mathfrak{m}}\} = \{\nu \in \mathcal{V}_x\ ;\ \nu \geq_x \nu_{\mathfrak{m}}\}.$$

Moreover, the partial orderings $\leq$ and $\leq_x$ agree on $\mathcal{V} \cap \mathcal{V}_x$.

Proof (Lemma 3.60). Write $\mu_i = N^{-1}\nu_i$. These are valuations normalized by $\mu_i(\mathfrak{m}) = 1$. The condition $\nu_1 \leq_x \nu_2$ is equivalent to $\mu_1(\phi)/\mu_1(x) \leq \mu_2(\phi)/\mu_2(x)$ for all ϕ. Here we may assume that $\phi \in \mathfrak{m}$ is irreducible and apply Proposition 3.25:

$$\frac{\mu_1(\phi)/\mu_1(x)}{\mu_2(\phi)/\mu_2(x)} = \frac{\alpha(\mu_1 \wedge \nu_\phi)\alpha(\mu_2 \wedge \nu_x)}{\alpha(\mu_2 \wedge \nu_\phi)\alpha(\mu_1 \wedge \nu_x)}. \tag{3.17}$$

First assume $\nu_1 \leq_x \nu_2$. We need to show that $[\nu_x, \mu_1] \subset [\nu_x, \mu_2]$. By choosing ϕ such that $\nu_\phi \wedge \mu_i = \nu_{\mathfrak{m}}$ for $i = 1, 2$ we get $\mu_1 \wedge \nu_x \geq \mu_2 \wedge \nu_x$ from (3.17). If equality holds, then $\mu_1(x) = \mu_2(x)$, which together with $\nu_1 \leq_x \nu_2$ gives $\mu_1 \leq \mu_2$. From this we easily conclude $[\nu_x, \mu_1] \subset [\nu_x, \mu_2]$. If instead $\mu_1 \wedge \nu_x > \mu_2 \wedge \nu_x$, then we claim that $\mu_1 \in [\nu_x, \mu_2 \wedge \nu_x]$. Indeed, otherwise pick $\phi \in \mathfrak{m}$ irreducible such that $\nu_\phi \geq \mu_1$. Then $\mu_1 \wedge \nu_\phi > \mu_1 \wedge \nu_x$ and $\mu_2 \wedge \nu_\phi = \mu_2 \wedge \nu_x$, which by (3.17) contradicts $\nu_1 \leq_x \nu_2$. Thus $\mu_1 \in [\nu_x, \mu_2 \wedge \nu_x]$, which implies $[\nu_x, \mu_1] \subset [\nu_x, \mu_2]$.

Conversely assume $[\nu_x, \mu_1] \subset [\nu_x, \mu_2]$. We want to prove $N\mu_1 \leq_x N\mu_2$ i.e. $\mu_1(\phi)/\mu_1(x) \leq \mu_2(\phi)/\mu_2(x)$ for all ϕ. The assumption implies that $\mu_1 \wedge \nu_x \geq \mu_2 \wedge \nu_x$. If equality holds, then $\mu_1(x) = \mu_2(x)$, and the assumption gives $\mu_1 \leq \mu_2$, so that $N\mu_1 \leq_x N\mu_2$. If instead $\mu_1 \wedge \nu_x > \mu_2 \wedge \nu_x$ i.e. $\mu_1(x) > \mu_2(x)$, then the assumption implies that $\mu_1 \in [\nu_{\mathfrak{m}}, \nu_x]$. Pick $\phi \in \mathfrak{m}$ irreducible. If $\mu_1 \wedge \nu_\phi \leq \mu_2 \wedge \nu_\phi$, then (3.17) immediately gives $\mu_1(\phi)/\mu_1(x) \leq \mu_2(\phi)/\mu_2(x)$. If instead $\mu_1 \wedge \nu_\phi > \mu_2 \wedge \nu_\phi$, then $\mu_1 \wedge \nu_\phi \geq \mu_1 \wedge \nu_x$, and $\mu_2 \wedge \nu_\phi = \mu_2 \wedge \nu_x$ so again (3.17) results in $\mu_1(\phi)/\mu_1(x) \leq \mu_2(\phi)/\mu_2(x)$. Thus $\nu_1 \leq_x \nu_2$, which concludes the proof. □

3.9.2 Relative Parameterizations

We now introduce natural parameterizations on $\mathcal{V}_x$, analogous to skewness and thinness on $\mathcal{V}$. The multiplicity $m(\phi) = \nu_{\mathfrak{m}}(\phi)$, which plays a key role on $\mathcal{V}$, is here replaced by the *relative multiplicity* $m_x(\phi) = \nu_x(\phi)$. By (3.14), this number also equals the intersection multiplicity between the curves $\{\phi = 0\}$ and $\{x = 0\}$. In particular $m_x(\phi)$ is an integer unless x divides ϕ in which case it is infinite.

Definition 3.64. *Pick $\nu \in \mathcal{V}_x$. We define*

- the relative skewness*: $\alpha_x(\nu) := \sup_{\phi\in\mathfrak{m}} \frac{\nu(\phi)}{m_x(\phi)} \in \overline{\mathbf{R}}_+$;*
- the relative multiplicity*: if ν is infinitely singular, then $m_x(\nu) := \infty$; otherwise $m_x(\nu) := \min\{m_x(\phi)\ ;\ \phi \in \mathfrak{m}$ irreducible, $N\nu_\phi \geq_x \nu\} \in \mathbf{N}$;*
- the relative thinness*: $A_x(\nu) := 1 + \int_{\nu_x}^{\nu} m_x(\mu)\, d\alpha_x(\mu) \in \overline{\mathbf{R}}_+$.*

Notice that $\alpha_x(\mathrm{div}_x) = 0$, $A_x(\mathrm{div}_x) = 1$ and $m_x(\mathrm{div}_x) = 1$.

Proposition 3.65. *Pick $\nu \in \mathcal{V} \setminus \{\nu_x\}$ and let $N\nu$ be its image in $\mathcal{V}_x$. Then*

$$\alpha_x(N\nu) = \alpha(\nu)/\nu(x)^2 \quad \text{and} \quad A_x(N\nu) = A(\nu)/\nu(x).$$

The relative multiplicity is given by

$$m_x(N\nu) = \begin{cases} 1 & \text{when } \nu \in [\nu_x, \nu_{\mathfrak{m}}], \\ m(\nu)\,\nu(x) & \text{otherwise.} \end{cases}$$

In particular, α_x and A_x are finite on quasimonomial valuations, and $m_x(\nu)$ is infinite iff ν is infinitely singular. We leave it to the reader to generalize Theorem 3.26 and Propositions 3.37, 3.46. In particular we have

(i) $A_x(\nu) \geq 1 + \alpha_x(\nu)$ with equality iff $m_x(\nu) = 1$, i.e. if ν is monomial in some local coordinates (x, y);
(ii) if $m_x(\nu) > 1$, then $A_x(\nu) < m_x(\nu)\alpha_x(\nu)$.

The main consequence of the preceding proposition is

Corollary 3.66. *The relative skewness defines parameterizations*

$$\alpha_x : \mathcal{V}_x \to [0, \infty] \quad \text{and} \quad \alpha_x : \mathcal{V}_{\mathrm{qm},x} \to [0, \infty[\,.$$

Similarly, the relative thinness gives parameterizations

$$A_x : \mathcal{V}_x \to [1, \infty] \quad \text{and} \quad A_x : \mathcal{V}_{\mathrm{qm},x} \to [1, \infty[\,.$$

As in the nonrelative case, we can use these parameterizations to define metrics on $\mathcal{V}_x$ and $\mathcal{V}_{\mathrm{qm},x}$.

Proof (Proposition 3.65). We fix $\nu \in \mathcal{V} \setminus \{\nu_x\}$. First consider the relative skewness. The supremum defining $\alpha_x(N\nu)$ can be taken over irreducible elements $\phi \in \mathfrak{m}$. For any such ϕ, Proposition 3.25 yields

$$\frac{(N\nu)(\phi)}{m_x(\phi)} = \frac{\nu(\phi)}{\nu(x)\nu_x(\phi)} = \frac{\alpha(\nu \wedge \nu_\phi)m(\phi)}{\nu(x)\alpha(\nu_\phi \wedge \nu_x)m(\phi)} = \frac{\alpha(\nu \wedge \nu_\phi)\alpha(\nu \wedge \nu_x)}{\alpha(\nu_\phi \wedge \nu_x)\nu(x)^2}. \tag{3.18}$$

If ν is divisorial or $\nu \notin]\nu_x, \nu_\mathfrak{m}]$, then we can choose $\phi \in \mathfrak{m}$ irreducible with $\nu_\phi \geq \nu$ and $\nu_\phi \wedge \nu_x = \nu \wedge \nu_x$. Then $\nu \wedge \nu_\phi = \nu$ so (3.18) immediately gives $(N\nu)(\phi)/m_x(\phi) \geq \alpha(\nu)/\nu(x)^2$, yielding the lower bound $\alpha_x(N\nu) \geq \alpha(\nu)/\nu(x)^2$. If $\nu \in]\nu_x, \nu_\mathfrak{m}[$ is irrational, then we get the same lower bound by a simple approximation argument. The corresponding upper bound also follows from (3.18). Indeed, we always have either $\nu \wedge \nu_x \leq \nu_\phi \wedge \nu_x$ or $\nu \wedge \nu_\phi \leq \nu_\phi \wedge \nu_x$, so $(N\nu)(\phi)/m_x(\phi) \leq \alpha(\nu)/\nu(x)^2$, implying $\alpha(N\nu) \leq \alpha(\nu)/\nu(x)^2$.

Let us now relate multiplicity to relative multiplicity. We may assume that ν is not infinitely singular, since otherwise $m_x(N\nu) = m(\nu) = \infty$.

If $\nu \in [\nu_x, \nu_\mathfrak{m}]$ then $N\nu \leq_x N\nu_y$ whenever (x, y) are local coordinates. Hence $m_x(N\nu) \leq m_x(y) = \nu_x(y) = 1$, so $m_x(N\nu) = 1$.

If $\nu \notin [\nu_x, \nu_\mathfrak{m}]$ then it follows from Proposition 3.61 that for $\phi \in \mathfrak{m}$ irreducible, the conditions $N\nu_\phi \geq_x N\nu$ and $\nu_\phi \geq \nu$ are equivalent. Moreover, they both imply $m_x(\phi) = m(\phi)\nu(x)$. Hence $m_x(N\nu) = m(\nu)\nu(x)$.

To relate thinness to relative thinness we make use of the preceding computations. Write $\nu' = \nu \wedge \nu_x$. Then

$$\begin{aligned} A_x(N\nu) &= 1 + \int_{\mathrm{div}_x}^{N\nu'} m_x(\tilde{\mu})\, d\alpha_x(\tilde{\mu}) + \int_{N\nu'}^{N\nu} m_x(\tilde{\mu})\, d\alpha_x(\tilde{\mu}) \\ &= 1 + \alpha_x(N\nu') + \int_{\nu'}^{\nu} m(\mu)\mu(x)\, \frac{d\alpha(\mu)}{\mu(x)^2} \\ &= 1 + \alpha(\nu')^{-1} + \frac{A(\nu) - A(\nu')}{\nu'(x)} = \frac{A(\nu)}{\nu'(x)} = \frac{A(\nu)}{\nu(x)}. \end{aligned}$$

This completes the proof. □

Proof (Corollary 3.66). It suffices to consider the full tree $\mathcal{V}_x$; the statements about $\mathcal{V}_{\mathrm{qm},x}$ will be direct consequences.

If $t \geq 1$, then $\alpha_x(N\nu_{x,t}) = t/t^2 = t^{-1}$. Hence the restriction of α_x to the segment $[\mathrm{div}_x, \nu_\mathfrak{m}]$ gives an order-preserving bijection onto the real interval $[0, 1]$. Any end in $\mathcal{V}_x$ different from div_x is of the form $N\nu$, where $\nu \neq \nu_x$ is an end in $\mathcal{V}$. Let us prove that α_x defines an order-preserving bijection of $[\mathrm{div}_x, N\nu]$ onto $[0, \alpha_x(N\nu)]$. Set $\nu' = \nu \wedge \nu_x$. By the above calculation, α_x gives an order-preserving bijection of $[\mathrm{div}_x, N\nu']$ onto $[0, \alpha_x(N\nu')]$. On the other hand, if $\mu \in [\nu', \nu]$, then $\mu(x) = \nu'(x)$ so $\alpha_x(N\mu) = \alpha(\mu)/\nu'(x)^2$. It hence follows from Theorem 3.26 that α_x gives an order-preserving mapping of $[N\nu', N\nu]$ onto $[\alpha_x(N\nu'), \alpha_x(N\nu)]$.

This shows that α_x defines a parameterization of the rooted nonmetric tree $(\mathcal{V}_x, \leq_x)$. The fact that relative thinness gives a parameterization is proved in the same way. Note that $A_x(N\nu_{x,t}) = 1 + t^{-1}$ for $t \geq 1$. $\square$

Finally we have the following calculation which will be needed in Chapter 4. The proof is left to the reader.

Lemma 3.67. *If $\nu, \nu' \in \mathcal{V}_x$, $\nu' <_x \nu$ and $m_x(\nu) = m_x(\nu')$, then $A_x(\nu) - A_x(\nu') = \nu(\phi) - \nu'(\phi)$ for any irreducible $\phi \in \mathfrak{m}$ such that $N\nu_\phi >_x \nu'$ and $m_x(\phi) = m_x(\nu')$.*

3.9.3 Balls of Curves

The relative valuative tree can also be understood in terms of balls of irreducible curves, just as in Section 3.8.

Definition 3.68. *As before, let $\mathcal{C}$ be the space of (local formal) irreducible curves. Define $\mathcal{C}_x = \mathcal{C} \setminus \{x = 0\}$. The* relative multiplicity *of a curve $C \in \mathcal{C}_x$ is defined by $m_x(C) = C \cdot \{x = 0\}$. Thus $m_x(\{\phi = 0\}) = m_x(\phi)$ for $\phi \in \mathfrak{m}$ irreducible. Given $C_1, C_2 \in \mathcal{C}_x$ set*

$$d_{\mathcal{C}_x}(C_1, C_2) := \frac{m_x(C_1) m_x(C_2)}{C_1 \cdot C_2}.$$

We leave to the reader to check the analogue of Lemma 3.56:

Lemma 3.69. *For any $C_1, C_2 \in \mathcal{C}_x$ we have*

$$d_{\mathcal{C}_x}(C_1, C_2) = \frac{1}{\alpha_x(N\nu_{C_1} \wedge_x N\nu_{C_2})}. \tag{3.19}$$

Moreover, $(\mathcal{C}_x, d_{\mathcal{C}_x})$ is an ultrametric space of infinite diameter.

As in the nonrelative case we can associate a parameterized tree $\mathcal{T}_{\mathcal{C}_x}$ to the ultrametric space $(\mathcal{C}_x, d_{\mathcal{C}_x})$. Let us review this construction, pointing out the differences that we encounter in the relative setting.

A point in $\mathcal{T}_{\mathcal{C}_x}$ is a closed ball in $\mathcal{C}_x$ of positive radius. The partial ordering is given by reverse inclusion. In order to have a minimal element in $\mathcal{T}_{\mathcal{C}_x}$ we also add the ball of infinite radius, i.e. the whole space $\mathcal{C}_x$. The parameterization on $\mathcal{T}_{\mathcal{C}_x}$ sends a ball of radius r to the real number r^{-1}.

There is a natural multiplicity function on $\mathcal{T}_{\mathcal{C}_x}$. Its value on a ball is the minimum relative multiplicity of any curve in the ball.

The completion of $\mathcal{T}_{\mathcal{C}_x}$ is taken in the sense of nonmetric trees. An element in the completion is represented by a decreasing sequence of balls. The multiplicity function and the parameterization on $\mathcal{T}_{\mathcal{C}_x}$ both extend naturally to the completion. An element in the completion, represented by a decreasing sequence of balls, has finite multiplicity iff the sequence has nonempty intersection in $\mathcal{C}_x$.

We can now define a map from $\mathcal{V}_{\mathrm{qm},x}$ to $\mathcal{T}_{\mathcal{C}_x}$ by sending a quasimonomial valuation ν to the ball $B_\nu := \{C \; ; \; N\nu_C >_x \nu\}$. (That this is a ball follows from (3.19).) Notice that div_x is sent to the whole space $\mathcal{C}_x$.

Similarly, $B \mapsto \nu_B := \wedge_x\{N\nu_C \; ; \; C \in B\}$ gives a map from $\mathcal{T}_{\mathcal{C}_x}$ to $\mathcal{V}_{\mathrm{qm},x}$. A straightforward adaptation of the proof of Theorem 3.57 gives

Theorem 3.70. *The mappings $\nu \mapsto B_\nu$ and $B \mapsto \nu_B$ preserve multiplicity and give inverse isomorphisms between the parameterized trees $(\mathcal{V}_{\mathrm{qm},x}, \alpha_x)$ and $(\mathcal{T}_{\mathcal{C}_x}, \mathrm{r}^{-1})$. They also extend uniquely to isomorphisms between $\mathcal{V}_x$ and the completion of $\mathcal{T}_{\mathcal{C}_x}$.*

3.9.4 Homogeneity

Recall that we defined $\tilde{\mathcal{V}}$ as the set of *all* centered valuations $\nu : R \to \overline{\mathbf{R}}_+$ without any normalization. This space $\tilde{\mathcal{V}}$ is naturally a bundle with fibers that are rays of the form $\mathbf{R}_+^*\nu$, $\nu \in \tilde{\mathcal{V}}$. The trees $\mathcal{V}$ and $\mathcal{V}_x$, consisting of elements $\nu \in \tilde{\mathcal{V}}$ satisfying $\nu(\mathfrak{m}) = 1$ and $\nu(x) = 1$, respectively, can then be viewed as (the images of) sections of this bundle.

Suppose we wish to define functions skewness and thinness on the full space $\tilde{\mathcal{V}}$ in such a way that the restrictions to $\mathcal{V}$ and $\mathcal{V}_x$ recover skewness (thinness) and relative skewness (relative thinness), respectively. We also want these functions to be homogeneous on each ray $\mathbf{R}_+^*\nu$. Proposition 3.65 indicates that we should then impose $\alpha(t\nu) = t^2\alpha(\nu)$, and $A(t\nu) = tA(\nu)$ for $\nu \in \tilde{\mathcal{V}}$ and $t > 0$. In other words, *skewness is homogeneous of degree two, and thinness is homogeneous of degree one.* Let us give further evidence that these are the correct degrees.

First, as noted in Remark 3.33, the skewness $\alpha(\nu)$ of a valuation $\nu \in \mathcal{V}$ is equal to the inverse of its volume $\mathrm{Vol}(\nu) := \lim_{c\to\infty} \frac{2}{c^2} \dim_{\mathbf{C}} R/\{\nu \geq c\}$. It is clear that $\mathrm{Vol}(t\nu) = t^{-2}\,\mathrm{Vol}(\nu)$. Hence keeping $\alpha = \mathrm{Vol}^{-1}$ on all of $\tilde{\mathcal{V}}$ implies the required homogeneity property.

Alternatively, we shall see in Section 7.12 that there is a natural intersection product on $\mathcal{V}$ given by $\nu \cdot \nu' := \alpha(\nu \wedge \nu')$. In particular $\alpha(\nu) = \nu \cdot \nu$. By extending this intersection product in a bilinear way, restricting to $\tilde{\mathcal{V}}$, and still requiring $\alpha(\nu) = \nu \cdot \nu$, we again get $\alpha(t\nu) = t\nu \cdot t\nu = t^2\alpha(\nu)$.

As for thinness, we rely on Remark 3.50. Consider ν divisorial. Let $\mathfrak{m}_\nu$ be the maximal ideal of the valuation ring R_ν, and let $J_{R_\nu/R}$ be the Jacobian ideal. If $\nu \in \mathcal{V}$ is normalized by $\nu(\mathfrak{m}) = 1$, we have seen that $A(\nu) = \nu(J_{R_\nu/R}\,\mathfrak{m}_\nu)$. Taking this as a definition for thinness, we get $A(t\nu) = tA(\nu)$. Indeed, multiplying ν by a positive constant does not change the ideals $\mathfrak{m}_\nu$ and $J_{R_\nu/R}$.

4

Valuations Through Puiseux Series

We now present an alternative approach, based on Puiseux series, for classifying valuations and obtaining the tree structure on the valuative tree $\mathcal{V}$, or more precisely the relative valuative tree $\mathcal{V}_x$ as defined in Section 3.9.

Fix a smooth formal curve at the origin, say $\{x = 0\}$. Then any local irreducible curve, save for $\{x = 0\}$, is represented by a (non-unique) Puiseux series in x. In the same spirit, we show that every valuation in $\mathcal{V}_x$ is represented by a valuation on the power series ring in one variable with coefficients that are Puiseux series in x. Moreover, the set $\widehat{\mathcal{V}}_x$ of all such (normalized) valuations has a natural tree structure and the restriction map from $\widehat{\mathcal{V}}_x$ to $\mathcal{V}_x$ is a tree morphism. In fact, $\mathcal{V}_x$ is naturally the orbit space of $\widehat{\mathcal{V}}_x$ under the action by the relevant Galois group.

This Puiseux approach is closely related to Berkovich's theory of analytic spaces and we show how the valuative tree naturally embeds (as a nonmetric tree) as the closure of a disk in the Berkovich projective line $\mathbb{P}^1(k)$ over the local field $k = \mathbf{C}((x))$. Moreover, $\mathcal{V}_{\mathrm{qm},x}$ embeds as a subtree of the Bruhat-Tits building of PGL_2. The latter space has a natural metric and we show that the induced metric on $\mathcal{V}_{\mathrm{qm},x}$ is exactly the relative thin metric.

As references for this chapter we point out [Te3] for Puiseux expansions, and [Be] and [BT] for Berkovich spaces and Bruhat-Tits buildings, respectively. In fact, many of the results presented here are contained, at least implicitly, in [Be] but are presented there in a different language and with less details. We also refer to [Ri] for a concrete exposition in a context similar to ours.

4.1 Puiseux Series and Valuations

We fix an element of R representing a smooth formal curve. In other words, fix $x \in \mathfrak{m}$ with multiplicity $m(x) = 1$. Think of x as a variable and let $k = \mathbf{C}((x))$ be the fraction field of $\mathbf{C}[[x]]$, i.e. the field of Laurent series in x. Let $\hat{k}$ be the algebraic closure of k: its elements are finite or infinite *Puiseux series* of the form

$$\hat{\phi} = \sum_{j \geq 1} a_j x^{\hat{\beta}_j} \quad \text{with} \quad a_j \in \mathbf{C}^*,\ \hat{\beta}_{j+1} > \hat{\beta}_j \in \mathbf{Q} \tag{4.1}$$

and where the rational numbers $\hat{\beta}_j$ have bounded denominators, i.e. $m\hat{\beta}_j \in \mathbf{Z}$ for all j for some integer m. Notice that this implies that if the series is infinite, then $\hat{\beta}_j \to \infty$ as $j \to \infty$.

We endow k and $\hat{k}$ with the valuation $\nu_\star$ defined by $\nu_\star|_{\mathbf{C}^*} = 0$, and $\nu_\star(x) = 1$, and we let $\bar{k}$ be the completion of $\hat{k}$ with respect to $\nu_\star$. The elements of $\bar{k}$ are finite or infinite series of the form (4.1), with the restriction that $\hat{\beta}_j \to \infty$ if the series is infinite, but without the restriction on the denominators. Shortly we will consider even more general series, dropping the condition that $\hat{\beta}_j \to \infty$. Clearly $\nu_\star$ extends to series of that type: we have $\nu(\phi) = \hat{\beta}_1$.

We define $\hat{k}_+$ to be the subset of $\hat{k}$ consisting of Puiseux series $\hat{\phi}$ with $\nu_\star(\hat{\phi}) > 0$, i.e. $\hat{\beta}_1 > 0$ in (4.1). Similarly we define k_+ and $\bar{k}_+$.

Fix $y \in \mathfrak{m}$ such that (x, y) are local formal coordinates at $\mathbf{C}^2$, i.e. $\nu_x \wedge \nu_y = \nu_\mathfrak{m}$. Think of y as a variable and consider the ring $\bar{k}[[y]]$ of formal power series with Puiseux series coefficients.

We are interested in valuations $\hat{\nu} : \bar{k}[[y]] \to \overline{\mathbf{R}}_+$ extending $\nu_\star$ on $\bar{k}$ and satisfying $\hat{\nu}(y) > 0$. As in Proposition 2.10, any such valuation is determined by its values on polynomials in y and—since $\bar{k}$ is algebraically closed—in fact on its values on linear polynomials $y - \hat{\psi}$, $\hat{\psi} \in \bar{k}$.

Proposition 4.1. *Any valuation $\hat{\nu} : \bar{k}[[y]] \to \overline{\mathbf{R}}_+$ extending $\nu_\star$ on $\bar{k}$ and satisfying $\hat{\nu}(y) > 0$ can be uniquely represented by a number $\hat{\beta} \in \overline{\mathbf{R}}_+$ and a finite or infinite series $\hat{\phi}$ of the form* (4.1) *(with no restriction on* $\lim \hat{\beta}_j$ *if the series is infinite) such that $\hat{\beta} > \hat{\beta}_j > 0$ for all j, and $\hat{\beta} = \lim \hat{\beta}_j$ if the series is infinite. More precisely, we have*

$$\hat{\nu}(y - \hat{\psi}) = \min\{\hat{\beta}, \nu_\star(\hat{\psi} - \hat{\phi})\} \quad \textit{for all } \hat{\psi} \in \bar{k}. \tag{4.2}$$

Conversely, if $\hat{\phi}$ and $\hat{\beta}$ satisfy these conditions then there exists a unique valuation $\hat{\nu}$ on $\bar{k}[[y]]$ satisfying (4.2) *and extending $\nu_\star$ on $\bar{k}$*

Definition 4.2. *We write* $\mathrm{val}[\hat{\phi}; \hat{\beta}]$ *for the valuation $\hat{\nu} : \bar{k}[[y]] \to \overline{\mathbf{R}}_+$ determined by the pair $(\hat{\phi}, \hat{\beta})$ as in Proposition 4.1. Moreover, we say that $\hat{\nu}$ is*

(i) of point type *if $\hat{\phi} \in \bar{k}$ and $\hat{\beta} = \infty$; we then write $\hat{\nu} = \hat{\nu}_{\hat{\phi}}$;*
(ii) of finite type *if $\hat{\phi} \in \hat{k}$ and $\hat{\beta} < \infty$;*
(iii) rational *if $\hat{\nu}$ is of finite type and $\hat{\beta}$ is rational;*
(iv) irrational *if $\hat{\nu}$ is of finite type and $\hat{\beta}$ is irrational;*
(v) of special type *if $\hat{\phi} \notin \bar{k}$.*

Remark 4.3. The proof below shows that if $\hat{\nu} = \mathrm{val}[\hat{\phi}; \hat{\beta}]$ is of point type or special type then

$$\hat{\nu}(y - \psi) = \nu_\star(\hat{\psi} - \hat{\phi}) \quad \text{for all } \hat{\psi} \in \bar{k}, \tag{4.3}$$

exhibiting the fact that $\hat{\beta}$ is determined by $\hat{\phi}$ in this case.

Proof (Proposition 4.1). Pick a valuation $\hat{\nu}$ on $\bar{k}[[y]]$ extending $\nu_\star$ and satisfying $\hat{\nu}(y) > 0$. Let us construct $\hat{\phi}$ and $\hat{\beta}$ such that (4.2) holds.

Set $\hat{\beta}_1 := \hat{\nu}(y) > 0$. Then $\hat{\nu}(y - \hat{\psi}) \geq \min\{\hat{\beta}_1, \nu_\star(\hat{\psi})\}$ for all $\hat{\psi} \in \bar{k}$. There are now two possibilities: either equality holds for all $\hat{\psi}$, or $\hat{\beta}_1 \in \mathbf{Q}$ and there exists a unique $\theta_1 \in \mathbf{C}^*$ such that $\hat{\nu}(y - \hat{\phi}_1) > \hat{\beta}_1$, where $\hat{\phi}_1 := a_1 x_1^{\hat{\beta}}$. In the first case we stop, setting $\hat{\phi} = \hat{\phi}_1$ and $\hat{\beta} = \hat{\beta}_1$. In the second case we let $\hat{\beta}_2 := \hat{\nu}(y - \hat{\phi}_1)$ and continue. Inductively we construct $\hat{\phi}_j = \hat{\phi}_{j-1} + \theta_j x^{\hat{\beta}_j}$ with $\theta_j \in \mathbf{C}^*$ and $\hat{\beta}_{j+1} > \hat{\beta}_j \in \mathbf{Q}$, such that $\hat{\nu}(y - \hat{\psi}) \geq \min\{\hat{\beta}_{j+1}, \nu_\star(\hat{\psi} - \hat{\phi}_j)\}$. If this process stops at some finite step j_0, then we set $\hat{\phi} = \hat{\phi}_{j_0} = \sum_1^{j_0} \theta_j x^{\hat{\beta}_j}$, $\hat{\beta} = \hat{\beta}_{j_0+1}$. Otherwise $(\hat{\beta}_j)_1^\infty$ is strictly increasing with limit $\hat{\beta}$ and $\hat{\phi}_j$ converges to a series $\hat{\phi}$ of the type (4.1), where $\lim \hat{\beta}_j = \hat{\beta}$ may or may not be finite. In fact $\hat{\phi} \in \bar{k}$ iff $\hat{\beta} = \infty$.

We claim that (4.2) holds. In the case when $\hat{\phi}$ is a finite series, i.e. the procedure above stops after finitely many steps, then this is clear by construction. If $\hat{\phi}$ is an infinite series, then for $\hat{\psi} \in \bar{k}$ we have $\hat{\nu}(y - \hat{\psi}) \geq \min\{\hat{\beta}_{j+1}, \nu_\star(\hat{\psi} - \hat{\phi}_j)\}$ for each j, with equality unless $\nu_\star(\hat{\psi} - \hat{\phi}_{j+1}) > \hat{\beta}_{j+1}$. Passing to the limit we obtain $\hat{\nu}(y - \hat{\psi}) \geq \min\{\hat{\beta}, \nu_\star(\hat{\psi} - \hat{\phi})\}$ and strict inequality would imply that $\nu_\star(\hat{\psi} - \hat{\phi}_j) > \hat{\beta}_j$ for all j. If $\hat{\phi} \in \bar{k}$, i.e. $\hat{\beta} = \infty$, then this gives $\nu_\star(\hat{\psi} - \hat{\phi}) = \infty$ so that $\hat{\psi} = \hat{\phi}$ and (4.2) holds. If $\hat{\phi} \notin \bar{k}$, i.e. $\hat{\beta} < \infty$, then $\nu_\star(\hat{\psi} - \hat{\phi}_j) > \hat{\beta}_j$ for all j leads to a contradiction, since $\hat{\psi} \in \bar{k}$ but $\hat{\phi} \notin \bar{k}$. Thus (4.2) holds also in this case.

Notice that $\hat{\beta}$ and $\hat{\phi}$ are unique. Indeed, it follows from (4.2) that the number $\hat{\beta}$ satisfies $\hat{\beta} = \sup\{\hat{\nu}(y - \hat{\psi})\ ;\ \hat{\psi} \in \bar{k}\}$ (compare Definition 4.7 below). Moreover, suppose that $\hat{\phi} = \sum a_j x^{\hat{\beta}_j}$ and $\hat{\phi}' = \sum a'_j x^{\hat{\beta}'_j}$ both satisfy (4.2) and that $\hat{\beta}_j, \hat{\beta}'_j < \hat{\beta}$ for all j. Then $\nu_\star(\hat{\phi} - \hat{\phi}') \geq \hat{\beta}$, so $\hat{\phi} = \hat{\phi}'$.

Finally we show that given $\hat{\phi}$ and $\hat{\beta}$ satisfying the conditions in the proposition, there exists a unique valuation $\hat{\nu}$ satisfying (4.2). Uniqueness is immediate from (4.2) since $\hat{\nu}$ is determined on its values on linear polynomials $y - \hat{\psi}$. As for existence, let us define $\hat{\nu} : \bar{k}[y] \to \overline{\mathbf{R}}_+$ using (4.2), $\hat{\nu}|_{\bar{k}} = \nu_\star$ and $\hat{\nu}(\prod(y - \hat{\psi}_i)) = \sum \hat{\nu}(y - \hat{\psi}_i)$. The nontrivial part is to show that $\hat{\nu}$ is actually a valuation on $\bar{k}[y]$, i.e. that

$$\hat{\nu}(P + Q) \geq \min\{\hat{\nu}(P), \hat{\nu}(Q)\} \quad \text{for all } P, Q \in \bar{k}[y]. \tag{4.4}$$

Once we have proved (4.4), the fact that $\hat{\nu}$ extends uniquely to a valuation on the power series ring $\bar{k}[[y]]$ is proved as in Proposition 2.10.

To prove (4.4) let us rephrase the defining property (4.2) somewhat. First assume that $\hat{\phi}$ is a finite Puiseux series and that $\hat{\beta} \in \mathbf{Q}$. For $a \in \mathbf{C}$ define $\hat{\phi}_a = \hat{\phi} + a x^{\hat{\beta}}$. Then $\hat{\phi}_a \in \bar{k}$ and from (4.2) it follows that there exists a subset $X(\psi)$ of $\mathbf{C}$ containing at most one element such that if $a \notin X(\psi)$, then $\hat{\nu}(y-\hat{\psi}) = \nu_\star(\hat{\psi}-\hat{\phi}_a)$. As a consequence, given $P \in \bar{k}[y]$ there exists a finite subset $X(P)$ such that $\hat{\nu}(P) = \nu_\star(P(\hat{\phi}_a))$ for $a \notin X(P)$. Here $P(\hat{\phi}_a)$ is the element of $\bar{k}$ obtained by substituting $\hat{\phi}_a$ for y in the polynomial $P \in \bar{k}[y]$. If $P, Q \in \bar{k}[y]$ and $a \notin X(P) \cup X(Q) \cup X(P+Q))$, we then get

$$\begin{aligned}\hat{\nu}(P+Q) = \nu_\star(P(\hat{\phi}_a)+Q(\hat{\phi}_a)) &\geq \min\{\nu_\star(P(\hat{\phi}_a)), \nu_\star(Q(\hat{\phi}_a))\}\\ &= \min\{\hat{\nu}(P), \hat{\nu}(Q)\},\end{aligned}$$

so that $\hat{\nu}$ is a valuation in this case.

The same argument works also if $\hat{\phi}$ is a finite Puiseux series and $\hat{\beta}$ is irrational: we simply replace $\hat{\phi}_a$ by $\hat{\phi}+x^{\hat{\beta}}$ in the argument above. Of course $\hat{\phi}+x^{\hat{\beta}}$ is not an element of $\bar{k}$, but the computations still make sense. If $\hat{\phi}$ is an infinite Puiseux series and $\hat{\beta} = \infty$ (so that $\hat{\phi} \in \bar{k}$), then (4.2) instead implies $\hat{\nu}(P) = \nu_\star(P(\hat{\phi}))$ for all $P \in \bar{k}[y]$, which gives (4.4) by the same argument as above. This method in fact also works if $\hat{\phi} \notin \bar{k}$ and $\hat{\beta} < \infty$.

Thus $\hat{\nu}$ is a valuation, completing the proof. □

4.2 Tree Structure

Let $\widehat{\mathcal{V}}_x$ be the set of valuations $\hat{\nu} : \bar{k}[[y]] \to \overline{\mathbf{R}}_+$ extending $\nu_\star$ on $\bar{k}$ and satisfying $\hat{\nu}(y) > 0$. To $\widehat{\mathcal{V}}_x$ we also add the valuation $\hat{\nu}_\star$ defined by $\hat{\nu}_\star(y-\hat{\phi}) = 0$ for all $\hat{\phi} \in \bar{k}$ with $\nu_\star(\phi) \geq 0$. Note that by definition this valuation is a rational valuation of finite type. Our objective now is to show that $\widehat{\mathcal{V}}_x$ carries a tree structure similar to that of the relative valuative tree $\mathcal{V}_x$ discussed in Section 3.9.

Recall that the main technical tool for obtaining the tree structure on $\mathcal{V}_x$ was the technique of SKP's. Here, in the case of $\widehat{\mathcal{V}}_x$, the situation is a bit simpler and Proposition 4.1 provides the technical result that we need.

4.2.1 Nonmetric Tree Structure

There is a natural partial ordering $\leq$ on $\widehat{\mathcal{V}}_x$: $\hat{\nu} \leq \hat{\mu}$ iff $\hat{\nu} \leq \hat{\mu}$ pointwise as functions on $\bar{k}[y]$. As $\bar{k}$ is algebraically closed, it suffices to check this condition on linear polynomials in y. Also notice that if $\hat{\nu} \in \mathcal{V}_x$, $\hat{\phi} \in \bar{k}$ and $\nu_\star(\hat{\phi}) \leq 0$, then $\hat{\nu}(y-\hat{\phi}) = \nu_\star(\hat{\phi})$. Hence

$$\hat{\nu} \leq \hat{\mu} \quad \text{iff} \quad \hat{\nu}(y-\hat{\psi}) \leq \hat{\mu}(y-\hat{\psi}) \text{ for every } \hat{\psi} \in \bar{k}_+. \tag{4.5}$$

Proposition 4.4. *The partial ordering $\leq$ gives $\widehat{\mathcal{V}}_x$ the structure of a complete nonmetric tree rooted at $\hat{\nu}_\star$. Its ends are the valuations of point type or special type. Its branch points are the rational valuations. Its regular points are the irrational valuations.*

Corollary 4.5. *Write $\widehat{\mathcal{V}}_{\text{fin}}$ for the set of valuations in $\widehat{\mathcal{V}}_x$ of finite type. Then $\widehat{\mathcal{V}}_{\text{fin}}$ is a rooted subtree of $\widehat{\mathcal{V}}_x$ containing none of its ends (apart from the root $\hat{\nu}_\star$).*

We denote by $\wedge$ the infimum in $\widehat{\mathcal{V}}_x$ with respect to the partial ordering above.

Proof (Proposition 4.4). The key to the proof is the following characterization of the partial ordering using the structure result above.

Lemma 4.6. *Consider $\hat{\nu}_1, \hat{\nu}_2 \in \widehat{\mathcal{V}}_x$ and write $\hat{\nu}_i = \text{val}[\hat{\phi}_i, \hat{\beta}_i]$ as in Definition 4.2. Then $\hat{\nu}_1 \leq \hat{\nu}_2$ iff $\hat{\beta}_2 \geq \hat{\beta}_1$ and $\nu_\star(\hat{\phi}_2 - \hat{\phi}_1) \geq \hat{\beta}_1$.*

We postpone the proof of the lemma. Let us show that $(\widehat{\mathcal{V}}_x, \leq)$ satisfies the axioms (T1)-(T3) for a nonmetric tree. As for (T1) this is easy as $\hat{\nu}_\star$ is clearly the unique minimal element of $\widehat{\mathcal{V}}_x$. Now consider (T2). Fix $\hat{\nu} \in \widehat{\mathcal{V}}_x$, $\hat{\nu} \neq \hat{\nu}_\star$. We have to show that $\{\hat{\mu} \in \widehat{\mathcal{V}}_x \ ; \ \hat{\mu} \leq \hat{\nu}\}$ is a totally ordered set isomorphic to a real interval. Write $\hat{\nu} = \text{val}[\hat{\phi}, \hat{\beta}]$ as in Definition 4.2. Thus $\hat{\beta} > 0$ and $\hat{\phi} = \sum_{j=1}^n a_j x^{\hat{\beta}_j}$, where $1 \leq n \leq \infty$. Write $\hat{\phi}_i = \sum_{j=1}^i a_j x^{\hat{\beta}_j}$ for $1 \leq i < n+1$. By Lemma 4.6 we have $\hat{\mu} \leq \hat{\nu}$ iff $\hat{\mu} = \text{val}[\hat{\phi}_i; \hat{\alpha}]$ for some i and $\hat{\beta}_i < \hat{\alpha} \leq \hat{\beta}_{i+1}$. Here we adopt the convention that $\hat{\beta}_0 = 0$ and that $\hat{\beta}_{n+1} = \hat{\beta}$ if $n < \infty$. Thus $\{\hat{\mu} \in \widehat{\mathcal{V}}_x \ ; \ \hat{\nu}_\star < \hat{\mu} \leq \hat{\nu}\}$ is isomorphic to the union of the intervals $]\hat{\beta}_i, \hat{\beta}_{i+1}]$, hence to the real interval $]0, \hat{\beta}]$. This proves (T2).

As for (T3), it is easier to prove the equivalent statement (T3') given in Remark 3.3. Moreover (T3') clearly follows if we can prove that every totally ordered subset of $\widehat{\mathcal{V}}_x$ has a majorant in $\widehat{\mathcal{V}}_x$. This will in fact also prove that $\widehat{\mathcal{V}}_x$ is a complete tree. Thus consider such a totally ordered subset $\mathcal{S} \subset \widehat{\mathcal{V}}_x$. By Lemma 4.6 the Puiseux series defining $\hat{\nu} \in \mathcal{S}$ has a length that is a nondecreasing function of ν. The parameter $\hat{\beta}$ is also an increasing function on $\mathcal{S}$. It follows easily from this and from Lemma 4.6 that there exists a valuation in $\widehat{\mathcal{V}}_x$ of point type or special type dominating all the $\hat{\nu} \in \mathcal{S}$. Thus $\widehat{\mathcal{V}}_x$ is a complete tree.

It is clear from Lemma 4.6 that valuations in $\widehat{\mathcal{V}}_x$ of point type or special type are maximal elements, hence ends. Similarly, a valuation of finite type, say $\hat{\nu} = \text{val}[\hat{\phi}, \hat{\beta}]$ with $\hat{\phi} \in \hat{k}$ and $\hat{\beta} < \infty$ cannot be an end as we obtain a valuation dominating it by increasing $\hat{\beta}$.

Finally, let us show that rational valuations are branch points and irrational valuations regular points in $\widehat{\mathcal{V}}_x$. First consider $\hat{\nu} = \text{val}[\hat{\phi}, \hat{\beta}]$ rational, say $\hat{\phi} = \sum_1^{j_0} a_j x^{\hat{\beta}_j}$ and $\mathbf{Q} \ni \hat{\beta} > \hat{\beta}_{j_0}$. For $a \in \mathbf{C}^*$ define $\phi_a = \phi + a x^{\hat{\beta}}$ and $\hat{\nu}_a = \text{val}[\phi_a; \infty]$. Then $\hat{\nu}_a > \hat{\nu}$ for all a but if $a \neq b$, then $\hat{\nu}_a \wedge \hat{\nu}_b = \hat{\nu}$. Thus $\hat{\nu}$ is a branch point. Now suppose $\hat{\nu}$ is irrational, with $\hat{\phi} = \sum_1^{j_0} a_j x^{\hat{\beta}_j}$ and $\mathbf{Q} \not\ni \hat{\beta} > \hat{\beta}_{j_0}$. If $\hat{\nu}_1, \hat{\nu}_2 > \hat{\nu}$, then $\hat{\nu}_i = \text{val}[\phi + \phi_i; \hat{\alpha}_i]$ where $\hat{\alpha}_i > \hat{\beta}$ and $\hat{\nu}_\star(\phi_i) > \hat{\beta}$, $i = 1, 2$. Pick $\hat{\alpha}$ such that $\hat{\alpha} > \hat{\beta}$ but $\hat{\alpha} < \hat{\alpha}_i$, and define $\hat{\mu} = \text{val}[\hat{\phi}; \hat{\alpha}]$. Then $\hat{\nu} < \hat{\mu} < \hat{\nu}_i$ for $i = 1, 2$ so that $\hat{\nu}_1$ and $\hat{\nu}_2$ define the same tree tangent vector at $\hat{\nu}$. □

Proof (Lemma 4.6). First suppose that $\hat{\beta}_2 \geq \hat{\beta}_1$ and $\nu_\star(\hat{\phi}_2 - \hat{\phi}_1) \geq \hat{\beta}_1$. Let us show that $\hat{\nu}_1 \leq \hat{\nu}_2$. It suffices to show that $\hat{\nu}_1(y - \hat{\psi}) \leq \hat{\nu}_2(y - \hat{\psi})$ for every $\hat{\psi} \in \bar{k}$. Fix $\hat{\psi} \in \bar{k}$. By definition, $\hat{\nu}_i(y - \hat{\psi}) = \min\{\hat{\beta}_i, \nu_\star(\hat{\phi}_i - \hat{\psi})\}$ for $i = 1, 2$. The assumptions now give $\hat{\beta}_2 \geq \min\{\hat{\beta}_1, \nu_\star(\hat{\phi}_1 - \hat{\psi})\}$ and

$$\nu_\star(\hat{\phi}_2 - \hat{\psi}) \geq \min\{\nu_\star(\hat{\phi}_1 - \hat{\phi}_2), \nu_\star(\hat{\phi}_1 - \hat{\psi})\} \geq \min\{\hat{\beta}_1, \nu_\star(\hat{\phi}_1 - \hat{\psi})\}$$

so that $\hat{\nu}_1(y - \hat{\psi}) \leq \hat{\nu}_2(y - \hat{\psi})$ as claimed.

For the converse, first suppose $\hat{\nu}_\star(\hat{\phi}_2 - \hat{\phi}_1) < \hat{\beta}_1$. Pick $\hat{\psi} = \hat{\phi}_1$. Then $\hat{\nu}_1(y - \hat{\psi}) = \hat{\beta}_1$ but $\hat{\nu}_2(y - \hat{\psi}) = \min\{\hat{\beta}_2, \hat{\nu}_\star(\hat{\phi}_2 - \hat{\phi}_1)\} < \hat{\beta}_1$ so $\hat{\nu}_1 \not\leq \hat{\nu}_2$. Next suppose $\hat{\nu}_\star(\hat{\phi}_2 - \hat{\phi}_1) \geq \hat{\beta}_1$ but $\hat{\beta}_1 > \hat{\beta}_2$. Pick $\hat{\psi} = \hat{\phi}_2$. Then $\hat{\nu}_1(y - \hat{\psi}) = \min\{\hat{\beta}_1, \hat{\nu}_\star(\hat{\phi}_2 - \hat{\phi}_1)\} = \hat{\beta}_1$ but $\hat{\nu}_2(y - \hat{\psi}) = \hat{\beta}_2$ so again $\hat{\nu}_1 \not\leq \hat{\nu}_2$. This completes the proof. □

4.2.2 Puiseux Parameterization

Next we parameterize $\widehat{\mathcal{V}}_x$.

Definition 4.7. *If $\hat{\nu} \in \widehat{\mathcal{V}}_x$, then define the* Puiseux parameter *of $\hat{\nu}$ by*

$$\hat{\beta}(\hat{\nu}) := \sup\{\hat{\nu}(y - \hat{\psi}) \; ; \; \hat{\psi} \in \bar{k})\}.$$

Proposition 4.8. *The Puiseux parameter defines a parameterization $\hat{\beta} : \widehat{\mathcal{V}}_x \to [0, \infty]$ of the nonmetric tree $\widehat{\mathcal{V}}_x$ rooted in the valuation $\hat{\nu}_\star$. Moreover:*

(i) if $\hat{\nu}$ is rational, then $\hat{\beta}(\hat{\nu})$ is rational;
(ii) if $\hat{\nu}$ is irrational, then $\hat{\beta}(\hat{\nu})$ is irrational;
(iii) if $\hat{\nu}$ is of point type, then $\hat{\beta}(\hat{\nu}) = \infty$;
(iv) if $\hat{\nu}$ is of special type, then $\hat{\beta}(\hat{\nu}) \in (0, \infty]$.

Corollary 4.9. *The Puiseux parameter also defines a parameterization of the tree $\widehat{\mathcal{V}}_{\text{fin}}$ with values in $[0, \infty)$.*

Definition 4.10. *Given $\hat{\phi} \in \bar{k}_+$ and $\hat{\beta} \geq 0$ we denote by* $\text{val}[\hat{\phi}; \hat{\beta}]$ *the unique valuation in the segment $[\hat{\nu}_\star, \hat{\nu}_{\hat{\phi}}]$ of Puiseux parameter $\hat{\beta}$.*

Proposition 4.11. *If $\hat{\nu} = \text{val}[\hat{\phi}; \hat{\beta}]$ in the sense of Definition 4.10 then* (4.2) *holds. Thus the notation $\hat{\nu} = \text{val}[\hat{\phi}; \hat{\beta}]$ is compatible with that of Definition 4.2.*

Proposition 4.12. *If $\hat{\nu} \in \widehat{\mathcal{V}}_x$ and $\hat{\phi} \in \bar{k}_+$, then $\hat{\nu}_{\hat{\phi}} \geq \hat{\nu}$ iff $\hat{\nu}(y - \hat{\phi}) = \hat{\beta}(\hat{\nu})$.*

Definition 4.13. *The* Puiseux metric *is the metric on the tree $\widehat{\mathcal{V}}_{\text{fin}}$ induced by the Puiseux parameterization, i.e. the unique tree metric such that the distance between $\hat{\nu}_\star$ and $\hat{\nu}$ equals $\hat{\beta}(\hat{\nu})$.*

Proof (Proposition 4.8). Pick $\hat{\nu} \in \widehat{\mathcal{V}}_x$ and write $\hat{\nu} = \mathrm{val}[\hat{\phi}; \hat{\beta}]$ as in Definition 4.2. It follows from (4.2) that $\hat{\beta}(\hat{\nu}) = \hat{\beta}$. The proof of Proposition 4.4 then immediately gives that the Puiseux parameter defines a parameterization of $\widehat{\mathcal{V}}_x$. Assertions (i)-(iv) are direct consequences of Definition 4.2. □

Proof (Proposition 4.11). Write $\hat{\nu} = \mathrm{val}[\phi'; \hat{\beta}']$ in the sense of Definition 4.2. It follows from (4.2) that $\hat{\beta}'$ is the Puiseux parameter of $\hat{\nu}$, hence $\hat{\beta}' = \hat{\beta}$. Moreover, we have $\hat{\nu}_{\hat{\phi}} = \mathrm{val}[\hat{\phi}; \infty]$ in the sense of Definition 4.2 so since $\hat{\nu}_{\hat{\phi}} \geq \hat{\nu}$, Lemma 4.6 gives $\nu_\star(\hat{\phi} - \hat{\phi}') \geq \hat{\beta}$. It is then easy to see that

$$\min\{\hat{\beta}, \nu_\star(\hat{\psi} - \hat{\phi})\} = \min\{\hat{\beta}, \nu_\star(\hat{\psi} - \hat{\phi}')\} = \hat{\nu}(y - \hat{\psi})$$

for any $\hat{\psi} \in \hat{k}$, which completes the proof. □

Proof (Proposition 4.12). The proof is similar to that of Proposition 4.11. Write $\hat{\beta} = \hat{\beta}(\hat{\nu})$. Then $\hat{\nu} = \mathrm{val}[\hat{\phi}'; \hat{\beta}]$ in the notation of Definition 4.2 for some $\hat{\phi}' \in \bar{k}_+$. Lemma 4.6 gives $\hat{\nu}_{\hat{\phi}} \geq \hat{\nu}$ iff $\nu_\star(\hat{\phi} - \hat{\phi}') \geq \hat{\beta}$. Since $\hat{\nu}(y - \hat{\phi}) = \min\{\hat{\beta}, \nu_\star(\hat{\phi} - \hat{\phi}')\}$, this is equivalent to $\hat{\nu}(y - \hat{\phi}) = \hat{\beta}$. □

4.2.3 Multiplicities

Consider a Puiseux series $\hat{\phi} \in \bar{k}$ written in the form (4.1). Write $\hat{\beta}_j = r_j/s_j$ with $\gcd(r_j, s_j) = 1$. We define the *multiplicity* of $\hat{\phi}$ by $m(\hat{\phi}) = \mathrm{lcm}(s_j)$. Thus $m(\hat{\phi}) \in \overline{\mathbf{N}}$ and $m(\hat{\phi}) < \infty$ iff $\hat{\phi} \in \hat{k}$.

As in the case of $\mathcal{V}_{\mathrm{qm}}$ we can extend the notion of multiplicities from elements in $\bar{k}$ to valuations in $\widehat{\mathcal{V}}_x$ of finite type, using the tree structure.

Definition 4.14. *If $\hat{\nu} \in \widehat{\mathcal{V}}_x$ is of point type or finite type, then the* multiplicity *of $\hat{\nu}$ is $m(\hat{\nu}) = \min\{m(\hat{\phi})\ ;\ \hat{\phi} \in \bar{k}_+, \hat{\nu}_{\hat{\phi}} \geq \hat{\nu}\}$. If $\hat{\nu}$ is of special type then we set $m(\hat{\nu}) = \infty$.*

Proposition 4.15. *If $\hat{\mu} \leq \hat{\nu}$ then $m(\hat{\mu})$ divides $m(\hat{\nu})$. Further:*

(i) $m(\hat{\nu}) = \infty$ iff $\hat{\nu}$ is of special type or of point type associated to an element not in $\hat{k}$;
(ii) $m(\hat{\nu}) = 1$ iff there exists a change of variables $z = y - \hat{\phi}$ with $\hat{\phi} \in k_+$, such that have $\hat{\nu}(z - \hat{\psi}) = \min\{\hat{\beta}(\hat{\nu}), \nu_\star(\hat{\psi})\}$ for every $\hat{\psi} \in \bar{k}$.

Proof. We first claim that if $\hat{\nu} \in \widehat{\mathcal{V}}_x$ of finite type or point type and $\hat{\nu} = \mathrm{val}[\hat{\phi}; \hat{\beta}]$ as in Definition 4.2, then $m(\hat{\nu}) = m(\hat{\phi})$. This is trivial if $\hat{\nu}$ is of point type. If $\hat{\nu}$ is of finite type and $\hat{\nu}_{\hat{\psi}} > \hat{\nu}$, then $\nu_\star(\hat{\psi} - \hat{\phi}) \geq \hat{\beta}$. This implies that either $\hat{\psi} = \hat{\phi}$, in which case $m(\hat{\psi}) = m(\hat{\phi})$, or $\hat{\psi} = \hat{\phi}$ plus higher order terms, in which case $m(\hat{\psi})$ is a multiple of $m(\hat{\phi})$.

This claim together with Lemma 4.6 implies that $\hat{\mu} \leq \hat{\nu}$ implies that $m(\hat{\mu})$ divides $m(\hat{\nu})$. It also immediately gives (i).

Finally we prove (ii). If the condition in (ii) holds then $\hat{\nu} \leq \hat{\nu}_{\hat{\phi}}$ by Proposition 4.12. Since $\hat{\phi} \in k$, $m(\hat{\phi}) = 1$ and so $m(\hat{\nu}) = 1$. Conversely, if $m(\hat{\nu}) = 1$, then we can find $\phi \in \hat{k}_+$ with $m(\hat{\nu})$ such that $\hat{\nu} \leq \hat{\nu}_{\hat{\phi}}$. Thus $\hat{\phi} \in k_+$ and $\hat{\nu} = \mathrm{val}[\hat{\phi}; \hat{\beta}(\hat{\nu})]$ so we are done by Proposition 4.11. □

4.3 Galois Action

The Galois group $G := \mathrm{Gal}(\hat{k}/k)$ acts on $\hat{k}$ and by duality on valuations in $\widehat{\mathcal{V}}_x$. Here we show that this action respects the tree structure so that the orbit space $\widehat{\mathcal{V}}_x/G$ is in itself a tree. Later we shall show that $\widehat{\mathcal{V}}_x/G$ is in fact isomorphic to the relative valuative tree $\mathcal{V}_x$.

4.3.1 The Galois Group

As the field $\hat{k}$ can be viewed as the injective limit of field extensions, the Galois group $\mathrm{Gal}(\hat{k}/k)$ is naturally a projective limit. Let us be more precise.

For any $m \geq 1$, set $k_m = \mathbf{C}((x^{1/m}))$. The map $x^{1/m} \mapsto (x^{1/mn})^n \in k_{mn}$ defines a field extension $k_m \to k_{mn}$ and the fields k_m form an injective system with injective limit $\hat{k}$.

The Galois group $G_m = \mathrm{Gal}(k_m/k)$ is isomorphic to the set of complex numbers ω with $\omega^m = 1$. The action of $\omega \in G_m$ on $k_m = \mathbf{C}((x^{1/m}))$ is denoted by ω^* (or ω_* which is the same as G is abelian) and satisfies $\omega^*(x^{1/m}) = \omega x^{1/m}$. With this identification, there is a group homomorphism from G_{mn} to G_m given by $\omega \mapsto \omega^n$. The groups G_m form a projective system whose projective limit is the Galois group $G = \mathrm{Gal}(\hat{k}/k)$.

If $\omega \in G$, then we write ω_m for its image in G_m. The action of $\omega \in G$ on a monomial $x^\beta \in \hat{k}$ is then given as follows. Write $\beta = p/q$ with $\gcd(p,q) = 1$ and pick m such that q divides m, say $m = qr$. Then $\omega^*(x^\beta) = \omega_m^{pr} x^\beta$. This does not depend on the choice of m. It follows that G also acts naturally on the completion $\bar{k}$ of $\hat{k}$.

It is clear that $\nu_\star(\omega^*\hat{\phi}) = \nu_\star(\hat{\phi})$ and $m(\omega^*\hat{\phi}) = m(\hat{\phi})$ for every $\hat{\phi} \in \bar{k}$. Moreover, if $m(\hat{\phi}) = 1$, i.e. if $\hat{\phi} \in k$, then $\omega^*\hat{\phi} = \hat{\phi}$.

We extend the action of the Galois group to $\bar{k}[y]$ by declaring $\omega^* y = y$ for every $\omega \in G$. It follows that the restriction of ω^* to $k[[y]] \supset R$ is the identity for every ω.

4.3.2 Action on $\widehat{\mathcal{V}}_x$

By duality, the Galois group G acts on valuations in $\widehat{\mathcal{V}}_x$. If $\omega \in G$, then the action of ω is denoted by ω_* and is given by $(\omega_*\hat{\nu})(\hat{\psi}) := \hat{\nu}(\omega^*\hat{\psi})$. We claim that ω_* maps $\widehat{\mathcal{V}}_x$ into itself. Indeed, if $\hat{\nu}(y) > 0$, then $(\omega_*\hat{\nu})(y) = \hat{\nu}(y) > 0$, and if $\hat{\nu}$ extends $\nu_\star$, then so does $\omega_*\hat{\nu}$ since $\nu_\star(\omega^*\hat{\phi}) = \nu_\star(\hat{\phi})$ for every $\hat{\phi} \in \bar{k}$.

This calculation in conjunction with (4.3) also shows that the action is well-behaved on valuations of point type: $\omega_* \hat{\nu}_{\hat{\phi}} = \hat{\nu}_{\eta^* \hat{\phi}}$ for every $\hat{\phi} \in \bar{k}_+$, where $\eta = \omega^{-1} \in G$.

Proposition 4.16. *If $\omega \in G$, then $\omega_* : (\widehat{\mathcal{V}}_x, \hat{\beta}) \to (\widehat{\mathcal{V}}_x, \hat{\beta})$ is an isomorphism of parameterized trees and preserves multiplicity.*

Proof. We have to show that ω_* is a bijection that preserves the partial ordering, the Puiseux parameter and the multiplicity.

That ω_* is a bijection is obvious since $\omega_*^{-1}\omega_* = \omega_*\omega_*^{-1} = \text{id}$. That it preserves the partial ordering, and the Puiseux parameter are simple consequence of the equation

$$(\omega_*\hat{\nu})(y - \hat{\psi}) = \hat{\nu}(y - \omega^*\hat{\psi}) \quad \text{for } \hat{\nu} \in \widehat{\mathcal{V}}_x \text{ and } \hat{\psi} \in \bar{k}. \tag{4.6}$$

Indeed, (4.5) and (4.6) immediately give that $\hat{\nu} \le \hat{\mu}$ iff $\omega_*\hat{\nu} \le \omega_*\hat{\mu}$. Similarly, Definition 4.7 and (4.6) yield $\hat{\beta}(\omega_*\hat{\nu}) = \hat{\beta}(\hat{\nu})$ for every $\hat{\nu} \in \widehat{\mathcal{V}}_x$.

Finally let us show that $m(\omega_*\hat{\nu}) = m(\hat{\nu})$ for $\hat{\nu} \in \widehat{\mathcal{V}}_x$. Consider $\hat{\phi} \in \hat{k}$ such that $\hat{\nu}_{\hat{\phi}} > \hat{\nu}$ and $m(\hat{\phi}) = m(\hat{\nu})$. Write $\hat{\psi} = (\omega^{-1})^*\hat{\phi}$. Then $m(\hat{\psi}) = m(\hat{\phi}) = m(\hat{\nu})$ and $\hat{\nu}_{\hat{\psi}} = \omega_*\hat{\nu}_{\hat{\phi}} > \omega_*\hat{\nu}$. Hence $m(\omega_*\hat{\nu}) \le m(\hat{\nu})$. The reverse inequality follows by applying the same argument to ω^{-1} rather than ω. □

4.3.3 The Orbit Tree

Proposition 4.16 implies that the orbit space $\widehat{\mathcal{V}}_x/G$ has a natural tree structure. Let us be more precise. Declare two elements $\hat{\nu}, \hat{\mu}$ to be equivalent if there exists $\omega \in G$ with $\hat{\mu} = \omega_*\hat{\nu}$. Let $[\hat{\nu}]$ be denote the equivalence class containing $\hat{\nu}$ and let $\widehat{\mathcal{V}}_x/G$ be the set of equivalence classes. We define a partial ordering on $\widehat{\mathcal{V}}_x/G$ by $[\hat{\nu}] \le [\hat{\mu}]$ iff $\hat{\nu} \le \hat{\mu}$ for some representatives $\hat{\nu}$, $\hat{\mu}$. Proposition 4.16 implies that $\widehat{\mathcal{V}}_x/G$ is a complete, rooted nonmetric tree under this partial ordering and that $\hat{\beta}$ defines a well-defined parameterization. Moreover, the multiplicity m is also well-defined.

4.4 A Tale of Two Trees

We now wish to relate the tree $\widehat{\mathcal{V}}_x$ to the valuative tree $\mathcal{V}_x$. The elements of $\widehat{\mathcal{V}}_x$ are valuations on the ring $\bar{k}[[y]] \supset \mathbf{C}[[x]][[y]] = R$. Any element in $\widehat{\mathcal{V}}_x$ restricts to $\nu_\star$ on $\mathbf{C}[[x]]$ so valuations in $\widehat{\mathcal{V}}_x$ restrict to valuations on R satisfying $\nu(x) = 1$, and we end up with a mapping $\Phi : \widehat{\mathcal{V}}_x \to \mathcal{V}_x$. Our first goal is to show that Φ respects the two tree structures in the strongest possible sense.

Notice that Φ is not injective: the valuations $\text{val}[y \pm x^{3/2}; 5/2]$ in $\widehat{\mathcal{V}}_x$ both map to the valuation $\nu_{y^2 - x^3, 2}$ in $\mathcal{V}_x$. Our second goal is to understand this lack of injectivity. In fact, we shall see that Φ factors through the action of the Galois group $G = \text{Gal}(\hat{k}/k)$, resulting in an isomorphism between the orbit tree $\widehat{\mathcal{V}}_x/G$ defined in Section 4.3.3 and the relative valuative tree $\mathcal{V}_x$.

4.4.1 Minimal Polynomials

The main tool in the study of the restriction map from $\widehat{\mathcal{V}}_x$ to $\mathcal{V}_x$ is the relation between Puiseux series in $\hat{k}$ and irreducible elements in $\mathfrak{m}$. Recall that if $\phi \in \hat{k}_+$, then the *minimal polynomial* of $\hat{\phi}$ over k is given by

$$\phi(x,y) = \prod_{j=0}^{m-1} (y - \hat{\phi}_j) \tag{4.7}$$

where $m = m(\hat{\phi})$ is the multiplicity of $\hat{\phi}$, where $\omega \in G$ is an element whose image in G_m generates G_m, and where $\hat{\phi}_j = (\omega^j)^*(\hat{\phi})$. Thus $\phi \in \mathfrak{m}$ is irreducible in R and if $\psi \in \mathfrak{m} \subset \bar{k}[[y]]$ is irreducible in R and $y - \hat{\phi}$ divides ψ in $\bar{k}[[y]]$, then $\psi = \phi\xi$, where ξ is a unit in R. Conversely, if $\psi \in \mathfrak{m}$ is irreducible in R, then there exists a unit $\xi \in R$ and $\hat{\phi} \in \hat{k}_+$ such that $\psi = \phi\xi$, where ϕ is the minimal polynomial of $\hat{\phi}$.

Finally, two elements $\hat{\phi}, \hat{\psi} \in \hat{k}_+$ have the same minimal polynomial iff there exists $\omega \in G$ such that $\hat{\psi} = \omega^*\hat{\phi}$.

4.4.2 The Morphism

Recall that we are working with the relative tree $\mathcal{V}_x$. Thus the elements of $\mathcal{V}_x$ are normalized by $\nu(x) = 1$, and the root of $\mathcal{V}_x$ is the valuation div_x defined by $\phi \mapsto \max\{m \; ; \; x^m \mid \phi\}$. Also recall that $\mathcal{V}_x$ is equipped with the relative partial ordering $\leq_x$, the parameterization A_x by relative thinness, and the relative multiplicity function m_x. (The relative skewness will be of lesser importance here and can anyway be recovered from the relative thinness and multiplicity.)

By Section 3.7, $\widehat{\mathcal{V}}_x$ has a tree structure, comprised of the partial ordering $\leq$, the Puiseux parameter $\hat{\beta}$ and the multiplicity function m. Let us denote by $1 + \hat{\beta}$ the parameterization of $\widehat{\mathcal{V}}_x$ given by $(1 + \hat{\beta})(\hat{\nu}) = 1 + \hat{\beta}(\hat{\nu})$ for all $\hat{\nu} \in \widehat{\mathcal{V}}_x$.

We define a mapping $\Phi : \widehat{\mathcal{V}}_x \to \mathcal{V}_x$ letting $\Phi(\hat{\nu})$ be the restriction of $\hat{\nu}$ from $\bar{k}[[y]]$ to $R \subset \bar{k}[[y]]$. The fact that $\hat{\nu}$ restricts to $\hat{\nu}_\star$ on $\bar{k}$ implies that $\Phi(\hat{\nu})(x) = 1$. In particular $\Phi(\hat{\nu}_\star) = \mathrm{div}_x$ with the normalization convention above.

Theorem 4.17. *The mapping*

$$\Phi : (\widehat{\mathcal{V}}_x, 1 + \hat{\beta}) \to (\mathcal{V}_x, A_x)$$

is a surjective morphism of parameterized trees and preserves multiplicity. Further, Φ factors through the projection $\widehat{\mathcal{V}}_x \to \widehat{\mathcal{V}}_x/G$, where $G = \mathrm{Gal}(\hat{k}/k)$, and the induced mapping

$$(\widehat{\mathcal{V}}_x/G, 1 + \hat{\beta}) \to (\mathcal{V}_x, A_x)$$

is a multiplicity-preserving isomorphism of parameterized trees.

Moreover, for $\hat{\nu} \in \widehat{\mathcal{V}}_x$ and $\nu = \Phi(\hat{\nu})$ we have:

(i) $\hat{\nu}$ *is of finite type iff* ν *is quasimonomial;*
(ii) $\hat{\nu}$ *is rational iff* ν *is divisorial;*
(iii) $\hat{\nu}$ *is irrational iff* ν *is irrational;*
(iv) $\hat{\nu}$ *is of point type and finite multiplicity iff* ν *is a curve valuation; in this case* $\hat{\nu} = \hat{\nu}_{\hat{\phi}}$ *and* $\nu = \nu_\phi$*, where* ϕ *is the minimal polynomial of* $\hat{\phi}$ *over* k*;*
(v) $\hat{\nu}$ *is of point type and infinite multiplicity iff* ν *is infinitely singular with infinite thinness;*
(vi) $\hat{\nu}$ *is of special type iff* ν *is infinitely singular with finite thinness.*

Recall that the relative and nonrelative tree structures coincide on the subtree $\{\nu \geq_x \nu_{\mathrm{m}}\} = \{\nu \wedge \mathrm{div}_x = \nu_{\mathrm{m}}\}$ of $\mathcal{V}_x$ by Proposition 3.61. Hence we have

Corollary 4.18. Φ *restricts to a surjective morphism*

$$(\{\hat{\nu} \geq \mathrm{val}[0;1]\}, 1 + \hat{\beta}) \to (\{\nu \geq_x \nu_{\mathrm{m}}\}, A)$$

of parameterized trees and preserves multiplicity.

The fact that (relative) thinness can be computed in terms of Puiseux series is very useful in practice and will play an important role in Chapter 6.

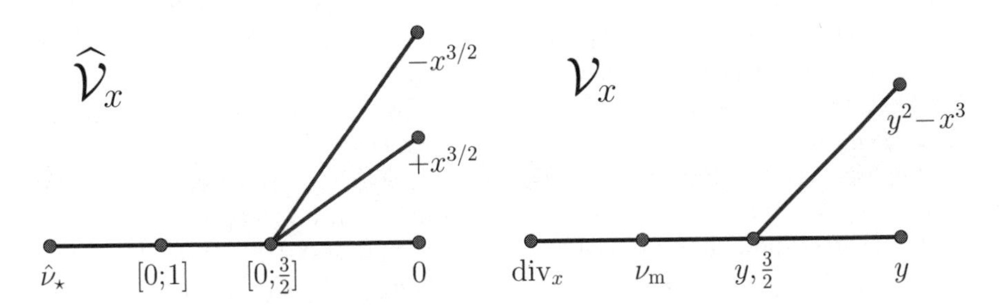

Fig. 4.1. The tree structures on the spaces $\widehat{\mathcal{V}}_x$ and $\mathcal{V}_x$ and their relation given by Theorem 4.17. The segment in $\widehat{\mathcal{V}}_x$ joining the root $\hat{\nu}_\star$ and the valuation $\hat{\nu}_0$ of point type is mapped onto the segment in $\mathcal{V}_x$ joining the root div_x and the curve valuation ν_y. The two segments in $\widehat{\mathcal{V}}_x$ joining $\hat{\nu}_\star$ and the valuations $\nu_{\pm x^{3/2}}$ of point type are both mapped onto the segment in $\mathcal{V}_x$ joining the root div_x and the curve valuation $\nu_{y^2-x^3}$.

4.4.3 Proof

We now turn to the proof of Theorem 4.17. Almost the whole proof boils down to a study of the factorization in (4.7).

First notice that Φ is strictly increasing. Indeed, if $\hat{\nu} \leq \hat{\mu}$ in $\widehat{\mathcal{V}}_x$, then $\hat{\nu} \leq \hat{\mu}$ pointwise on $\bar{k}[y]$ so by restriction $\Phi(\hat{\nu}) \leq \Phi(\hat{\mu})$ on R, whence $\hat{\nu} \leq_x \hat{\mu}$. If moreover $\nu < \mu$, then there exists $\hat{\phi} \in \hat{k}_+$ such that $\hat{\nu}(y - \hat{\phi}) < \hat{\mu}(y - \hat{\phi})$. This implies $\Phi(\hat{\nu})(\phi) < \Phi(\hat{\mu})(\phi)$, where $\phi \in \hat{k}[y] \subset \mathfrak{m}$ is the minimal polynomial of $\hat{\phi}$ over k. Thus $\Phi(\nu) <_x \Phi(\mu)$.

Let us study the restriction of Φ to valuations of point type and finite multiplicity. Consider $\hat{\phi} \in \hat{k}_+$ and let ϕ be the minimal polynomial of $\hat{\phi}$ over k. It is given by (4.7). Then $\hat{\nu}_{\hat{\phi}}(y - \hat{\phi}) = \infty$, so $\Phi(\hat{\nu}_{\hat{\phi}})(\phi) = \infty$, which implies $\Phi(\hat{\nu}_{\hat{\phi}}) = \nu_\phi$ (normalized by $\nu_\phi(x) = 1$)). See the discussion following Lemma 1.5. Essentially the same argument shows that if $\hat{\nu} \in \widehat{\mathcal{V}}_x$ and $\Phi(\hat{\nu}) = \nu_\phi$ is a curve valuation, then $\hat{\nu} = \hat{\nu}_{\hat{\phi}}$ as above. This proves (iv). Also notice that (4.7) shows that the multiplicity of $\hat{\phi}$ equals the relative multiplicity of ϕ (they are both equal to m).

The next step is to study the restriction of Φ to $\widehat{\mathcal{V}}_{\text{fin}}$, the set of valuations of finite type. Generally speaking, a valuation of finite type can be understood through the valuations of point type (say of finite multiplicity) that dominate it. Thus the previous paragraph, together with the fact that Φ is order-preserving, allows us to understand the restriction of Φ to $\widehat{\mathcal{V}}_{\text{fin}}$. Of course we have to make this precise.

We claim that if $\hat{\nu} \in \widehat{\mathcal{V}}_{\text{fin}}$, and $\phi \in \mathfrak{m}$ is irreducible, then $\Phi(\hat{\nu}) <_x \nu_\phi$ iff there exists $\hat{\phi} \in \hat{k}$ such that ϕ is the minimal polynomial of $\hat{\phi}$ over k (up to a unit) and $\hat{\nu} < \hat{\nu}_{\hat{\phi}}$. Indeed, write $\phi = \prod_j (y - \hat{\phi}_j)$ as in (4.7), and write $\hat{\nu}_j = \hat{\nu}_{\hat{\phi}_j}$. Suppose $\Phi(\hat{\nu}) <_x \nu_\phi$ but $\hat{\nu} \not< \hat{\nu}_j$ for all j. Then we may pick $\hat{\nu}' < \hat{\nu}$ such that $\hat{\nu}' \not< \hat{\nu}_j$ for all j. Since Φ is order-preserving we get $\Phi(\hat{\nu}') <_x \Phi(\hat{\nu}) <_x \nu_\phi$. But $\Phi(\hat{\nu}')(\phi) = \Phi(\hat{\nu})(\phi)$ since $\phi = \prod (y - \hat{\phi}_j)$, so this is a contradiction. The other implication is easy.

Next we wish to show that the restriction of Φ to $\widehat{\mathcal{V}}_{\text{fin}}$ preserves multiplicity and transports the (adjusted) Puiseux parameterization to relative thinness.

Let us first show that if $\hat{\nu}$ is of finite type, then $m_x(\Phi(\hat{\nu})) = m(\hat{\nu})$. Consider $\hat{\phi} \in \hat{k}_+$ with $\hat{\nu}_{\hat{\phi}} > \hat{\nu}$ and $m(\hat{\phi}) = m(\hat{\nu})$. Then $\nu_\phi >_x \Phi(\hat{\nu})$, where ϕ is the minimal polynomial of $\hat{\phi}$ over k. As noted above, (4.7) yields $m_x(\phi) = m(\hat{\phi})$. We infer $m_x(\Phi(\hat{\nu})) \le m(\hat{\nu})$. On the other hand, we saw above that if $\phi \in \mathfrak{m}$ is irreducible and $\nu_\phi > \nu$, then ϕ is, up to a unit in R, the minimal polynomial over k of some $\hat{\phi} \in \hat{k}_+$ such that $\hat{\nu}_{\hat{\phi}} > \hat{\nu}$. This gives the reverse inequality.

To relate the Puiseux parameterization to relative thinness, we first consider $\hat{\nu}, \hat{\nu}' \in \widehat{\mathcal{V}}_x$ of finite type with $\hat{\nu}' < \hat{\nu}$ such that the multiplicity is constant, say equal to m, on the segment $]\hat{\nu}', \hat{\nu}]$ in $\widehat{\mathcal{V}}_x$. Pick $\hat{\phi}_0 \in \hat{k}_+$ with $\hat{\nu}_{\hat{\phi}_0} > \hat{\nu}$ and $m(\hat{\phi}_0) = m$. Write $\nu = \Phi(\hat{\nu})$ and $\nu' = \Phi(\hat{\nu}')$. Let $\phi = \prod_j (y - \hat{\phi}_j)$ be the minimal polynomial of $\hat{\phi}_0$ over k. Our assumptions imply that $\nu_\star(\hat{\phi}_i - \hat{\phi}_j) < \hat{\beta}(\hat{\nu}')$ for $i \neq j$. Hence $\hat{\nu}(y - \hat{\phi}_j) = \hat{\nu}'(y - \hat{\phi}_j)$ for $j \neq 0$ in view of (4.2). Moreover, $\hat{\nu}(y - \hat{\phi}_0) = \hat{\beta}(\hat{\nu})$ and $\hat{\nu}'(y - \hat{\phi}_0) = \hat{\beta}(\hat{\nu}')$. This leads to

$$A_x(\nu) - A_x(\nu') = \nu(\phi) - \nu'(\phi) = \sum_0^{m-1} \left(\hat{\nu}(y - \hat{\phi}_j) - \hat{\nu}'(y - \hat{\phi}_j) \right) = \hat{\beta}(\hat{\nu}) - \hat{\beta}(\hat{\nu}'),$$

where we have used Lemma 3.67. We also have $\Phi(\hat{\nu}_\star) = \mathrm{div}_x$ and $A_x(\mathrm{div}_x) = 1 = 1 + \hat{\beta}(\hat{\nu}_\star)$. Given $\hat{\nu} \in \widehat{\mathcal{V}}_{\mathrm{fin}}$ we can break up the segment $[\hat{\nu}_\star, \hat{\nu}]$ into finitely many pieces on each of which the multiplicity is constant. By applying the equation above to each piece we get $A_x(\Phi(\hat{\nu})) = 1 + \hat{\beta}(\hat{\nu})$.

We have shown that $\Phi : \widehat{\mathcal{V}}_x \to \mathcal{V}_x$ is order-preserving and that $A_x \circ \Phi = 1 + \hat{\beta}$ on $\widehat{\mathcal{V}}_{\mathrm{fin}}$. Since $\widehat{\mathcal{V}}_x$ is the completion of $\widehat{\mathcal{V}}_{\mathrm{fin}}$, this immediately implies that $A_x \circ \Phi = 1 + \hat{\beta}$ on all of $\widehat{\mathcal{V}}_x$, so that Φ gives a morphism of parameterized trees.

We also know that $m_x(\Phi(\hat{\nu})) = m(\hat{\nu})$ for $\hat{\nu}$ of finite type and for $\hat{\nu}$ of point type of finite multiplicity. If $\hat{\nu}$ is either of special type or of point type with infinite multiplicity, then $m(\hat{\nu}) = \infty$. Thus $m(\hat{\mu}) \to \infty$ as $\hat{\mu} \to \hat{\nu}$ along the segment $[\hat{\nu}_\star, \hat{\nu}[$. The valuations in this segment are of finite type, so $m_x(\Phi(\hat{\mu})) \to \infty$. As Φ is order-preserving, $m_x(\Phi(\hat{\mu}))$ increases to $m_x(\Phi(\hat{\nu}))$ so $m_x(\Phi(\hat{\nu})) = \infty$. Thus Φ preserves multiplicity on all of $\widehat{\mathcal{V}}_x$.

Statements (i)-(vi) now follow from the characterizations of the elements in $\widehat{\mathcal{V}}_x$ and $\mathcal{V}_x$ in terms of the tree structure.

Finally, to show that Φ induces an isomorphism of the orbit tree $\widehat{\mathcal{V}}_x/G$ onto $\mathcal{V}_x$ it suffices to prove that if $\hat{\nu}, \hat{\mu} \in \widehat{\mathcal{V}}_x$, then $\Phi(\hat{\nu}) = \Phi(\hat{\mu})$ iff there exists $\omega \in G$ such that $\hat{\mu} = \omega_* \hat{\nu}$. One of these implications is trivial since the restriction ω^* to R is the identity for any $\omega \in G$.

For the other implication, suppose $\Phi(\hat{\nu}) = \Phi(\hat{\mu}) =: \nu$ for some $\hat{\nu}, \hat{\mu} \in \widehat{\mathcal{V}}_x$. Then $\hat{\nu}$ and $\hat{\mu}$ share the same Puiseux parameter and multiplicity. First assume the multiplicity is finite. Then $m_x(\nu) < \infty$ so we can find $\phi \in \mathfrak{m}$ irreducible with $\nu_\phi \geq \nu$. As we showed above, this implies that there exist $\hat{\phi}, \hat{\psi} \in \hat{k}$ both having ϕ as minimal polynomial over k and such that $\hat{\nu}_{\hat{\phi}} \geq \hat{\nu}$ and $\hat{\nu}_{\hat{\psi}} \geq \hat{\mu}$. Then $\eta^* \hat{\psi} = \hat{\phi}$ for some $\eta \in G$, hence $\omega_* \hat{\nu}_{\hat{\psi}} = \hat{\nu}_{\hat{\phi}}$, where $\omega = \eta^{-1} \in G$. By Proposition 4.16 we infer that ω_* maps the segment $[\hat{\nu}_\star, \hat{\nu}_{\hat{\psi}}]$ onto $[\hat{\nu}_\star, \hat{\nu}_{\hat{\phi}}]$, preserving the Puiseux parameter. Since $\hat{\nu}$ and $\hat{\mu}$ have the same Puiseux parameter, this implies $\omega_* \hat{\mu} = \hat{\nu}$.

Now suppose $\Phi(\hat{\nu}) = \Phi(\hat{\mu}) = \nu$ and $m(\hat{\nu}) = m(\hat{\mu}) = \infty$. Pick an increasing sequence $(\hat{\nu}_n)_1^\infty$ of rational valuations in the segment $[\hat{\nu}_\star, \hat{\nu}[$ such that $m_n := m(\hat{\nu}_n)$ is strictly increasing and $\hat{\nu}_n \to \hat{\nu}$ as $n \to \infty$. Also pick $\hat{\phi}_n \in \hat{k}$ such that $m(\hat{\phi}_n) = m_n$, and $\hat{\nu}_{\hat{\phi}_n} \wedge \hat{\nu} = \hat{\nu}_n$. Write $\nu_n = \Phi(\hat{\nu}_n)$ and let $\phi_n \in \mathfrak{m}$ be the minimal polynomial of $\hat{\phi}_n$ over k. Then ν_n increases to ν, $m_x(\nu_n) = m_n$, $m_x(\phi_n) = m_n$ and $\nu_{\phi_n} \wedge \nu = \nu_n$. By what precedes there is a unique preimage $\hat{\mu}_n$ under Φ of ν_n in the segment $[\hat{\nu}_\star, \hat{\mu}[$ and we have $m(\hat{\mu}_n) = m_n$. All preimages of ν_{ϕ_n} under Φ are of the form $\hat{\nu}_{\hat{\psi}_n}$ where $\hat{\psi}_n \in \hat{k}$ and the minimal polynomial of $\hat{\psi}_n$ over k is ϕ_n. We automatically have $m(\hat{\psi}_n) = m_n$ and we can pick $\hat{\psi}_n$ such that $\hat{\nu}_{\hat{\psi}_n} \wedge \hat{\mu} = \hat{\mu}_n$.

Since $\hat{\psi}_n$ and $\hat{\phi}_n$ have the same minimal polynomial over k there exists $\eta_n \in G$ such that $\eta_n^* \hat{\psi}_n = \hat{\phi}_n$. Notice while there are many such η_n, they all have the same image in G_{m_n}. Thus there exists a unique $\eta \in G_{m_n}$ whose image in G_{m_n} agrees with that of η_n. This η satisfies $\eta^*\hat{\psi}_n = \hat{\phi}_n$ for all n. As a consequence $\omega_*\hat{\nu}_{\hat{\psi}_n} = \hat{\nu}_{\phi_n}$ for all n, where $\omega = \eta^{-1} \in G$. Since ω_* preserves the Puiseux parameter, $\omega_*\hat{\mu}_n = \hat{\nu}_n$ for all n, so finally $\omega_*\hat{\mu} = \hat{\nu}$ as desired.

This concludes the proof of Theorem 4.17.

4.5 The Berkovich Projective Line

We next indicate how the valuative spaces $\mathcal{V}_x$ and $\widehat{\mathcal{V}}_x$ are naturally embedded (as nonmetric trees) in the Berkovich projective lines over the local fields k and $\bar{k}$, respectively.

The Berkovich affine line $\mathbb{A}^1(k)$ is defined [Be, p.19] to be the set of valuations $\nu : k[y] \to (-\infty, +\infty]$ extending $\nu_\star$. The line $\mathbb{A}^1(\bar{k})$ is defined analogously. We will write ν for elements of $\mathbb{A}^1(k)$ and $\hat{\nu}$ for elements of $\mathbb{A}^1(\bar{k})$.

Both lines are endowed with the weak topology, defined in terms of pointwise convergence. By definition, $\widehat{\mathcal{V}}_x$ is a subset of $\mathbb{A}^1(\bar{k})$. Berkovich defines [Be, p.18] the open disk in $\mathbb{A}^1(\bar{k})$ with center $\hat{\phi} \in \bar{k}$ and radius $r > 0$ to be $D(\hat{\phi}, r) = \{\hat{\nu} \; ; \; \hat{\nu}(y - \hat{\phi}) > \log r\}$. It follows that *$\widehat{\mathcal{V}}_x$ is the closure of the open disk in $\mathbb{A}^1(\bar{k})$ of radius 1 centered at zero.* Disks in $\mathbb{A}^1(k)$ are defined as images of disks under the restriction map $\mathbb{A}^1(\bar{k}) \to \mathbb{A}^1(k)$. It follows that the disk $D(0,1)$ in $\mathbb{A}^1(k)$ is the set of valuations $\nu : k[y] \to (-\infty, \infty]$ extending $\nu_\star$ and satisfying $\nu(y) > 0$. Any such valuation is an element of $\mathcal{V}_x$, if we use the normalization $\nu(x) = 1$. Conversely, any element of $\mathcal{V}_x$, save for div_x, can be obtained in this way. Hence *$\mathcal{V}_x$ is homeomorphic to the closure of the open disk in $\mathbb{A}^1(\bar{k})$ of radius 1 centered at zero.*

The Berkovich projective line $\mathbb{P}^1(k)$ is defined as $\mathbb{A}^1(k) \cup \{\nu_\infty\}$ where $\nu_\infty : k[y] \to [-\infty, \infty]$ equals $\nu_\star$ on k and $-\infty$ elsewhere. The line decomposes as

$$\mathbb{P}^1(k) = \{\hat{\nu}_\star\} \sqcup \bigsqcup_{\alpha \in \mathbf{P}^1(\mathbf{C})} D_\alpha,$$

where $D_\alpha := D(y - \alpha, 1)$ for $\alpha \in \mathbf{C}$ and $D_\infty := \{\nu \in \mathbb{A}^1(k) \; ; \; \nu(y) < 0\} \cup \{\nu_\infty\}$ and $\hat{\nu}_\star \in \mathbb{A}^1(k)$ denotes the valuation $\hat{\nu}_\star(\sum_j a_j y^j) = \min_j \nu_\star(a_j)$.

Any translation $y \mapsto y + \tau$, $\tau \in \mathbf{C}$, induces an isomorphism of $\mathbb{P}^1(k)$ sending D_α homeomorphically to $D_{\alpha+\tau}$. Similarly, the inversion $y \mapsto y^{-1}$ induces a homeomorphism of D_∞ onto D_0. The open disks D_α are hence all homeomorphic, and their closures are $\overline{D_\alpha} = D_\alpha \cup \{\hat{\nu}_\star\}$.

On any disk $\overline{D_\alpha}$ we can put the nonmetric tree structure induced by the relative tree structure on $\mathcal{V}_x$ (rooted at div_x, see Section 3.9). For instance, on $\overline{D_0} \simeq \mathcal{V}_x$, the tree structure is given by $\nu \leq \mu$ iff $\nu(\phi) \leq \mu(\phi)$ for all $\phi \in k[y]$. Patched together, these partial orderings endow $\mathbb{P}^1(k)$ with a natural nonmetric tree structure rooted at $\nu_\star$ (see [Be, Theorem 4.2.1]). As we shall see below (Theorem 5.1), the weak tree topology on $\mathbb{P}^1(k)$ coincides with the weak topology.

This discussion goes through, essentially verbatim, with k replaced by $\bar{k}$.

4.6 The Bruhat-Tits Metric

The standard projective line $\mathbf{P}^1(\bar{k})$ embeds naturally in the Berkovich line $\mathbb{P}^1(\bar{k})$: $\hat{\phi} \in \bar{k}$ corresponds to the valuation $\hat{\nu}_{\hat{\phi}}$ with $\hat{\nu}_{\hat{\phi}}(y - \hat{\phi}) = \infty$ and ∞ to $\hat{\nu}_\infty$. The points in $\mathbf{P}^1(\bar{k})$ are ends of the nonmetric tree $\mathbb{P}^1(\bar{k})$; hence $\mathbb{H} := \mathbb{P}^1(\bar{k}) \setminus \mathbf{P}^1(\bar{k})$ has a nonmetric tree structure rooted at $\hat{\nu}_\star$. The group $\mathrm{PGL}_2(\bar{k})$ of Möbius transformations acts on $\mathbf{P}^1(\bar{k})$ and this action extends to $\mathbb{P}^1(\bar{k})$ by $(M_*\hat{\nu})(y - \hat{\phi}) = \hat{\nu}(My - \phi)$.

Berkovich noted that $\mathbb{H}$ is isomorphic (as a non-metric tree) to the Bruhat-Tits building of $\mathrm{PGL}_2(\bar{k})$. We will not define this building nor the isomorphism here. Suffice it to say that the building is a metric tree on which $\mathrm{PGL}_2(\bar{k})$ acts by *isometries*. This last condition in fact defines the tree metric on $\mathbb{H}$ up to a constant. To see this, consider the segment in $\mathbb{H}$ parameterized by $\mathrm{val}[0;t]$, $0 \le t < \infty$. Fix any rational $t_0 > 0$ and pick $\hat{\phi}_0$ with $\hat{\nu}_\star(\hat{\phi}_0) = t_0$. Then $M \in \mathrm{PGL}_2(\bar{k})$ defined by $My = y\hat{\phi}_0$ gives $M_* \mathrm{val}[0;t] = \mathrm{val}[0; t + t_0]$. But the only translation invariant metric on $\mathbf{R}$ is the Euclidean metric (up to a constant), so after normalizing we get that $[0, \infty[\ni t \mapsto \mathrm{val}[0,t] \in \mathbb{H}$ is an isometry. Now if $\hat{\phi} \in \bar{k}$, then $M(y) := y - \hat{\phi}$ induces an isometry of the segment $[\hat{\nu}_\star, \hat{\nu}_{\hat{\phi}}[$ onto $[\hat{\nu}_\star, \hat{\nu}_0[$. Similarly, $M(y) = 1/y$ gives an isometry of $[\hat{\nu}_\star, \infty[$ onto $[\hat{\nu}_\star, \hat{\nu}_0[$. Thus the metric on $\mathbb{H}$ is uniquely defined.

In particular we see that if $\hat{\phi} \in \bar{k}_+$, then $[0, \infty[\ni t \mapsto \mathrm{val}[\hat{\phi}, t] \in \mathbb{H}$ is an isometry. Thus *the Puiseux metric on* $\widehat{\mathcal{V}}_{\mathrm{fin}}$ *is induced by the Bruhat-Tits metric on* $\mathbb{H} \supset \widehat{\mathcal{V}}_x$. Passing to $\mathcal{V}_x$ we conclude using Theorem 4.17 than *the relative thin metric on* $\mathcal{V}_{\mathrm{qm},x}$ *is induced by the Bruhat-Tits metric.*

4.7 Dictionary

<table>
<tr><th colspan="3">Valuations in $\mathcal{V}_x$</th><th colspan="2">Valuations in $\widehat{\mathcal{V}}_x$</th><th>Berkovich</th></tr>
<tr><td rowspan="2">Quasim.</td><td colspan="2">Divisorial</td><td rowspan="2">Finite type</td><td>Rational</td><td>Type 2</td></tr>
<tr><td colspan="2">Irrational</td><td>Irrational</td><td>Type 3</td></tr>
<tr><td rowspan="3">Not quasi-monomial</td><td colspan="2">Curve</td><td rowspan="2">Point type</td><td>$m < \infty$</td><td rowspan="2">Type 1</td></tr>
<tr><td rowspan="2">Inf sing</td><td>$A = \infty$</td><td>$m = \infty$</td></tr>
<tr><td>$A < \infty$</td><td colspan="2">Special type</td><td>Type 4</td></tr>
</table>

Table 4.1. Terminology

We end this chapter with a dictionary between valuations in $\mathcal{V}_x$ i.e. on $k[y]$, their preimages in $\widehat{\mathcal{V}}_x$ (i.e. on $\bar{k}[y]$) under the restriction map (see Theorem 4.17), and Berkovich's terminology [Be, p.18]. See Appendix C for a more extensive dictionary.

Note that except for $\nu = \mathrm{div}_x$, the terminology for a valuation $\nu \in \mathcal{V}_x$ and for its nonrelative counterpart $\nu/\nu(\mathfrak{m}) \in \mathcal{V}$ coincide. The valuation div_x in $\mathcal{V}_x$ is not centered at the origin, and is of divisorial type. It corresponds to ν_x which is a curve valuation in $\mathcal{V}$. The preimage of div_x in $\widehat{\mathcal{V}}_x$ is $\hat{\nu}_\star$ which is of rational type in our terminology and of type 2 in Berkovich's terminology. With our convention, div_x also fits into the table above.

5

Topologies

Our objective now is to introduce, analyze and compare different topologies on the valuative tree $\mathcal{V}$.

There are at least three natural ways of defining topologies. First, we can exploit the tree structure of $\mathcal{V}$. This leads to one topology defined using the nonmetric tree structure, and several others defined in terms of metric tree structures.

Second, recall that $\mathcal{V}$ is by definition a collection of (normalized) functions from R to $\overline{\mathbf{R}}_+$. Hence we can define topologies using pointwise or (suitably normalized) uniform convergence.

A third type of topologies can be defined on the set $\mathcal{V}_K$ of all equivalence classes of centered Krull valuations, and then passing to $\mathcal{V}$. As a Krull valuation takes values in an abstract ordered group, pointwise or uniform convergence a priori does not makes sense. However, we can always check whether the value is positive, zero or negative. Different topologies on the set $\{+, 0, -\}$ then lead to different topologies on $\mathcal{V}_K$. These can then be turned into topologies on the valuative tree.

The chapter is organized as follows. We start in Section 5.1 by considering the *weak topology*. As we show, it can be defined equivalently either in terms of pointwise convergence or in terms of the nonmetric tree structure.

Then in Sections 5.2 and 5.3 we study the *strong topology*. More precisely, there is one strong topology on $\mathcal{V}$ and one on the subtree $\mathcal{V}_{\mathrm{qm}}$ of quasimonomial valuations. They both arise from the parameterization of $\mathcal{V}$ by skewness. Alternatively, they can be defined in terms of normalized uniform convergence.

As a slight variation, Section 5.4 contains a brief discussion of the *thin topology*, which also comes in two versions: one on $\mathcal{V}$ and one on $\mathcal{V}_{\mathrm{qm}}$. Both versions arise from the parameterization of $\mathcal{V}$ by thinness.

After that we turn to the *Zariski topology*. It is defined on the set $\mathcal{V}_K$ of (equivalence classes of) centered Krull valuations. We analyze it in Section 5.5. The main result is that if we start with $\mathcal{V}_K$ endowed with the Zariski topology and identify exceptional curve valuations with their associated divisorial valuations, then we recover the valuative tree with the weak topology.

A refinement of the Zariski topology is the *Hausdorff-Zariski topology*, considered in Section 5.6. We show that it also has a natural interpretation as a weak tree topology, but now given in terms of a $\overline{\mathbf{N}}$-tree structure on $\mathcal{V}_K$.

Finally we compare the different topologies on $\mathcal{V}$ in Section 5.7.

5.1 The Weak Topology

At this stage we have two candidates for the weak topology on the valuative tree $\mathcal{V}$: the weak topology defined by pointwise convergence and the weak tree topology induced by the nonmetric tree structure on $\mathcal{V}$ (see Section 3.1.4). In this section we show that these two topologies are in fact identical. We also study its main properties.

5.1.1 The Equivalence

We defined the weak topology on $\mathcal{V}$ by describing converging sequences: $\nu_k \to \nu$ iff $\nu_k(\phi) \to \nu(\phi)$ for every $\phi \in \mathfrak{m}$. Equivalently, a basis of open sets is given by $\{\nu \in \mathcal{V} \; ; \; t < \nu(\phi) < t'\}$ over $\phi \in \mathfrak{m}$ irreducible and $t' > t \geq 1$. The weak tree topology is generated by $U(\vec{v})$, where $\vec{v}$ runs over all tangent vectors in $\mathcal{V}$.

Theorem 5.1. *The weak topology on $\mathcal{V}$ coincides with the weak tree topology.*

Proof. We use Proposition 3.25 repeatedly. If $\vec{v} \in T\nu$ is a tangent vector not represented by $\nu_{\mathfrak{m}}$ (true e.g. if $\nu = \nu_{\mathfrak{m}}$), then $U(\vec{v}) = \{\mu \; ; \; \mu(\phi) > \alpha\, m(\phi)\}$, where $\alpha = \alpha(\nu)$ and ν_ϕ represents $\vec{v}$. If instead $\vec{v}$ is represented by $\nu_{\mathfrak{m}}$, then $U(\vec{v}) = \{\mu \in \mathcal{V} \; ; \; \mu(\phi) < \alpha\, m(\phi)\}$, where $\nu_\phi \geq \nu$. In both cases $U(\vec{v})$ is open in the weak topology.

Conversely, every nontrivial set of the form $\{\nu(\phi) > t\}$ or $\{\nu(\phi) < t\}$ with $t \geq 1$ and $\phi \in \mathfrak{m}$ irreducible is of the form $U(\vec{v})$. This completes the proof. □

5.1.2 Properties

We now investigate the weak topology further.

Proposition 5.2. *The weak topology is compact but not metrizable.*

Proof. Compactness follows from Theorem 5.1 and the fact that any complete, parameterizable tree is weakly compact (see Theorem 3.26). Alternatively, $\mathcal{V}$ is naturally embedded as a closed subspace of the compact product space $[0, \infty]^R$.

To prove that $\mathcal{V}$ is not metrizable it suffices to show that $\nu_{\mathfrak{m}} \in \mathcal{V}$ has no countable basis of open neighborhoods. Fix local coordinates (x, y). For $\theta \in \mathbf{C}^*$, let $\nu_\theta := \nu_{y+\theta x, 2}$ (see Definition 3.27) and $U_\theta := \{\nu \not\geq \nu_\theta\}$. Any U_θ is an open neighborhood of $\nu_{\mathfrak{m}}$. Suppose $\{V_k\}_{k\geq 1}$ is a countable basis of open neighborhoods of $\nu_{\mathfrak{m}}$. Then for any θ there exists $k = k(\theta) \geq 0$ with $V_k \subset U_\theta$. As $\mathbf{C}^*$ is uncountable, one can find k and a sequence θ_i with $\theta_i \neq \theta_j$ for $i \neq j$ such that $V_k \subset U_{\theta_i}$. Now $\nu_{\theta_i} \to \nu_{\mathfrak{m}}$, so $\nu_{\theta_i} \in V_k$ for $i \gg 1$. But $\nu_{\theta_i} \notin U_{\theta_i}$. This is a contradiction. □

Proposition 5.3. *The four subsets of $\mathcal{V}$ consisting of divisorial, irrational, infinitely singular and curve valuations are all weakly dense in $\mathcal{V}$.*

Proof. Instead of proving all of this directly we will appeal to Proposition 5.9 which asserts that the divisorial, irrational and infinitely singular valuations are all (individually) dense in $\mathcal{V}$ in a stronger topology that the weak topology. Hence it suffices to show that, say, any divisorial valuation can be weakly approximated by curve valuations. But if ν is divisorial, then by Proposition 3.20 there exists a sequence ϕ_n of irreducible elements in $\mathfrak{m}$ such that the associated curve valuations ν_{ϕ_n} represent distinct tangent vectors at ν. Then $\nu_{\phi_n} \to \nu$ as $n \to \infty$. □

Remark 5.4. In Section 3.9 we considered the relative valuative tree $\mathcal{V}_x$ consisting of centered valuations ν on R normalized by $\nu(x) = 1$ (and with the valuation div_x added). We could define two relative weak topologies on $\mathcal{V}_x$: one using the relative nonmetric tree structure and one using pointwise convergence. The proof of Theorem 5.1 could easily be adjusted to show that these relative weak topologies are in fact the same.

5.2 The Strong Topology on $\mathcal{V}$

As with the weak topology, we have two candidates for the strong topology on $\mathcal{V}$: one defined using uniform sequential convergence (see (5.2) below) and one induced by the tree metric (3.5) on $\mathcal{V}$. We now show that these two coincide and analyze the properties of the strong topology.

5.2.1 Strong Topology I

A general metric tree admits two natural topologies, the weak topology as a nonmetric tree and the topology as a metric space. We emphasize that the latter topology depends on the choice of metric.

The valuative tree carries several interesting metric tree structures. For the moment we are mainly interested in the tree metric d induced by skewness and defined in (3.5):

$$d(\mu, \nu) = \left(\alpha(\mu \wedge \nu)^{-1} - \alpha(\mu)^{-1}\right) + \left(\alpha(\mu \wedge \nu)^{-1} - \alpha(\nu)^{-1}\right). \tag{5.1}$$

We shall refer to the induced topology as the *strong tree topology* on $\mathcal{V}$.

Before analyzing the strong topology in detail, let us note the following general result.

Proposition 5.5. *The topology on a metric tree $(\mathcal{T}, d)$ induced by the metric is at least as strong as the weak topology.*

Proof. Let σ_n be a sequence of points in the metric tree $(\mathcal{T}, d)$ that converges to $\sigma_\infty \in \mathcal{T}$ in the metric. Use σ_∞ as a root. If $\sigma_n \not\to \sigma_\infty$ weakly, then there would exist $\sigma > \sigma_\infty$ such that $\sigma_n \geq \sigma$ for infinitely many n. Thus $d(\sigma_n, \sigma_\infty) \geq d(\sigma, \sigma_\infty) > 0$ for these n, a contradiction. □

Remark 5.6. If $\mathcal{T}$ is as in Example 3.6 with X infinite and $g(x) \equiv 1$, then the two topologies are not equivalent: if $(x_n)_{n\geq 0}$ are distinct elements in X then $(x_n, 1)$ converges weakly to $(x, 0)$, but $d((x_n, 1), (x, 0)) = 1$.

5.2.2 Strong Topology II

Another strong topology on $\mathcal{V}$ can be defined in a quite general setting (i.e. for other rings than R) in terms of the metric

$$d^{\mathrm{str}}_{\mathcal{V}}(\nu_1, \nu_2) = \sup_{\phi \in \mathfrak{m} \text{ irreducible}} \left| \frac{m(\phi)}{\nu_1(\phi)} - \frac{m(\phi)}{\nu_2(\phi)} \right| \tag{5.2}$$

This topology is stronger than the weak topology and any formal invertible mapping $f : (\mathbf{C}^2, 0) \to (\mathbf{C}^2, 0)$ induces an isometry $f_* : \mathcal{V} \to \mathcal{V}$ for $d^{\mathrm{str}}_{\mathcal{V}}$.

5.2.3 The Equivalence

We now show that the metric $d^{\mathrm{str}}_{\mathcal{V}}$ is compatible with the tree metric d given by (5.1). As a consequence, the two strong topologies on $\mathcal{V}$ defined in Sections 5.2.1 and 5.2.2 are really the same.

Theorem 5.7. *The strong topology on $\mathcal{V}$ is identical to the strong tree topology. More precisely, if d and d^{str} are the metrics on $\mathcal{V}$ given by (5.1) and (5.2), respectively, then, for $\nu_1, \nu_2 \in \mathcal{V}$:*

$$d^{\mathrm{str}}_{\mathcal{V}}(\nu_1, \nu_2) \leq d(\nu_1, \nu_2) \leq 2 d^{\mathrm{str}}_{\mathcal{V}}(\nu_1, \nu_2) \tag{5.3}$$

Proof. Let us consider $\phi \in \mathfrak{m}$ irreducible. Set $\nu = \nu_1 \wedge \nu_2$. We first show that

$$\left|\nu_1^{-1}(\phi) - \nu_2^{-1}(\phi)\right| = \max_{i=1,2} \left\{\nu^{-1}(\phi) - \nu_i^{-1}(\phi)\right\}. \tag{5.4}$$

To see this, notice that Proposition 3.25 implies $\mu(\phi) = (\mu \wedge \nu_\phi)(\phi)$ for any $\mu \in \mathcal{V}$. Thus we may replace ν_i by $\nu_i \wedge \nu_\phi$, so that $\nu \leq \nu_i \leq \nu_\phi$, $i = 1, 2$. But this means that $\nu = \nu_1 \leq \nu_2 \leq \nu_\phi$ or $\nu = \nu_2 \leq \nu_1 \leq \nu_\phi$ and then (5.4) is immediate.

Multiplying (5.4) by $m(\phi)$ and taking the supremum over $\phi \in \mathfrak{m}$ we get $d^{\mathrm{str}}_{\mathcal{V}}(\nu_1, \nu_2) = \max_i d(\nu_i, \nu)$, which implies (5.3). □

5.2.4 Properties

We now investigate the strong topology further.

Proposition 5.8. *The strong topology on $\mathcal{V}$ is strictly stronger than the weak topology. It is not locally compact.*

Proof. The first assertion follows from Proposition 5.2, or from the last assertion as $\mathcal{V}$ is weakly compact Consider $\nu \in \mathcal{V}$ divisorial and pick $\phi_n \in \mathfrak{m}$ irreducible with $\nu_{\phi_n} > \nu$, such that ν_{ϕ_n} represent distinct tangent vectors at ν. For fixed $\varepsilon > 0$, set $\nu_n = \nu_{\phi_n,\alpha+\varepsilon}$, where α is the skewness of ν. Any ν_n is at distance $\varepsilon/\alpha(\alpha+\varepsilon)$ from ν. Further, $\nu_n \to \nu$ weakly, so ν_n has no strong accumulation point. If ν had a compact strong neighborhood, it would contain a ball of positive radius, say bigger than ε. Then ν_n would have a strongly convergent subsequence. This is impossible. □

Proposition 5.9. *The three subsets of $\mathcal{V}$ consisting of divisorial, irrational and infinitely singular valuations are all strongly dense in $\mathcal{V}$.*

The strong closure of the set $\mathcal{C}$ of curve valuations is the set of valuations of infinite skewness. In particular, $\overline{\mathcal{C}} \setminus \mathcal{C}$ contains only infinitely singular valuations.

Proof. For the first assertion, we first prove that any valuation $\nu \in \mathcal{V}$ can be strongly approximated by divisorial and irrational valuations. If ν is not infinitely singular, then $\nu = \nu_{\phi,t}$ for some irreducible $\phi \in \mathfrak{m}$ and $t \in [1, \infty]$. Then $\mu := \nu_{\phi,s}$ converges strongly to ν as $s \to t$ and μ is divisorial (irrational) if s is rational (irrational). If ν is infinitely singular, then $\nu = \lim \nu_n$ for an increasing sequence ν_n, where ν_n can be chosen to be all divisorial or all irrational.

The fact that any divisorial valuation can be approximated by an infinitely singular valuation follows from Lemma 5.16 below and $\alpha(\mu) - \alpha(\nu) < A(\mu) - A(\nu)$ when $\mu > \nu$. This completes the proof of the first assertion.

For the second assertion, first notice that skewness defines a strongly continuous function $\alpha : \mathcal{V} \to [1, \infty]$, so since every curve valuation has infinite skewness, so does every valuation in $\overline{\mathcal{C}}$. Conversely, suppose ν is an infinitely singular valuation with infinite skewness and let $(\nu_n)_0^\infty$ be its approximating sequence. Then $\alpha(\nu_n) \to \infty$ as $n \to \infty$. For each n, let μ_n be a curve valuation with $\mu_n > \nu_n$ and $m(\mu_n) = m(\nu_n) < b(\nu_n)$. We get

$$d(\nu_n, \nu) = \left(\frac{1}{\alpha(\nu \wedge \mu_n)} - \frac{1}{\alpha(\nu)}\right) + \left(\frac{1}{\alpha(\nu \wedge \mu_n)} - \frac{1}{\alpha(\mu_n)}\right) = \frac{2}{\alpha(\nu_n)} \to 0,$$

so $\mu_n \to \nu$ as $n \to \infty$. This completes the proof. □

Remark 5.10. It is also possible to define a *relative strong topology* on $\mathcal{V}_x$, based on the parameterization by relative skewness. Suitable versions of all the results above continue to hold in the relative setting. In Proposition 5.9

we should then use the convention that div_x is a divisorial valuation, and in particular not an element of $\mathcal{C}$. Also notice that a valuation in $\mathcal{V}_x \setminus \{\mathrm{div}_x\}$ has infinite relative skewness iff the corresponding valuation in $\mathcal{V}$ has infinite (nonrelative) skewness.

5.3 The Strong Topology on $\mathcal{V}_{\mathrm{qm}}$

In many instances it is natural and convenient to work on the tree $\mathcal{V}_{\mathrm{qm}}$ consisting of quasimonomial valuations. As described in Section 3.3.4 we can use the parameterization by skewness to define a metric on $\mathcal{V}_{\mathrm{qm}}$: for $\nu, \mu \in \mathcal{V}_{\mathrm{qm}}$ we have:

$$d_{\mathrm{qm}}(\mu, \nu) = (\alpha(\mu) - \alpha(\mu \wedge \nu)) + (\alpha(\nu) - \alpha(\mu \wedge \nu)). \tag{5.5}$$

We refer to the resulting topology as the *strong tree topology* on $\mathcal{V}_{\mathrm{qm}}$.

On the other hand, we could also consider the following strong metric:

$$d_{\mathrm{qm}}^{\mathrm{str}}(\mu, \nu) = \sup_{\phi \in \mathfrak{m} \text{ irreducible}} \left| \frac{\nu_1(\phi)}{m(\phi)} - \frac{\nu_2(\phi)}{m(\phi)} \right|$$

(It is in fact not hard to see that the supremum can be taken over all $\phi \in \mathfrak{m}$, not just ϕ irreducible.) Following the proof of Theorem 5.7, we infer

Theorem 5.11. *The strong topology coincides with the strong tree topology on $\mathcal{V}_{\mathrm{qm}}$. More precisely, for $\nu_1, \nu_2 \in \mathcal{V}_{\mathrm{qm}}$ we have*

$$d_{\mathrm{qm}}^{\mathrm{str}}(\nu_1, \nu_2) \leq d_{\mathrm{qm}}(\nu_1, \nu_2) \leq 2 d_{\mathrm{qm}}^{\mathrm{str}}(\nu_1, \nu_2). \tag{5.6}$$

Proposition 5.12. *The completion of $(\mathcal{V}_{\mathrm{qm}}, d_{\mathrm{qm}})$ is a tree naturally isomorphic to the union of $\mathcal{V}_{\mathrm{qm}}$ and all infinitely singular valuations with finite skewness.*

Proof. The subset $\mathcal{V}'$ of $\mathcal{V}$ consisting of valuations with finite skewness is a strongly open subtree that contains all quasimonomial valuations. The metric d_{qm} extends naturally to $\mathcal{V}'$ as a metric $d_{\mathcal{V}'}$. Let us show that $(\mathcal{V}', d_{\mathcal{V}'})$ is complete. Any $d_{\mathcal{V}'}$-Cauchy sequence ν_n in $\mathcal{V}'$ is $d_{\mathcal{V}}$-Cauchy, hence $d_{\mathcal{V}}(\nu_n, \nu) \to 0$ for some $\nu \in \mathcal{V}$ by completeness of $d_{\mathcal{V}}$. But it is easy to see that ν_n must have uniformly bounded skewness, so $\alpha(\nu) < \infty$ and then $d_{\mathcal{V}'}(\nu_n, \nu) \to 0$ as well. □

Remark 5.13. We can also equip the tree $\mathcal{V}_{\mathrm{qm},x}$ with a relative strong topology defined in terms of the parameterization by relative skewness. See Section 3.9 and compare Remark 5.10.

5.4 Thin Topologies

Instead of skewness we can use thinness to parameterize the valuative tree. As noted in Section 3.6 we can use this parameterization to define metrics D on $\mathcal{V}$ and D_{qm} :

$$D(\mu,\nu) = \left(A(\mu \wedge \nu)^{-1} - A(\mu)^{-1}\right) + \left(A(\mu \wedge \nu)^{-1} - A(\nu)^{-1}\right) \tag{5.7}$$

$$D_{\mathrm{qm}}(\mu,\nu) = (A(\mu) - A(\mu \wedge \nu)) + (A(\nu) - A(\mu \wedge \nu)). \tag{5.8}$$

We refer to D and D_{qm} as the *thin metric* on $\mathcal{V}$ and $\mathcal{V}_{\mathrm{qm}}$, respectively, and to the induced topologies as the *thin topologies.*

The thin metrics share a lot of similarities with the previously defined metrics d and d_{qm} defined in terms of skewness. We summarize them in the following propositions.

Proposition 5.14. *The metric space $(\mathcal{V}, D)$ is complete, and not locally compact. The three subsets consisting of divisorial, irrational and infinitely singular valuations are dense in $(\mathcal{V}, D)$. The closure of the set of curve valuations is the set of valuations of infinite thinness.*

Proposition 5.15. *The completion of $(\mathcal{V}_{\mathrm{qm}}, D_{\mathrm{qm}})$ is naturally isomorphic to the union of $\mathcal{V}_{\mathrm{qm}}$ and all infinitely singular valuations with finite thinness.*

Proof (Propositions 5.14 and 5.15). The proofs of these propositions are completely analogous to those of Propositions 5.8, 5.9 and 5.12. The only point which is not clear is the fact that we may approximate any divisorial valuation by a sequence of infinitely singular valuations. It is an immediate consequence of the lemma below. □

Lemma 5.16. *For any divisorial valuation ν and $\varepsilon > 0$, there exists an infinitely singular valuation $\mu > \nu$, with $A(\mu) < A(\nu) + \varepsilon$.*

Proof. Let us extend the approximating sequence $(\nu_i)_0^g$ of ν to an infinite approximating sequence $(\nu_i)_0^\infty$ as follows. First suppose $b(\nu) > m(\nu)$. By convention, $\nu_{g+1} = \nu$. Pick a curve valuation $\mu_{g+2} > \nu$ with $m(\mu_{g+2}) = b(\nu)$. Consider a divisorial valuation ν_{g+2} in the segment $]\nu, \mu_{g+2}[$. It follows from (3.13) that if ν_{g+2} is close enough to ν, then $b(\nu_{g+2}) > m(\nu)$. We may therefore pick ν_{g+2} in this segment such that $b(\nu_{g+2}) > m(\nu_{g+2}) = b(\nu)$ and $\alpha(\nu_{g+2}) - \alpha(\nu) < \varepsilon/(2b(\nu))$.

Inductively, given $(\nu_i)_1^{g+k}$ we construct divisorial valuations ν_{g+k+1} satisfying $\nu_{g+k+1} > \nu_{g+k}$, $b(\nu_{g+k+1}) > m(\nu_{g+k+1}) = b(\nu_{g+k})$ and $\alpha(\nu_{g+k+1}) - \alpha(\nu_{g+k}) < \varepsilon/(2^k b(\nu_{g+k}))$. Then $(\nu_i)_0^\infty$ defines an approximating sequence for an infinitely singular valuation μ, satisfying $\mu > \nu$ and $A(\mu) < A(\nu) + \varepsilon$.

If $b(\nu) = m(\nu)$ then we redefine ν_{g+1} to be a divisorial valuation with $\nu_{g+1} > \nu$, $b(\nu_{g+1}) > m(\nu_{g+1}) = m(\nu)$ and $\alpha(\nu_{g+1}) - \alpha(\nu) < \varepsilon/b(\nu)$. We may then continue as before and construct an infinite approximating sequence $(\nu_i)_0^\infty$ of an infinitely singular valuation μ with $\mu > \nu$ and $A(\mu) < A(\nu) + 2\varepsilon$. □

Remark 5.17. Using the relative thinness defined in Section 3.9 we can define relative thin topologies on the trees $\mathcal{V}_x$ and $\mathcal{V}_{\mathrm{qm},x}$. Compare Remarks 5.10 and 5.13.

Remark 5.18. Although we shall not pursue this further here, the analysis in Chapter 4 implies that the (relative) thin topologies can be defined in terms of (suitably normalized) uniform convergence when the valuations are extended to the ring of power series in one variable with coefficients that are Puiseux series.

5.5 The Zariski Topology

We now turn to a classical, but quite different construction, the Zariski topology. This topology, which is defined on the set $\mathcal{V}_K$, of equivalence classes of Krull valuations, is a not Hausdorff topology since divisorial valuations do not define closed points. We show how to make it Hausdorff by identifying a divisorial valuation with the valuations in its closure. The latter valuations are exactly the exceptional curve valuations, or, equivalently, the elements of $\mathcal{V}_K \setminus \mathcal{V}$. It is a remarkable fact that this procedure recovers $\mathcal{V}$ endowed with the weak topology.

5.5.1 Definition

The Zariski topology is defined on the set $\mathcal{V}_K$ of equivalence classes of centered Krull valuations on R (not necessarily $\overline{\mathbf{R}}_+$-valued, see Section 1.3). A Krull valuation ν is determined, up to equivalence, by its valuation ring $R_\nu = \{\nu \geq 0\} \subset K$, so an open set in $\mathcal{V}_K$ can also be viewed as a set of valuation rings satisfying certain conditions.

Definition 5.19. *A basis for the Zariski topology on $\mathcal{V}_K$ is given by*

$$V(z) := \{\nu \ ; \ \nu(z) \geq 0\} = \{\nu \ ; \ z \in R_\nu\}$$

over $z \in K$. In other words, an arbitrary open set in the Zariski topology is a union of finite intersections of sets of the type $V(z)$.

Remark 5.20. The topological space $\mathcal{V}_K$ endowed with the Zariski topology is called the *Riemann-Zariski variety* of R (see [ZS2, p.110], [Va]). This variety can actually be defined in a quite general context. When R is the ring of an algebraic variety V (of any dimension and over any field), one can show that the Riemann-Zariski variety is the projective limit of all birational models of V, and that the topology defined above coincides with the projective limit of the usual Zariski topology on each model.

Proposition 5.21. *$\mathcal{V}_K$ is quasi-compact but not Hausdorff. Moreover,*

(i) *Any non-divisorial valuation is a closed point.*
(ii) *The closure of a divisorial valuation associated to an exceptional component E is the set of exceptional curve valuations $\nu_{E,p}$ whose centers lie on E.*

Proof. That $\mathcal{V}$ is not Hausdorff follows from (ii). For the quasi-compactness of $\mathcal{V}_K$, see Theorem 5.26 below. Both (i) and (ii) are consequences of Lemma 1.5 since a valuation μ lies in the closure of another valuation $\nu \neq \mu$ iff $R_\mu \subsetneq R_\nu$. □

Remark 5.22. $\mathcal{V} \subset \mathcal{V}_K$ is neither open nor closed for the Zariski topology. Indeed, Example 5.31 shows that a sequence of divisorial valuations can converge to an exceptional curve valuation. Conversely, if E is an irreducible component of $\pi^{-1}(0)$ for some composition of blowups π, and $(p_j)_1^\infty$ is a sequence of distinct points on E, then the exceptional curve valuations ν_{E,p_j} converge to the divisorial valuation ν_E. See Appendix B.

Remark 5.23. The Zariski topology can be geometrically described as follows: a basis is given by $V(\pi, C)$, where π ranges over compositions of finitely many blowups and C over Zariski open subsets of the exceptional divisor $\pi^{-1}(0)$. Here C is the complement in $\pi^{-1}(0)$ of finitely many irreducible components and finitely many points, and $V(\pi, C)$ is the set of all $\nu \in \mathcal{V}_K$ whose associated sequence of infinitely near points $\Pi[\nu] = (p_j)$ are eventually in C.

5.5.2 Recovering $\mathcal{V}$ from $\mathcal{V}_K$

We can try to turn $\mathcal{V}_K$ into a Hausdorff space $\widetilde{\mathcal{W}}_K$ by identifying ν and ν' if they both belong to the Zariski closure of the same Krull valuation μ. By Proposition 5.21 this amounts to identifying exceptional curve valuations with their associated divisorial valuation. Let $p : \mathcal{V}_K \to \widetilde{\mathcal{W}}_K$ be the natural projection and endow $\widetilde{\mathcal{W}}_K$ with the quotient topology. Then $\widetilde{\mathcal{W}}_K$ is quasi-compact, being the image of a quasi-compact space by a continuous map.

Consider the natural injection $\imath : \mathcal{V} \to \mathcal{V}_K$, where $\mathcal{V}$ carries the weak topology.

Theorem 5.24. *The composition $p \circ \imath : \mathcal{V} \to \widetilde{\mathcal{V}}_K$ is a homeomorphism.*

Proof. Since $\mathcal{V}$ contains all divisorial valuations but no exceptional curve valuations, injectivity and surjectivity of $p \circ \imath$ follow from Proposition 5.21. Since $\mathcal{V}$ is Hausdorff and $\widetilde{\mathcal{V}}_K$ is quasi-compact, continuity of $(p \circ \imath)^{-1}$ will imply that $p \circ \imath$ is in fact a homeomorphism.

Therefore, let us show that $(p \circ \imath)^{-1}$ is continuous. A basis for the weak (tree) topology is given by the open sets $U(\vec{v})$, over tangent vectors $\vec{v}$. In fact, it suffices to take $\vec{v} \in T\nu$ with ν divisorial. For $\phi, \psi \in \mathfrak{m}$ irreducible and $t > 0$ define

$$V(\phi, \psi, t) := \left\{ \mu \in \mathcal{V} \; ; \; \frac{\mu(\phi)}{m(\phi)} < t \frac{\mu(\psi)}{m(\psi)} \right\}.$$

Proposition 3.25 implies that if $\vec{v} \in T\nu$ is not represented by ν_{m} (true e.g. if $\nu = \nu_{\mathrm{m}}$), then $U(\vec{v}) = V(\phi, \psi, 1)$, where ν_ψ represents $\vec{v}$ and $\nu_\phi \wedge \nu_\psi = \nu$. If instead $\vec{v}$ is represented by ν_{m}, then $U(\vec{v}) = V(\phi, \psi, \alpha(\nu))$, where $\nu_\phi \geq \nu$ and $\nu_\psi \wedge \nu_\phi = \nu_{\mathrm{m}}$.

Hence it suffices to show that $(p \circ \imath)(V)$ is open in $\widetilde{\mathcal{V}_K}$ for any $V = V(\phi, \psi, t)$. This amounts to $p^{-1}(p \circ \imath)(V)$ being open in $\mathcal{V}_K$. Define

$$W = W(\phi, \psi, t) := \bigcup_{p/q > t} \left\{ \nu \in \mathcal{V}_K \ ; \ \nu \left(\psi^{pm(\phi)} / \phi^{qm(\psi)} \right) \geq 0 \right\}.$$

Then W is open in $\mathcal{V}_K$ and $\imath(V) = \imath(\mathcal{V}) \cap W$. We claim that if $\nu \in \mathcal{V}_K$ is divisorial and $\nu' \in \overline{\{\nu\}}$ an exceptional curve valuation, then $\nu \in W$ iff $\nu' \in W$. This claim easily implies that $p^{-1}(p \circ \imath)(V) = W$ is open, completing the proof.

As for the claim, $\nu \in W$ implies $t' := \nu(\psi)m(\phi)/\nu(\phi)m(\psi) > t$. Pick $t'' \in (t, t')$ rational, $t'' = p/q$. Then ν belongs to the closed set $\{\mu \ ; \ \mu(\psi^{pm(\phi)}/\phi^{qm(\psi)}) > 0\} \subset W$, hence so does ν'. Conversely, if $\nu' \in W$ then $\nu \in W$ as $R_{\nu'} \subset R_\nu$. □

5.6 The Hausdorff-Zariski Topology

We now recall a natural refinement of the Zariski topology, the Hausdorff-Zariski (or simply HZ) topology. It is still defined on the set $\mathcal{V}_K$ of equivalence classes of Krull valuations. As we show, the HZ topology is the weak tree topology for a natural $\overline{\mathbf{N}}$-tree structure on $\mathcal{V}_K$.

5.6.1 Definition

Let $\mathcal{F} = \{0, +, -\}^K$ be the set of functions from K to $\{0, +, -\}$. Any Krull valuation defines an element in $\mathcal{F}$ by setting $\nu(\phi) = +, -, 0$ iff $\nu(\phi) > 0$, $\nu(\phi) = 0$ and $\nu(\phi) < 0$, respectively. Further, two Krull valuations are equivalent iff they define the same element of $\mathcal{F}$. (This is essentially equivalent to the fact that equivalence classes of Krull valuations are in 1-1 correspondence with valuation rings in K.)

Hence we can consider $\mathcal{V}_K$ as a subset of $\mathcal{F}$. It is easy to see that Zariski topology on $\mathcal{V}_K$ is exactly the topology induced from the product topology on $\mathcal{F}$ associated to the topology on $\{0, +, -\}$ whose open sets are given by $\emptyset$, $\{0, +\}$ and $\{0, +, -\}$. We define the *Hausdorff-Zariski topology* or simply *HZ topology* on $\mathcal{V}_K$ to be the topology induced by the product topology associated to the discrete one on $\{0, +, -\}$. The HZ topology is Hausdorff by construction and we have

Lemma 5.25. *A sequence $\nu_n \in \mathcal{V}_K$ converges towards ν in the Hausdorff-Zariski topology iff for all $\phi \in K$ with $\nu(\phi) > 0$, $\nu(\phi) = 0$ and $\nu(\phi) < 0$ one has $\nu_n(\phi) > 0$, $\nu_n(\phi)$ and $\nu_n(\phi) < 0$, respectively, for sufficiently large n.*

From Tychonov's theorem, and the fact that $\mathcal{V}_K$ is closed in $\mathcal{F}$ (see [ZS2, p.114]) one deduces the following fundamental result.

Theorem 5.26 ([ZS2, p.113]). *The space $\mathcal{V}_K$ endowed with the Hausdorff-Zariski topology is compact. Thus $\mathcal{V}_K$ is quasi-compact in the Zariski topology.*

5.6.2 The $\overline{\mathbf{N}}$-tree Structure on $\mathcal{V}_K$

We now introduce a tree structure on $\mathcal{V}_K$ which plays the same role for the HZ topology as the $\mathbf{R}$-tree structure plays for the weak topology on $\mathcal{V}$. The new tree will be modeled on the totally ordered set $\overline{\mathbf{N}} = \mathbf{N} \cup \{\infty\}$ (see Section 3.1.2).

Definition 5.27. *Define a partial ordering $\trianglelefteq$ on $\mathcal{V}_K$ by*

$$\nu_1 \trianglelefteq \nu_2 \text{ iff the sequence of blowups } \Pi[\nu_2] \text{ contains } \Pi[\nu_1].$$

Proposition 5.28. *The space $(\mathcal{V}_K, \trianglelefteq)$ is a $\overline{\mathbf{N}}$-tree rooted at $\nu_{\mathfrak{m}}$. Any divisorial valuation is a branch point, with tangent space in bijection with $\mathbf{P}^1$. The ends of the tree are exactly the non-divisorial valuations.*

Proof. That $(\mathcal{V}_K, \trianglelefteq)$ is a $\overline{\mathbf{N}}$-tree is a straightforward consequence of the definition, as is the fact that the set of ends coincides with the set of non-divisorial valuations.

A tangent vector at a divisorial valuation ν is given by a point on the exceptional component E defining ν. Hence the tangent space at ν is in bijection with $E \simeq \mathbf{P}^1$. This completes the proof. □

Proposition 5.29. *The Hausdorff-Zariski topology coincides with the weak tree topology induced by $\trianglelefteq$.*

Proof. Pick a tangent vector $\vec{v}$ at a divisorial valuation ν_0, and consider the weak open set $U(\vec{v})$. If ν_0 corresponds to the exceptional curve E, and $\vec{v}$ to the point $p \in E$, $U(\vec{v})$ coincides with the set of valuations that are centered at p. Choose $\phi, \phi' \in K$ defining two smooth transversal curves at p disjoint from E. Then $U(\vec{v}) = \{\nu \; ; \; \nu(\phi), \nu(\phi') > 0\}$ is a weak tree open set. Hence the identity map from $\mathcal{V}_K$ endowed with the HZ topology onto $\mathcal{V}_K$ endowed with the weak tree topology associated to $\triangleleft$ is continuous. As $\mathcal{V}$ is compact in the HZ topology and Hausdorff in the weak tree topology, it follows that the identity map is a homeomorphism. □

In Chapter 6 we shall use sequences of infinitely near points in a different way and actually recover the valuative tree. The analysis in that chapter can be used to explain the precise relation between the $\overline{\mathbf{N}}$-tree and $\mathbf{R}$-tree structures on $\mathcal{V}$ associated to $\trianglelefteq$ and $\leq$, respectively. We will contend ourselves with the following illustrative example. Fix local coordinates (x, y) and consider the segment $I = [\nu_{\mathfrak{m}}, \nu_y[$ in $\mathcal{V}$, consisting of monomial valuations in these coordinates satisfying $1 = \nu(x) \leq \nu(y)$. The restriction of $\leq$ to I coincides with the natural order on $[1, \infty[$. On the other hand, $\trianglelefteq$ gives the lexicographic order on the continued fractions expansions of elements in $[1, \infty[$.

5.7 Comparison of Topologies

We now have several topologies on the trees $\mathcal{V}$ and $\mathcal{V}_{\mathrm{qm}}$ and on the space $\mathcal{V}_K$. We present here some comparisons between them. Our objective is not to prove, or even state, all possible results relating the different topologies, but to illustrate a few of the connections.

We split our analysis into two parts. In the first part we describe the relationships between the weak, the strong, the thin and the HZ topology on $\mathcal{V}$. In the second part we compare the strong and thin metrics on $\mathcal{V}$ and $\mathcal{V}_{\mathrm{qm}}$.

5.7.1 Topologies

By the HZ topology on $\mathcal{V}$ we mean the topology of $\mathcal{V}$ as a subset of $(\mathcal{V}_K, HZ)$.

Theorem 5.30. *The strong, the thin and the HZ topology on $\mathcal{V}$ are all stronger than the weak topology. Moreover:*

(i) if ν has infinite skewness and $\nu_n \to \nu$ weakly, then $\nu_n \to \nu$ strongly;
(ii) if ν has infinite thinness and $\nu_n \to \nu$ weakly, then $\nu_n \to \nu$ thinly;
(iii) if ν is non-divisorial and $\nu_n \to \nu$ weakly, then $\nu_n \to \nu$ in the HZ topology.

Essentially no other implications hold as the following examples indicate.

Example 5.31. Let $\nu_n := \nu_{y,1+n^{-1}}$. Then $\nu_n \to \nu_{\mathfrak{m}}$ strongly and thinly (hence weakly). But $\nu_n(y/x) = 1/n > 0$ and $\nu_{\mathfrak{m}}(y/x) = 0$, so $\nu_n \not\to \nu$ in the HZ topology. In fact, it converges to the exceptional curve valuation $\mathrm{val}[(x,y);((1,0),(1,1))]$.

In this example, the HZ limit valuation is in the closure of $\nu_{\mathfrak{m}}$ in $\mathcal{V}_K$. This is a general fact for limits of sequences in the weak and HZ topology.

Example 5.32. Pick p_n, q_n relatively prime with $p_n/q_n \to \sqrt{2}$. Set $\phi_n = y^{q_n} - x^{p_n}$, $\nu_n = \nu_{\phi_n}$ and $\nu = \nu_{y,\sqrt{2}}$. Then $\nu_n \to \nu$ weakly (hence in the HZ topology since ν is irrational) but $d(\nu_n,\nu) \to 1/\sqrt{2}$ and $D(\nu_n,\nu) \to 1/(1+\sqrt{2})$ so ν_n does not converge to ν neither strongly nor thinly.

Example 5.33. Let ν be an infinitely singular valuation with finite skewness but infinite thinness. See Remark A.4 for how to construct ν. Pick a sequence $(\mu_n)_1^\infty$ of divisorial valuations increasing to ν and for each n pick a curve valuation ν_n with $\nu_n \wedge \nu = \mu_n$. Then $D(\nu_n,\nu) = 2A(\mu_n)^{-1} \to 0$ so $\nu_n \to \nu$ thinly. However, $d(\nu_n,\nu) \to \alpha(\nu)^{-1} > 0$ so $\nu_n \not\to \nu$ strongly.

Example 5.34. It is possible to construct a sequence $(\nu_n)_1^\infty$ of infinitely singular (or even divisorial) valuations such that $A(\nu_n) \geq 3$ but $\alpha(\nu_n) \to 1$ as $n \to \infty$. See Remark A.5. Then $\nu_n \to \nu_{\mathfrak{m}}$ strongly but not thinly.

Proof (Theorem 5.30). We know from Proposition 5.5 that the strong and thin topologies are both stronger than the weak topology. To compare the HZ and weak topologies we consider the natural injection $\jmath : (\mathcal{V}, \mathrm{HZ}) \to (\mathcal{V}_K, \mathrm{Z})$, which is continuous as the HZ topology is stronger than the Zariski topology. By Theorem 5.24 there is a continuous mapping $q : (\mathcal{V}_K, \mathrm{Z}) \to (\mathcal{V}, \text{weak})$ which is the identity on $\mathcal{V} \subset \mathcal{V}_K$. Since $q \circ \jmath = \mathrm{id}$, we see that $\mathrm{id} : (\mathcal{V}, \mathrm{HZ}) \to (\mathcal{V}, \text{weak})$ is continuous. Hence the HZ topology is stronger than the weak topology.

For (i), suppose $\nu_n \to \nu$ weakly. As $\alpha(\nu) = \infty$ we may find $\phi_k \in \mathfrak{m}$ such that $\nu(\phi_k) \geq (k+1)m(\phi_k)$. For $n \gg 1$, $\nu_n(\phi_k) \geq k\, m(\phi_k)$. Thus $\nu, \nu_n \geq \nu_{\phi_k,k}$, so

$$d(\nu_n, \nu) \leq d(\nu_n, \nu_{\phi_k,k}) + d(\nu_{\phi_k,k}, \nu) \leq k^{-1} + k^{-1}.$$

We let $k \to \infty$ and conclude that ν_n converges strongly towards ν.

The proof of (ii) is essentially identical. As for (iii), suppose $\nu_n \to \nu$ weakly. Consider $\phi \in K$. If $\nu(\phi) > 0$ (< 0), then $\nu_n(\phi) > 0$ (< 0) for $n \gg 1$. If $\nu(\phi) = 0$, then as the residue field k_ν is isomorphic to $\mathbf{C}$, we may write $\phi = \lambda + \psi$, where $\lambda \in \mathbf{C}^*$ and $\nu(\psi) > 0$. For $n \gg 1$ $\nu_n(\psi) > 0$ so that $\nu_n(\phi) = 0$. Thus $\nu_n \to \nu$ in the HZ topology. □

5.7.2 Metrics

We have seen above that the strong and thin topologies on $\mathcal{V}$ are distinct. Nevertheless, we can compare the thin and the strong metrics whenever we have a bound on the multiplicity.

Proposition 5.35. *Fix $m \geq 1$. Then the inequalities*

$$d_{\mathrm{qm}} \leq D_{\mathrm{qm}} \leq m\, d_{\mathrm{qm}}$$

hold on the subtree $\{\nu \in \mathcal{V}_{\mathrm{qm}} \,;\, m(\nu) \leq m\}$ of $\mathcal{V}_{\mathrm{qm}}$ and these estimates are sharp. The identity map $(\mathcal{V}_{\mathrm{qm}}, D_{\mathrm{qm}}) \to (\mathcal{V}_{\mathrm{qm}}, d_{\mathrm{qm}})$ is continuous, but its inverse is not.

Proposition 5.36. *Fix $m \geq 1$. Then the inequalities*

$$m^{-1} D \leq d \leq 2(m+1)\, D,$$

hold on the subtree $\{\nu \in \mathcal{V} \,;\, m(\nu) \leq m\}$ of $\mathcal{V}$. The identity map $(\mathcal{V}, D) \to (\mathcal{V}, d)$ is not continuous, and neither is its inverse.

Proof (Proposition 5.35). The estimate is an immediate consequence of the definition of d_{qm} and D_{qm} and of (3.7). That the identity map $(\mathcal{V}_{\mathrm{qm}}, D_{\mathrm{qm}}) \to (\mathcal{V}_{\mathrm{qm}}, d_{\mathrm{qm}})$ is continuous follows immediately. However, its inverse is not continuous in view of Example 5.34. □

Proof (Proposition 5.36). As d and D are both tree metrics, it suffices to consider $\nu_1 \leq \nu_2$ with $m(\nu_1) \leq m$ and compare $d(\nu_1, \nu_2)$ with $D(\nu_1, \nu_2)$. Write

$A_i = A(\nu_i)$, $\alpha_i = \alpha(\nu_i)$ and $m_i = m(\nu_i)$ for $i = 1, 2$. Also write $d_{12} = d(\nu_1, \nu_2)$ and $D_{12} = D(\nu_1, \nu_2)$. Then $d_{12} = \alpha_1^{-1} - \alpha_2^{-1} = (\alpha_2 - \alpha_1)/(\alpha_1\alpha_2)$ and $D_{12} = (A_2 - A_1)/(A_1 A_2)$. Moreover, (3.7) gives $m_1(\alpha_2 - \alpha_1) \le A_2 - A_1 \le m(\alpha_2 - \alpha_2)$ and Proposition 3.48 implies $A_i \le m_i\alpha_i + 1$. Hence

$$D_{12} \le \frac{m(\alpha_2 - \alpha_1)}{A_1 A_2} \le m \frac{\alpha_2 - \alpha_1}{\alpha_1 \alpha_2} = m\, d_{12}$$

and

$$D_{12} \ge \frac{m_1(\alpha_2 - \alpha_1)}{(m_1\alpha_1 + 1)(m_2\alpha_2 + 1)} \ge \frac{m_1\alpha_1}{m_1\alpha_1 + 1} \frac{\alpha_2}{m_2\alpha_2 + 1} d_{12} \ge \frac{d_{12}}{2(m+1)}.$$

where we used $m_1, m_2 \le m$ and $\alpha_1, \alpha_2 \ge 1$.

That the identity maps $(\mathcal{V}, D) \to (\mathcal{V}, d)$ and $(\mathcal{V}, d) \to (\mathcal{V}, D)$ are both discontinuous follows from Example 5.33 and Example 5.34. □

6

The Universal Dual Graph

We have already described several different approaches to the valuative tree. They all fundamentally derive from the definition of a valuation as a function on the ring R. On the other hand, valuations can also be viewed geometrically as sequences of infinitely near points. It is therefore natural to ask whether the valuative tree can be recovered through a purely geometric construction.

In this chapter we show that this is indeed possible. The construction goes as follows. To any composition of (point) blowups we associate the dual graph of the exceptional divisor. The set of vertices of this graph is naturally a poset and the collection of all such posets forms an injective system whose limit is a nonmetric tree Γ^* modeled on the rational numbers. By filling in the irrational points and adding all the ends we obtain a nonmetric tree Γ modeled on the real line. We call Γ the *universal dual graph*. Its points are encoded by sequences of infinitely near points above the origin. We show how to equip Γ with a natural *Farey parameterization* as well as an integer valued *multiplicity* function.

The main result is then that there exists a natural isomorphism from the universal dual graph Γ to the valuative tree $\mathcal{V}$. Its inverse maps a valuation to its associated sequence of infinitely near points as defined in Section 1.7.

The chapter is organized as follows. We start out by defining Γ as a nonmetric tree, then equip it with the Farey parameterization and multiplicity function, respectively. Along the way we show that sequences of infinitely near points correspond uniquely to either points in Γ, or tangent vectors at branch points in Γ. We then state and prove the main result, namely the isomorphism between Γ and $\mathcal{V}$. From this we deduce a number of applications, and show how the dual graph of the minimal desingularization of a reduced curve can be described inside the universal dual graph. Finally we discuss a relative version of the universal dual graph, analogous to the relative valuative tree, and explain the self-similar structure of these objects.

6.1 Nonmetric Tree Structure

We first define the universal dual graph as a nonmetric tree. Later we will equip it with a natural parameterization and multiplicity function.

6.1.1 Compositions of Blowups

Let us denote by $\mathfrak{B}$ the set of all modifications above the origin. This means that each element of $\mathfrak{B}$ is a mapping $\pi : X_\pi \to (\mathbf{C}^2, 0)$ where X_π is a smooth complex surface and π is a proper map which is a bijection outside the exceptional divisor $\pi^{-1}(0)$. It is well-known that each such π is a composition of (point) blowups (see [La, Theorem 5.7] for instance). We emphasize that these blowups are not necessarily associated to a sequence of infinitely near points.

The set $\mathfrak{B}$ has a natural partial ordering: $\pi \trianglerighteq \pi'$ iff $\pi = \pi' \circ \tilde{\pi}$ for some composition of blowups (see Section 5.6.2). We shall use repeatedly

Lemma 6.1. *The set $\mathfrak{B}$ is naturally an inverse system: any finite subset of $\mathfrak{B}$ admits a supremum; and every nonempty subset of $\mathfrak{B}$ admits an infimum.*

Proof. To prove that any finite subset admits a supremum, it is sufficient to consider the case of two elements. If π, π' belong to $\mathfrak{B}$, the identity map lifts to a bimeromorphic map $\mathrm{id} : X_\pi \to X_{\pi'}$. It is a classical fact that for any bimeromorphic map of surfaces there exist compositions of point blowups $\varpi : Y \to X_\pi$, $\varpi' : Y \to X_{\pi'}$ such that $\pi \circ \varpi = \pi' \circ \varpi'$. (Take a resolution of singularities of the graph of id, and apply [La, Theorem 5.7] for instance). This proves that any finite set admits a supremum.

In the discussion above, one can be more precise. By taking the minimal desingularization (see [La, Theorem 5.9]) of the graph of id, one can show that Y is "minimal" in the sense that any other surface Y' dominating both X_π and $X_{\pi'}$ dominates Y. We call the natural map sending Y to the base $(\mathbf{C}^2, 0)$ the *join* of π and π'.

Now take any non-empty subset $B \subset \mathfrak{B}$, and define $M := \{\pi \ ; \ \pi \trianglelefteq \varpi \text{ for all } \varpi \in B\}$. The set M contains the blowup of the origin and is hence non-empty. It is also a finite set, as any map $\pi \in \mathfrak{B}$ dominates at most finitely many other elements in $\mathfrak{B}$. Then the join of the collection of all $\pi \in M$ is dominated by all elements in B. This is the infimum of B. □

6.1.2 Dual Graphs

To any $\pi \in \mathfrak{B}$ we may attach a *dual graph* Γ_π: vertices are in bijection with irreducible components of the exceptional divisor $\pi^{-1}(0)$ (we shall refer to these as *exceptional components*), and edges with intersection points between two components. By decomposing an element $\pi \in \mathfrak{B}$ into a composition of point blowups, we see that the dual graph Γ_π can be obtained inductively as follows.

If π is a single blowup of the origin, then Γ_π is a single point that we denote by E_0. Otherwise we may write $\pi = \pi' \circ \tilde{\pi}$, where $\tilde{\pi}$ is the blowup of a point p on the exceptional divisor $(\pi')^{-1}(0)$, resulting in a new exceptional component E_p. The dual graph of π can then be obtained from the dual graph $\Gamma_{\pi'}$ of π' through an *elementary modification*. There are two kinds of elementary modifications, depending on the location of p. We say that p is a *free* (*satellite*) point if p is a regular (singular) point on $(\pi')^{-1}(0)$.

If p is free, i.e. $p \in E$ for a unique irreducible component $E \subset (\pi')^{-1}(0)$, then the elementary modification consists of adding to $\Gamma_{\pi'}$ a segment joining E and the new vertex E_p. We say that this modification is of the *first kind.* See Figure 6.1.

If p is a satellite point, it is the intersection of two irreducible components E and E'. The elementary modification now consists of adding a new vertex E_p to the segment between E and E'; we say that the modification is of the *second kind.* Again see Figure 6.1.

Fig. 6.1. Elementary modifications on the dual graph. The first kind of modification is illustrated to the left, the second kind to the right.

From this description, it follows that Γ_π is a finite simplicial tree. We shall denote by E_0 the vertex associated to the proper transform in Γ_π of the exceptional divisor obtained by blowing up the origin.

6.1.3 The Q-tree

Following Section 3.1.7 we can view the simplicial trees Γ_π as $\mathbf{N}$-trees. This goes as follows. For $\pi \in \mathfrak{B}$, consider the set Γ_π^* of vertices of Γ_π. It is a finite set, with a natural partial ordering $\le_\pi$ derived from the simplicial tree structure on Γ_π, rooted at E_0: $E_1 \le_\pi E_2$ iff $[E_0, E_1] \subset [E_0, E_2]$ as segments in Γ_π. The poset Γ_π^* is an $\mathbf{N}$-tree in the sense of Chapter 3.

If $\pi, \pi' \in \mathfrak{B}$ and $\pi \trianglerighteq \pi'$, then the dual graph Γ_π^* is obtained from $\Gamma_{\pi'}^*$ by performing a sequence of elementary modifications. In particular, there is a natural injective map $\imath_{\pi\pi'} : \Gamma_{\pi'}^* \to \Gamma_\pi^*$, and if $\pi'' \trianglelefteq \pi' \trianglelefteq \pi$, then $\imath_{\pi\pi'} \circ \imath_{\pi'\pi''} = \imath_{\pi\pi''}$

We claim that each map $\imath_{\pi\pi'}$ is order-preserving. By induction it suffices to show this when $\pi = \pi' \circ \tilde{\pi}$, where $\tilde{\pi}$ is the blowup of a point $p \in (\pi')^{-1}(0)$. Then $\imath_{\pi\pi'}$ is given by an elementary modification of the first or second kind: see Figure 6.1. If the elementary modification is of the first kind, i.e. if $E_p \in$

$\Gamma^*_\pi \setminus \imath_{\pi\pi'}(\Gamma^*_{\pi'})$ is obtained by blowing up a free point p on $E \in \Gamma^*_{\pi'}$, then $E_p > \imath_{\pi\pi'}(E)$, E_p is an end in Γ^*_π, and the segment $]\imath_{\pi\pi'}(E), E_p[$ in Γ^*_π is empty. If the elementary modification is of the second kind, i.e. if E_p is instead obtained by blowing up the intersection point of two elements $E, E' \in \Gamma^*_{\pi'}$, say with $E < E'$, then $\imath_{\pi\pi'}(E) < E_p < \imath_{\pi\pi'}(E')$ and the segments $]\imath_{\pi\pi'}(E), E_p[$ and $]E_p, \imath_{\pi\pi'}(E')[$ in Γ^*_π are empty.

Thus $\imath_{\pi\pi'}$ is order-preserving whenever $\pi' \trianglelefteq \pi$. Slightly abusively, we will consider $\Gamma^*_{\pi'}$ to be a subset of Γ^*_π and ignore the map $\imath_{\pi\pi'}$.

The set $\mathfrak{B}$ defines an inverse system, and $(\Gamma^*_\pi, \leq_\pi)_{\pi \in \mathfrak{B}}$ forms an injective system. We may therefore define the *universal dual graph* Γ^* as the injective limit

$$(\Gamma^*, \leq) = \operatorname{inj\,lim}_{\pi \in \mathfrak{B}} (\Gamma^*_\pi, \leq_\pi)$$

over all sequences of blowups π above the origin. Again we consider (slightly abusively) Γ^*_π as a subset of Γ^* for all π. Then Γ^* is the union of all the Γ^*_π's. It is important to note that any finite subset of Γ^* is contained in Γ^*_π for some π.

The partial ordering on Γ^* is determined as follows. If $E, E' \in \Gamma^*$, then $E, E' \in \Gamma^*_\pi$ for some $\pi \in \mathfrak{B}$ and $E \leq E'$ in Γ^* iff $E \leq E'$ in Γ^*_π.

Proposition 6.2. *The universal dual graph $(\Gamma^*, \leq)$ is a nonmetric* **Q**-*tree rooted at E_0. All its points are branch points.*

We shall describe the tangent spaces in Proposition 6.3 below.

Proof. Let us verify that the partial ordering on Γ^* satisfies the axioms (T1)-(T3) for a rooted nonmetric **Q**-tree on page 44. As for (T1), this is clear: E_0 is the unique minimal element of Γ^*.

Next consider (T2). Fix $E \in \Gamma^*$. In view of Lemma 3.11 it suffices to show that $[E_0, E] := \{F \in \Gamma^* \ ; \ F \leq E\}$ is totally ordered, countable, and has no gaps. It is totally ordered as the intersection with any Γ^*_π is. It cannot have any gaps, since if $E_1 < E_2 \leq E$ and there was no $E' \in \Gamma^*$ with $E_1 < E' < E_2$, then E_1 and E_2 would be adjacent vertices in some Γ_π. We could then blow up the intersection point between E_1 and E_2 and obtain $E' \in \Gamma^*$ with $E_1 < E' < E_2$, a contradiction. Thus the set $[E_0, E]$ has no gaps. The hardest part is to show that it is countable. Consider the minimal $\pi \in \mathfrak{B}$ such that $E \in \Gamma^*_\pi$. We shall describe the structure of Γ^*_π in Section 6.2, but for now it suffices to notice that any $F \in \Gamma^*$ belonging to the segment $[E_0, E]$ can be obtained by performing finitely many satellite blowups at intersection points of elements of $[E_0, E] \cap \Gamma^*_\pi$. More precisely, we define a sequence of proper birational morphisms $\pi = \pi_0 \trianglerighteq \pi_1 \trianglerighteq \dots$ inductively through their associated posets $\Gamma^*_n = \Gamma^*_{\pi_n}$ as follows: the elements of $\Gamma^*_{n+1} - \Gamma^*_n$ are obtained by blowing up all intersection points of elements of $[E_0, E] \cap \Gamma^*_n$. Then each Γ^*_n is a finite poset and $[E_0, E] = \bigcup_n [E_0, E] \cap \Gamma^*_n$. Thus $[E_0, E]$ is countable and (T2) holds.

Instead of proving (T3) we prove the equivalent statement (T3') in Remark 3.3. Thus consider an unbounded, totally ordered subset $\mathcal{S}$ of Γ^*. We will prove that $\mathcal{S}$ admits an increasing sequence in $\mathcal{S}$ without majorant in Γ^*. In doing so we will make use of the fact that if $E \in \Gamma^*$, then there exists $n = n(E) < \infty$ such that if $F \in \Gamma^*$ with $F \le E$, then the minimal $\pi \in \mathfrak{B}$ for which $F \in \Gamma^*_\pi$ is a composition of blowups, at most n of which are free (the number of satellite blowups can be arbitrarily high). Now consider $\mathcal{S}$ as above. Pick any $E_1 \in \mathcal{S}$, $E_1 > E_0$. Let $\pi_1 \in \mathfrak{B}$ be minimal such that $E_1 \in \Gamma^*_1 := \Gamma^*_{\pi_1}$. After replacing E_1 by $\max \Gamma^*_1 \cap \mathcal{S}$ we may assume that $E_1 = \max \Gamma^*_1 \cap \mathcal{S}$. Since $\mathcal{S}$ is unbounded, there exists $E_2 \in \mathcal{S}$ with $E_2 > E_1$. Notice that the assumption $E_1 = \max \Gamma^*_1 \cap \mathcal{S}$ implies that E_2 is obtained by blowing up a free point on E_1 followed by finitely many (free or satellite) blowups. As before we may assume that $E_2 = \max \Gamma^*_2 \cap \mathcal{S}$, where $\Gamma^*_2 = \Gamma^*_{\pi_2}$ and $\pi_2 \ge \pi_1$ is the minimal element of $\mathfrak{B}$ such that $E_2 \in \Gamma^*_2$. Inductively we construct an increasing sequence $(E_n)_1^\infty$ in $\mathcal{S}$ with the property that E_{n+1} is obtained by blowing up a free point on E_n followed by finitely many (free or satellite) blowups. The remark above then implies that the sequence (E_n) is unbounded in Γ^*, i.e. there is no $E \in \Gamma^*$ with $E_n < E$ for all n. This proves (T3') and hence (T3).

Thus Γ^* is a nonmetric $\mathbf{Q}$-tree rooted in E_0. Finally notice that any point $E \in \Gamma^*$ is a branch point since it becomes a branch point in some Γ^*_π after blowing up two or more free points on E. This concludes the proof. □

6.1.4 Tangent Spaces

The following geometric interpretation of the tangent spaces in Γ^* will play an important role in the sequel.

Proposition 6.3. *Let $E \in \Gamma^*$ and pick $\pi \in \mathfrak{B}$ such that $E \in \Gamma^*_\pi$. For $p \in E$, let $E_p \in \Gamma^*$ be the exceptional component obtained by blowing up p, and denote by $\vec{v}_p$ the tangent vector represented by E_p at E.*

Then the map $p \to \vec{v}_p$ induces a bijection between the set of points in E and the tangent space at E in Γ^*.

Proof. Pick a tangent vector $\vec{v}$ at E. By construction, $\vec{v}$ is represented by some point in Γ^*. We may hence find $\pi' \in \mathfrak{B}$ such that $\pi' \trianglerighteq \pi$, $E \in \Gamma^*_{\pi'}$, and $\vec{v}$ is represented by another component $F \in \Gamma^*_{\pi'}$. There is then a unique component E' intersecting E and lying in the segment $[E, F]$. This component represents $\vec{v}$ at E. This is also the case for the exceptional component E_p obtained by blowing up $p = E \cap E'$. If $\pi' = \pi \circ \varpi$, $\vec{v}$ is also represented by E_q where $q := \varpi(p)$. Whence $p \to \vec{v}_p$ is surjective.

To show that $p \to \vec{v}_p$ is injective, pick $p \ne q$. Then in the dual graph of the composition of π with the blowups at p and q, the component E_p and E_q represent different tangent vectors. This dual graph embeds in Γ^* by construction, hence $\vec{v}_p \ne \vec{v}_q$. □

From the proof we deduce the following consequence:

Corollary 6.4. *Let π, E and p be as above, and consider a proper birational morphism $\pi' : X_{\pi'} \to (\mathbf{C}^2, 0)$ dominating π, i.e. $\pi' = \pi \circ \varpi$. Assume that ϖ*

is an isomorphism above $X_\pi \setminus \{p\}$. Then all points in $\Gamma^*_{\pi'} \setminus \Gamma^*_\pi$ represent the tangent vector $\vec{v}_p$ at E.

6.1.5 The R-tree

In view of Section 3.1.8 there is a canonical rooted, nonmetric **R**-tree (or simply rooted, nonmetric tree) Γ^o associated to the rooted, nonmetric **Q**-tree Γ^*. The points in $\Gamma^o \setminus \Gamma^*$ can be viewed as decreasing sequences of closed segments in Γ^* with empty intersection in Γ^*. We will somewhat abusively refer also to Γ^o, and even its completion Γ, as the universal dual graph. In Section 6.2 we shall interpret the elements of Γ as sequences of infinitely near points.

Proposition 6.5. *The universal dual graph Γ is a complete nonmetric tree rooted in E_0 whose branch points are exactly the points in Γ^*.*

Proof. Immediate consequence of Proposition 3.12 and the definition of Γ as the completion of Γ^o. □

We shall denote the infimum in Γ by "$\wedge$", just as in the valuative tree.

Remark 6.6. The tangent space in the **R**-tree Γ at a branch point $E \in \Gamma^*$ is by construction canonically identified with the tangent space at E in the **Q**-tree Γ^*. The latter tangent space is described in Proposition 6.3: see also Theorem B.1.

6.2 Infinitely Near Points

Next we show that the points in Γ are encoded by sequences of infinitely near points. Recall that sequences of infinitely near points are in bijection with Krull valuations centered at $\mathfrak{m}$ by Theorem 1.10. We shall later see that Γ is indeed isomorphic to the valuative tree (Theorem 6.22).

6.2.1 Definitions and Main Results

In general we can classify sequences of infinitely near points into five categories. The terminology below is essentially in accordance with that of Spivakovsky [Sp].

Definition 6.7. *Let $\bar{p} = (p_j)_0^n$, $0 \le n \le \infty$ be a finite or infinite sequence of infinitely near points. We say that $\bar{p}$ is of*

- Type 0 *if $\bar{p}$ is finite;*
- Type 1 *if $\bar{p}$ is infinite and contains infinitely many free and infinitely many satellite points;*

- Type 2 *if $\bar{p}$ is infinite, contains only finitely many free points, and is not of Type 3;*
- Type 3 *if $\bar{p}$ is infinite, contains only finitely many free points, and has the following property: there exists (a unique) $j_0 \geq 1$ such that if $j > j_0$, then p_{j+1} is the satellite point defined by the intersection of E_j and the strict transform of E_{j_0};*
- Type 4 *if $\bar{p}$ is infinite and contains only finitely many satellite points.*

We describe in Figures 6.2 to 6.6 below the structure of the dual graph appearing in the successive blowups associated to a sequence of infinitely near points. The notation is explained in Section 6.2.2.

Definition 6.8. *Fix a finite sequence $(p_j)_0^n$ of infinitely near points (i.e. of Type 0). Write $\pi_{\bar{p}} \in \mathfrak{B}$ for the composition of blowups at all points $p_0, \dots, p_n$, and define $\gamma(\bar{p}) \in \Gamma^*$ to be the exceptional divisor of the blowup at p_n.*

Theorem 6.9. *The map $\bar{p} \to \gamma(\bar{p})$ gives a bijection between sequences of infinitely near points of Type 0, and Γ^*.*

Theorem 6.10. *Let $\bar{p} = (p_j)_0^\infty$ be an infinite sequence of infinitely near points not of Type 3, and define the truncation $\bar{p}_n = (p_j)_0^n$ for $n < \infty$. Then the sequence $\gamma(\bar{p}_n) \in \Gamma^*$ converges weakly*[1] *in Γ to an element $\gamma(\bar{p})$. We have:*

- *$\gamma(\bar{p})$ is an end in Γ if $\bar{p}$ is of Type 1 or 4;*
- *$\gamma(\bar{p})$ is a regular point if $\bar{p}$ is of Type 2.*

The map $\bar{p} \to \gamma(\bar{p})$ gives a bijection between sequences of infinitely near points not of Type 3, and Γ.

Theorem 6.11. *Let $\bar{p} = (p_j)_0^\infty$ be a sequence of infinitely near points of Type 3. In the notation of the previous theorem, the sequence $\gamma(\bar{p}_n) \in \Gamma^*$ converges weakly in Γ to an element $\gamma(\bar{p}) \in \Gamma^*$. For n large enough we have $[\gamma(\bar{p}), \gamma(\bar{p}_{n+1})] \subset [\gamma(\bar{p}), \gamma(\bar{p}_n)]$. In particular, the points $\gamma(\bar{p}_n)$ define the same tangent vector $\vec{v}(\bar{p})$ at $\gamma(\bar{p})$.*

Moreover, the map $\bar{p} \to \vec{v}(\bar{p})$ gives a bijection between infinitely near points of Type 3 and tangent vectors at points in Γ^*, i.e. at branch points in Γ.

Proposition 6.12. *The sequence $\bar{p}$ of infinitely near points associated to an irreducible curve C at the origin (see Example 1.13) is of Type 4. Conversely, every sequence of infinitely near points of Type 4 is associated to a unique irreducible curve.*

In view of this proposition and Theorem 6.10 we see that an irreducible curve C naturally defines an end in the universal dual graph Γ. We shall use this fact repeatedly in the sequel.

The following two results are consequences of Corollary 6.4 (see also Proposition 6.3 and Remark 6.6):

[1] We here use the weak topology induced by the tree structure on Γ.

Corollary 6.13. *If $\bar{p}' = (p_j)_0^{n'}$ is a sequence of infinitely near points not of Type 3, $0 \leq n < n' \leq \infty$ and $\bar{p} = (p_j)_0^n$, then the tangent vector at $E := \gamma(\bar{p}) \in \Gamma^*$ represented by $\gamma(\bar{p}') \in \Gamma$ is given by the point p_{n+1} on the exceptional component E.*

Corollary 6.14. *Consider any irreducible curve C and view C as an end in Γ. Let $\bar{p} = (p_j)_0^\infty$ be the sequence of infinitely near points (of Type 4) associated to C. Fix $n \geq 0$, set $E_n := \gamma((p_j)_0^n) \in \Gamma^*$ and let π_n be the composition of blowups at $p_0, \dots, p_n$. Then the tangent vector at E_n represented by C is given by the intersection of E_n with the strict transform of C under π_n.*

6.2.2 Proofs

Let us now present the proofs of the results above, starting with Theorem 6.9. If $\bar{p}$ is of Type 0, then $\gamma(\bar{p})$ is clearly an element in Γ^*. Moreover, it is clear that different sequences $\bar{p}$ give rise to different exceptional components $\gamma(\bar{p})$. To prove surjectivity, consider any $E \in \Gamma^*$. By Lemma 6.1 there is a minimal $\pi \in \mathfrak{B}$ for which $E \in \Gamma_\pi^*$. It is straightforward to verify that this E is associated to a sequence of infinitely near points. This completes the proof of Theorem 6.9.

Before turning to the proof of the next theorem, we introduce some notation which shall be used throughout the proofs below. Consider a finite sequence $(p_j)_0^n$ of infinitely near points. Recall that p_0 is the origin. Therefore, p_1 is always free (assuming $n \geq 1$). Define indices

$$0 = n_0 < \bar{n}_1 < n_1 < \bar{n}_2 < n_2 < \cdots < \bar{n}_g < n_g < \bar{n}_{g+1} \leq n \tag{6.1}$$

as follows: p_j is free for $n_i < j \leq \bar{n}_{i+1}$, $0 \leq i \leq g$, and satellite for $\bar{n}_i < j \leq n_i$, $1 \leq i \leq g$. Let $\pi = \pi_{\bar{p}}$ be the composition of blowups at $p_0, \dots, p_n$ and let E'_j be the strict transform of the blowup of p_j in the total space X_π.[2] Identify E'_j with its image in Γ_π. Write $E_i = E'_{n_i}$ for $0 \leq i \leq g$ $\bar{E}_i = E'_{\bar{n}_i}$ for $1 \leq i \leq g+1$. Finally set $E = E'_n = \gamma(\bar{p})$. The dual graph of $\pi_{\bar{p}}$ then looks as in Figure 6.2.

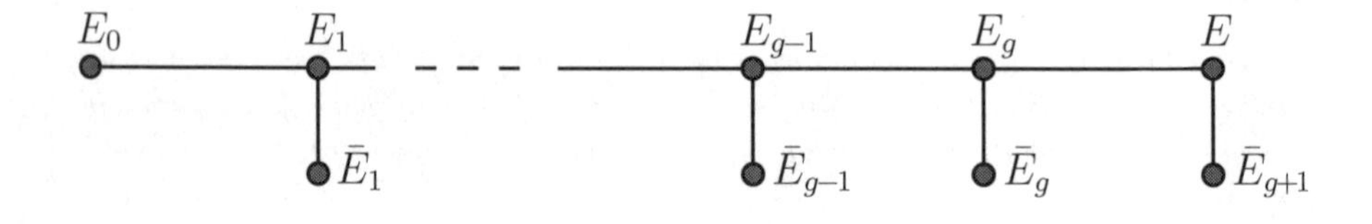

Fig. 6.2. The dual graph associated to a sequence of infinitely near points of Type 0. Only the branch points and ends are labeled. Compare Figure (5.7) in [Sp].

[2] The notation differs marginally from that of Section 1.7 where we would have written E_j instead of E'_j.

Let us now prove Theorems 6.10 and 6.11. Thus, pick an infinite sequence $\bar{p}$ of infinitely near points (i.e. of Type 1, 2, 3, or 4). We may define indices n_i, $0 \le i \le g$ and $\bar{n}_i$, $1 \le i \le g+1$ as in (6.1), allowing for the possibility that $g = \infty$. Consider the truncations $\bar{p}_n = (p_j)_0^n$ defined above. Let us show that $\gamma(\bar{p}_n)$ converges weakly to a point in Γ.

If $\bar{p}$ is of Type 1, then $g = \infty$ and the sequence $\gamma(\bar{p}_{n_i}) = E_i$ forms a strictly increasing sequence in Γ^*. In fact, this sequence is unbounded in Γ^*. This follows from the fact the sequence $\bar{p}$ contains infinitely many free blowups: see the proof of (T3') in Section 6.1.3. Thus $E_i := \gamma(\bar{p}_{n_i})$ converges to an end $\gamma(\bar{p})$ in Γ as $i \to \infty$. As $\gamma(\bar{p}_{n_i+k}) \ge \gamma(\bar{p}_{n_i})$ for all $k \ge 0$, $\lim_{j\to\infty} \gamma(\bar{p}_j) = \gamma(\bar{p})$. See Figure 6.3.

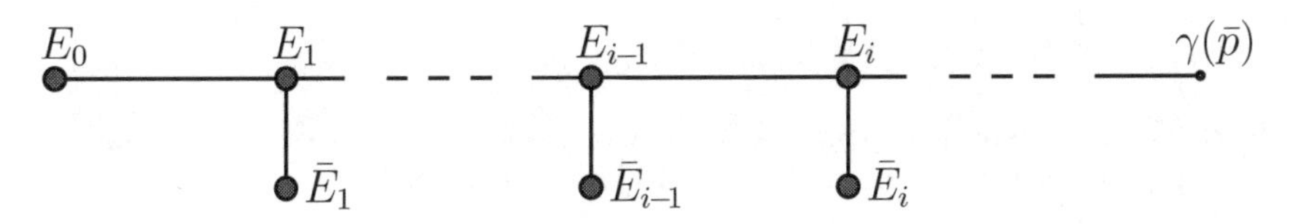

Fig. 6.3. Infinitely near points of Type 1 corresponding to an end in Γ (of infinite multiplicity). Compare Figure (9.1) in [Sp].

If $\bar{p}$ is of Type 4, then $g < \infty$ and the sequence $\gamma(\bar{p}_n) = E'_n$ for $n > n_g$ forms a strictly increasing sequence in Γ^*. As before it is unbounded, and therefore defines an end $E_{\bar{p}} = \gamma(\bar{p})$ in Γ. See Figure 6.4.

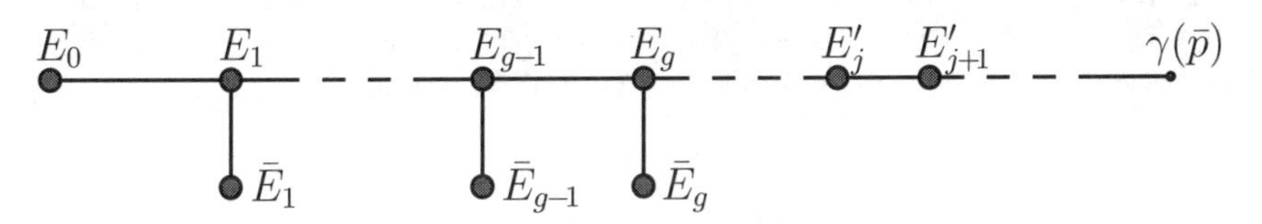

Fig. 6.4. Infinitely near points of Type 4 corresponding to an end in Γ (of finite multiplicity). Compare Figure (9.6) in [Sp].

If $\bar{p}$ is of Type 2 or 3, then $g < \infty$ and for $j > \bar{n}_{g+1}$, p_{j+1} is a satellite point, intersection of two exceptional divisors, one of which is E'_j and the other we denote by F_j. The segment $[E'_j, F_j]$ is then equal to either $[E'_j, E'_{j-1}]$ or to $[E'_j, F_{j-1}]$, hence the sequence $[E'_j, F_j]$ is decreasing. To conclude that E_j converges in Γ, we need to show that the intersection $\bigcap [E'_j, F_j]$ contains at most one point in Γ^*.

Thus pick $E \in \Gamma^*$ with $E \in [E'_j, F_j]$ for all j. If $E = E'_{j_0}$ for some j_0, then $\bar{p}$ is of Type 3, and p_{j+1} is the intersection of E with E'_j for $j > j_0$. Otherwise $E \ne E'_j$ for all j, in which case we may find j_0 sufficiently large such that E can be obtained from $p_{j_0+1} = E'_{j_0} \cap F_{j_0}$ by a sequence of blowups of satellite points. Whence $E = E'_j$, or $E \notin [E'_j, F_j]$ for some j. This is a contradiction, which shows that $\bigcap [E'_j, F_j]$ is empty for $\bar{p}$ of Type 2, and is reduced to $E = E_{j_0} \in \Gamma^*$ when $\bar{p}$ is of Type 3.

This concludes the proof that $\bar{p} \to \gamma(\bar{p})$ is well-defined for infinite sequences $\bar{p}$. See Figures 6.5 and 6.6. Note that we also showed that $\gamma(\bar{p})$ belongs to Γ^* if $\bar{p}$ is of Type 3; is an end in Γ if $\bar{p}$ is of Type 1 or 4; and is regular point if $\bar{p}$ is of Type 2.

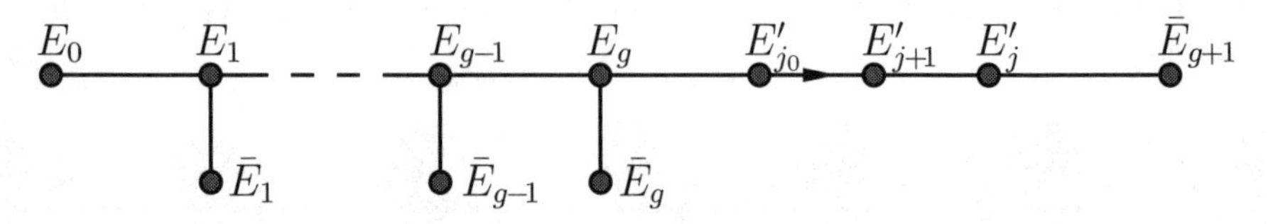

Fig. 6.5. Infinitely near points of Type 3 corresponding to a tangent vector at a branch point in Γ (here E'_{j_0}). Compare Figure (9.4) in [Sp].

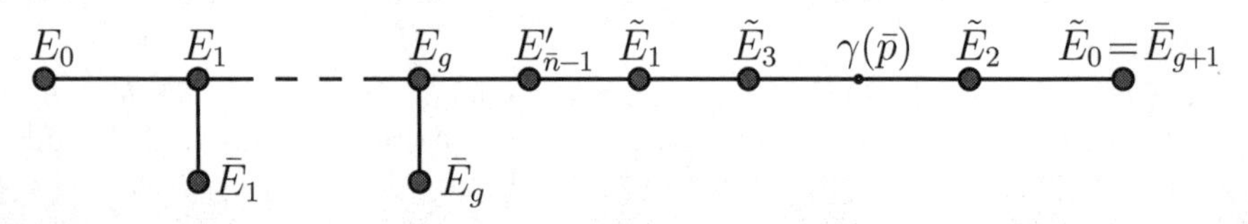

Fig. 6.6. Infinitely near points of Type 2 corresponding to a regular point in Γ. Here $\tilde{E}_k = E_{\tilde{n}_k}$ for some increasing sequence $(\tilde{n}_k)_0^\infty$ such that $\tilde{E}_{2k}$ ($\tilde{E}_{2k+1}$) decreases (increases) to $\gamma(\bar{p})$ as $k \to \infty$. Compare Figure (9.2) in [Sp].

To conclude the proof of Theorem 6.10 we only need to show that γ is bijective from the set of infinitely near points of Types 1, 2 and 4, to the set $\Gamma \setminus \Gamma^*$.

First consider injectivity. Pick $\bar{p} = (p_j)_0^\infty$ and $\bar{p}' = (p'_j)_0^\infty$ of Types 1, 2 or 4 with $\gamma(\bar{p}) = \gamma(\bar{p}')$ and $\bar{p} \neq \bar{p}'$. Write $\gamma_n := \gamma(\bar{p}_n)$, $\gamma'_n := \gamma(\bar{p}'_n)$. Pick n maximal such that $p_j = p'_j$ for all $j \leq n$. Thus $p_{n+1} \neq p'_{n+1}$.

Suppose p_{n+2} is a free point on γ_{n+1}. Then $\gamma_{n+2} > \gamma_{n+1}$, and $[\gamma_{n+k}, \gamma'_{n+1}]$ contains both γ_{n+1} and γ_n for all $k \geq 2$. This is impossible, hence p_{n+2} is a satellite point. Inductively, we see that p_{n+k+1} is necessarily the intersection of γ_{n+k} and γ_n (interpreted as exceptional components), and $\gamma_{n+k} \to \gamma_n$. Thus $\bar{p}$ is of Type 3, a contradiction. Hence $\bar{p} \to \gamma(\bar{p})$ is injective on the set of infinitely near points of Types 1,2 and 4.

Let us prove this map is also surjective. Pick $\gamma \in \Gamma \setminus \Gamma^*$. Let p_0 be the origin, and E'_0 the exceptional component obtained by blowing up p_0. The tangent vector $\vec{v}_1$ at E'_0 represented by γ is associated to a unique point $p_1 \in E'_0$ by Proposition 6.3. We construct inductively an infinite sequence $\bar{p} = (p_j)_0^\infty$ of infinitely near points, so that E'_j is the exceptional divisor of the blowup at p_j, and $p_{j+1} \in E'_j$ is the point associated to the tangent vector represented by γ at E'_j.

When $\bar{p}$ is of Type 1 or 4, p_{j_k} is a free point for an increasing subsequence j_k. But $\gamma \geq E'_{j_k}$, and $\gamma(\bar{p})$ is an end, hence $\gamma = \gamma(\bar{p})$. When $\bar{p}$ is of Type 2, there is a subsequence $(\tilde{n}_k)_0^\infty$ such that $\tilde{E}_{2k} := E_{\tilde{n}_{2k}}$ $(\tilde{E}_{2k+1} := E_{\tilde{n}_{2k+1}})$ decreases (increases) to $\gamma(\bar{p})$ as $k \to \infty$: see Figure 6.6. It is not difficult to see that $\gamma \geq \tilde{E}_{2k+1}$ and $\gamma \wedge \tilde{E}_{2k} < \gamma(\bar{p})$ for all k. But $\gamma(\bar{p})$ is a regular point in Γ, hence $\gamma = \gamma(\bar{p})$.

This concludes the proof of Theorem 6.10.

As for Theorem 6.11, consider $\bar{p} = (p_j)_0^\infty$ of Type 3 and pick j_0 as in the definition of Type 3. Then $[\gamma(\bar{p}), \gamma(\bar{p}_{j+1})] \subset [\gamma(\bar{p}), \gamma(\bar{p}_j)]$ for all $j > j_0$. In fact, we saw above that the segments $[\gamma(\bar{p}), \gamma(\bar{p}_j)]$ intersect only in $\gamma(\bar{p})$. This proves the first few assertions in Theorem 6.11: in particular $\vec{v}(\bar{p})$ is a well defined tangent vector at $\gamma(\bar{p})$. Notice that the pair $(\gamma(\bar{p}), \vec{v}(\bar{p}))$ is uniquely determined by the pair $((p_j)_0^{j_0}, p_{j_0+1})$. Using Proposition 6.3 (see also Remark 6.6), this easily implies that $\bar{p} \to \vec{v}(\bar{p})$ gives a bijection between Type 3 infinitely near points and tangent vectors at points in Γ^*. The proof is complete.

Proposition 6.12 is well-known so we shall only outline a proof. First, to any irreducible curve we associate a sequence of infinitely near points in terms of strict transforms as in Example 1.13. As the curve can be desingularized, all but finitely many points in the sequence are free, i.e. the sequence is of Type 4. Conversely, consider a sequence $\bar{p} = (p_j)_0^\infty$ of infinitely near points of Type 4. Let us show that $\bar{p}$ is associated to a unique irreducible curve. By lifting the situation, we may assume that all p_j are free. Let us use the notation $\tilde{\pi}_j$ above. Fix arbitrary preliminary coordinates (z', w') at the origin. As p_1 is free, we may write $\tilde{\pi}_1(z'_0, w'_0) = (z'_0, z'_0(\theta_0 + w'_0))$ for suitable coordinates $(z'_0 . w'_0)$ at p_1 and $\theta_0 \in \mathbf{C}^*$. Inductively, we get coordinates (z'_j, w'_j) at p'_{j+1} such that $\tilde{\pi}_j(z'_j, w'_j) = (z'_j, z'_j(\theta_j + w'_j))$ with $\theta_j \in \mathbf{C}^*$. Now define new coordinates (z, w) at the origin and (z_j, w_j) at p_{j+1} as follows: $z = z'$, $z_j = z'_j$, $w = w' - \sum_0^\infty \theta_k (z')^{k+1}$ and $w_j = w'_j - \sum_1^\infty \theta_{k+j}(z'_j)^k$. In these coordinates, $\tilde{\pi}_j(z_j, w_j) = (z_j, z_j w_j)$ for $j \geq 0$. It is then straightforward to verify that $\{w = 0\}$ is the unique irreducible curve associated to the sequence $\bar{p}$.

For the record we notice that all the maps $\tilde{\pi}_j$ are monomial in the coordinates that we constructed.

6.3 Parameterization and Multiplicity

After having described the connection between Γ and sequences of infinitely near points, we proceed to equip Γ with a parameterization and multiplicity function.

6.3.1 Farey Weights and Parameters

Let us associate to each element of Γ^* a vector $(a, b) \in (\mathbf{N}^*)^2$, called its *Farey weight*. It is defined inductively through the following combinatorial procedure (see [HP] for the toric case). If π is a single blowup of the origin, then Γ_π is a single point E_0 whose weight is defined to be $(2, 1)$. Otherwise we may write $\pi = \pi' \circ \tilde{\pi}$, where $\tilde{\pi}$ is the blowup of a point p on the exceptional divisor $(\pi')^{-1}(0)$. The Farey weights of the vertices in Γ_π that are strict transforms of vertices in $\Gamma_{\pi'}$ inherit their weights from the latter graph. The only other vertex in Γ_π is the exceptional divisor E_p obtained by blowing up p. The weight of E_p is determined as follows. When p is free, i.e. $p \in E$ for a unique exceptional divisor $E \subset (\pi')^{-1}(0)$, then the weight of E_p is defined to be $(a + 1, b)$, where (a, b) is the weight of E. When p is a satellite point, it is the intersection of two components E and E' whose weights are (a, b) and (a', b'), respectively. The weight of E_p is then $(a + a', b + b')$. See Figure 6.7 and compare with Figure 6.1.

Fig. 6.7. Farey weights under elementary modifications. The first kind of modification is illustrated to the left, the second kind to the right. Compare Figure 6.1.

The *Farey parameter* of a point in Γ^* with weight (a, b) is defined to be the rational number $A = a/b$. We shall later see that A defines a parameterization of Γ^*. Although, this can be proved directly (in roughly the same way that we verified condition (T2) in Section 6.1.3), we shall only prove

Lemma 6.15. *The Farey parameter A is a strictly increasing function on Γ^*.*

Proof. It is equivalent to prove that the restriction of A to the set of vertices Γ^*_π for any $\pi \in \mathfrak{B}$ is strictly increasing. We proceed by induction on the number of point blowup necessary to decompose π. When this number equals 1, π is the blowup of the origin, Γ^*_π is reduced to one point, and the claim is obvious. Otherwise, pick $\pi \in \mathfrak{B}$, and consider elements $E, E' \in \Gamma^*$ that are adjacent vertices in Γ_π. Assume $E < E'$. There are two cases. In the first case, E' is obtained from E by blowing up a free point and $a'/b' = (a + 1)/b > a/b$. In the second case, E' is obtained by blowing up the intersection point between two irreducible components E and E''. Then $E < E' < E''$ so the inductive assumption gives $a/b < a''/b''$, from which it is elementary to see that $a'/b' = (a + a'')/(b + b'') \in]a/b, a''/b''[$. This concludes the proof. $\square$

6.3.2 Multiplicities

We can also use the Farey weights to define *multiplicities* in the universal dual graph. Namely, if $E \in \Gamma^o$, then we set

$$m(E) := \min\{b(F) \ ; \ F \in \Gamma^*, F \geq E\},$$

where $(a(F), b(F))$ denotes the Farey weight of F.

By definition, m is integer-valued and increasing on Γ^o. It hence has a minimal extension to a function on Γ with values in $\overline{\mathbf{N}}$. Let us describe its basic structure. The key to such a description is given by

Lemma 6.16. *Let $E \in \Gamma^*$ and suppose F is obtained from E by blowing up a* free *point. Then $m(F) = b(F) = b(E)$. Moreover, the multiplicity is constant equal to $b(E)$ in the segment $]E, F]$.*

Proof. Let (a, b) be the Farey weight of E. As F is obtained by blowing a free point p on E, its Farey weight is by definition $(a+1, b)$. Whence $b(E) = b(F)$. We always have $m(F) \leq b(F)$. To prove the converse inequality, observe that if $F' \geq F$, then the sequence of infinitely near points associated to F' starts with the one of F followed by the blowup at a free point on F. An easy induction shows that the Farey weight (a', b') of F' satisfies $b' \geq b(F)$. Thus $m(F) \geq b(F)$. This shows $m(F) = b(F) = b(E)$.

Any F' representing the same tangent vector as F at E is obtained by a sequence of blowups starting with p, hence $b(F') \geq b = b(E)$. In particular the multiplicity of any element in the segment $]E, F]$ is at least $b(E)$. On the other hand, this multiplicity cannot exceed $m(F) = b(E)$. Thus $m \equiv b(E)$ on $]E, F]$. □

A direct consequence of this lemma is that, when $E \in \Gamma^*$ and its associated dual graph is as in Figure 6.2, $m(E) = b(E_g) = b(\bar{E}_{g+1})$. Define $m_i = b(E_{i-1})$ for $1 \leq i \leq g+1$. Then $1 = m_1 < m_2 < \cdots < m_g < m_{g+1}$. By Lemma 6.16 the multiplicity equals m_i on the segment $]E_i, \bar{E}_{i+1}]$. This is illustrated in Figure 6.8. In analogy with the situation on the valuative tree, we shall refer

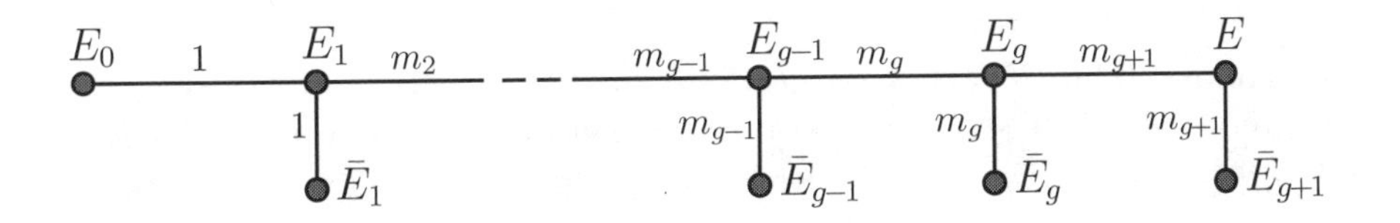

Fig. 6.8. Multiplicities.

to $E_1, \ldots, E_g$ as the *approximating sequence* of E.

Given $E \in \Gamma^*$ and a tangent vector $\vec{v}$ at E, define the multiplicity of $\vec{v}$ by $m(\vec{v}) = m(E)$ if $\vec{v}$ is represented by the root E_0, and

$$m(\vec{v}) = \min\{m(F) \ ; \ F \text{ represents } \vec{v}\}$$

otherwise.

Let $E \in \Gamma^*$ and $\pi \in \mathfrak{B}$ be minimal such that $E \subset \pi^{-1}(0)$. Its dual graph is given by Figure 6.8. It is thus clear that the two tangent vectors represented by E_0 and $\bar{E}_{g+1}$ (if $\bar{E}_{g+1} \neq E$) satisfy $m(\vec{v}) = m(E)$. Note that these tangent vectors correspond exactly to the intersection points of E with the other components of the exceptional divisor. Fix any other tangent vector $\vec{v}$, and $F \in \Gamma^*$ representing $\vec{v}$. Then F is obtained by first blowing up a free point p on E. Thus $b(F) \geq b(E)$, i.e. $m(\vec{v}) \geq b(E)$. On the other hand, Lemma 6.3 shows that $\vec{v}$ is represented by the exceptional component F obtained by blowing up p, and Lemma 6.16 gives $b(F) = b(E)$. Whence $m(\vec{v}) = b(F)$, and we obtain the following two corollaries.

Corollary 6.17. *We have $m(\vec{v}) = b(E)$ for all but at most two tangent vectors $\vec{v}$ at E. The exceptional tangent vectors have multiplicity $m(E)$.*

We shall call $b(E)$ the *generic multiplicity* of E.

Corollary 6.18. *The multiplicity of a tangent vector $\vec{v}$ at $E \in \Gamma^*$ is equal to the generic multiplicity iff there exists $\pi \in \mathfrak{B}$ such that $E \in \mathfrak{B}$ and the point on E defined by $\vec{v}$ (see Proposition 6.3) is free.*

We shall call such a tangent vector a *generic tangent vector*.

Note that when $m(\vec{v}) = b(E)$, π may be chosen to be the composition of blowup at the sequence of infinitely near points associated to E. Moreover, when blowing up at a free point, the generic multiplicity does not increase. Therefore we have

Corollary 6.19. *An element of Γ has infinite multiplicity iff it is an end associated to a sequence of infinitely near points of Type 1.*

Recall that any end of Γ corresponding to a sequence of infinitely near points of Type 4 can be viewed as an irreducible curve C. Write $m_\Gamma(C)$ for the multiplicity of C as an element of Γ. (We shall later show that $m_\Gamma(C)$ coincides with the usual multiplicity $m(C)$ of C.)

Corollary 6.20. *If C is an irreducible curve, $\pi \in \Gamma^*$ and the strict transform C' of C by π intersects $\pi^{-1}(0)$ in a free point $p \in E$, then $m_\Gamma(C) \geq b(E)$ with equality iff C' is smooth and transverse to E at p.*

Proof. Let $(q_j)_0^\infty$ be the sequence of infinitely near points associated to C' (note that $q_0 = p$) and let $F_j \in \Gamma^*$ be the exceptional divisor obtained by blowing up q_j. Then $m_\Gamma(C) = \lim b(F_j)$. By induction, $b(F_j) \geq b(E)$ with equality iff $(q_k)_0^j$ are all free. Thus $m_\Gamma(C) \geq b(E)$ with equality iff q_j is free for all j. The latter condition is equivalent to C' being smooth and transverse to E at p. □

Corollary 6.21. *For any $E \in \Gamma^o$ we have $m(E) = \min\{m_\Gamma(C) \; ; \; C > E\}$.*

Proof. The inequality $m(E) \leq \min\{m_\Gamma(C) \; ; \; C > E\}$ is obvious. To prove the other inequality it suffices to show that any $E \in \Gamma^*$ can be dominated by a curve C with $m_\Gamma(C) = b(E)$. For this, pick $\pi \in \mathfrak{B}$ such that $E \in \Gamma^*_\pi$, pick any free point p, and let C' be a smooth curve at p, transverse to E. By Corollary 6.20 the choice $C = \pi(C')$ works. □

6.4 The Isomorphism

We are now ready to formulate the main result of this chapter. It states that the universal dual graph, whose construction is purely geometric, is isomorphic—in the strongest possible sense—to the valuative tree.

The definition of the isomorphism goes as follows. A point $E \in \Gamma^*$ is an exceptional component of some $\pi \in \mathfrak{B}$. It hence defines a normalized divisorial valuation $\nu_E \in \mathcal{V}$: $\nu_E(\phi)$ is proportional to the order of vanishing of $\pi^*\phi$ along E. In other words, $\nu_E = b^{-1}\pi_* \operatorname{div}_E$ for some constant $b > 0$ (we will later see that b is indeed the generic multiplicity of E). This defines a mapping $\Phi : \Gamma^* \to \mathcal{V}_{\mathrm{div}}$.

Theorem 6.22. *The map $\Phi : \Gamma^* \to \mathcal{V}_{\mathrm{div}}$ extends uniquely to an isomorphism of parameterized trees $\Phi : (\Gamma, A) \to (\mathcal{V}, A)$. Here A denotes the Farey parameter on Γ and thinness on $\mathcal{V}$. Further, Φ preserves multiplicity.*

Before starting the proof of Theorem 6.22, let us state some immediate consequences and remarks. We refer to Section 6.6 below for more applications of the theorem. The first consequence was in fact already implicitly used above.

Corollary 6.23. *The Farey parameter induces a parameterization of the rooted, nonmetric tree Γ. The points in Γ^* are exactly the points in Γ^o having rational Farey parameter.*

Proof. The Farey parameter is indeed a parameterization, as thinness gives a parameterization of $\mathcal{V}$. As $\mathcal{V}_{\mathrm{div}}$ is the set of valuations in $\mathcal{V}_{\mathrm{qm}}$ having rational thinness, the points in Γ^* must be exactly the ones in Γ^o having rational Farey parameter. □

Our definition of the isomorphism $\Phi : \Gamma \to \mathcal{V}$ is quite indirect. In particular we do not explicitly define the value of Φ on elements of $\Gamma \setminus \Gamma^*$. In fact, we have

Corollary 6.24. *Let $\bar{p}$ be a sequence of infinitely near points, $\nu(\bar{p})$ be its associated Krull valuations as in Section 1.6, and $\gamma(\bar{p})$ be its associated point in Γ as in Section 6.2.*

When $\bar{p}$ is not of Type 3, $\Phi(\gamma(\bar{p})) = \nu(\bar{p})$. When $\bar{p}$ is of Type 3, $\nu(\bar{p})$ is an exceptional curve valuation, and $\Phi(\gamma(\bar{p})) = \nu'(\bar{p})$ where $\nu'(\bar{p})$ is the divisorial valuation associated to $\nu(\bar{p})$.

A direct consequence is:

Corollary 6.25. *For any irreducible curve C, viewed as an element of Γ, we have $\Phi(C) = \nu_C$, where ν_C is the curve valuation associated to C.*

Proof (Corollary 6.24). If $\bar{p}$ is finite (i.e. of Type 0), then the statement follows immediately from the definitions. Thus suppose $\bar{p} = (p_j)_0^\infty$ is infinite and consider the truncations $\bar{p}_n = (p_j)_0^n$. By the definition of $\gamma(\bar{p})$, $\gamma(\bar{p}_n) \in \Gamma^*$ tends to $\gamma(\bar{p})$ in the weak tree topology of Γ. On the other hand, it is also clear that $\nu(\bar{p}_n) \in \mathcal{V}_{\text{div}}$ converges to $\nu(\bar{p})$ in the Zariski topology on the set $\mathcal{V}_K$ of Krull valuations (see Section 5.5). Denote by $P(\nu(\bar{p}))$ the valuation in $\mathcal{V}$ corresponding to $\nu(\bar{p})$. We refer to the discussion in Section 1.6. When $\bar{p}$ is of Type 3, $P(\nu(\bar{p}))$ is the divisorial valuation associated to $\nu(\bar{p})$. By Theorem 5.24, $P(\nu(\bar{p}_n))$ converges weakly to $P(\nu(\bar{p}))$, hence in the weak tree topology of $\mathcal{V}$. As Φ is a tree isomorphism, $P(\nu(\bar{p})) = \Phi(\gamma(\bar{p}))$. □

6.5 Proof of the Isomorphism

To prove Theorem 6.22, we shall consider Φ as a mapping of Γ^* into $\mathcal{V}_{\text{div}}$ and proceed in four steps: first, Φ is a bijection onto $\mathcal{V}_{\text{div}}$; second, the image of the Farey parameter under Φ is equal to the thinness; third, Φ and its inverse are order preserving; and fourth, Φ preserves multiplicity.

This will show that $\Phi : \Gamma^* \to \mathcal{V}_{\text{div}}$ is an isomorphism of rooted, nonmetric $\mathbf{Q}$-trees. As thinness A is a parameterization of $\mathcal{V}_{\text{div}}$, the Farey parameter A is then a parameterization of Γ^* and $\Phi : (\Gamma^*, A) \to (\mathcal{V}_{\text{div}}, A)$ is an isomorphism of parameterized $\mathbf{Q}$-trees. It follows immediately that Φ extends to an isomorphism $\Phi : (\Gamma, A) \to (\mathcal{V}, A)$ of parameterized $\mathbf{R}$-trees. Finally, as m is lower-semicontinuous on Γ and $\mathcal{V}$, and $m = m \circ \Phi$ on Γ^*, we see that $m = m \circ \Phi$ on Γ, so that m preserves multiplicity.

6.5.1 Step 1: $\Phi : \Gamma^* \to \mathcal{V}_{\text{div}}$ is bijective

That Φ is surjective is automatic: any divisorial valuation ν is equivalent to $\pi_* \operatorname{div}_E$ for some $\pi \in \mathfrak{B}$ and some exceptional component E (see Proposition 1.12).

Let us give two arguments to prove that Φ is injective. Pick $E, E' \in \Gamma^*$ and $\pi \in \mathfrak{B}$ such that $E, E' \in \Gamma_\pi^*$. Suppose $E \neq E'$. By construction the center of the valuation $\Phi(E)$ in the total space X_π is irreducible and contains E. It is hence equal to E. Similarly, the center of $\Phi(E')$ is equal to E' so that $\Phi(E) \neq \Phi(E')$. We may also use a more geometric argument. Embed X_π in some projective space $\mathbf{P}^k$, $k \geq 2$. The curves E, E' are algebraic so we may find a polynomial $P = P(z_0, \ldots, z_k)$ homogeneous of degree d, such that $E \subset \{P = 0\}$, $E' \not\subset \{P = 0\}$. Choose a coordinate axis z_i such that $E \cup E' \not\subset \{z_i = 0\}$. Then the rational function $h = P/z_i^d$ satisfies $\operatorname{div}_E(h) > 0$ whereas $\operatorname{div}_{E'}(h) = 0$. Whence $\Phi(E) \neq \Phi(E')$.

6.5.2 Step 2: $A \circ \Phi = A$

We first need a preliminary result that gives geometric interpretations of the components a and b of the Farey weight.

Proposition 6.26. *Let $\pi \in \mathfrak{B}$, and E be an exceptional component with Farey weight (a,b). Then*

$$\begin{aligned} \operatorname{div}_E(\pi^*\mathfrak{m}) &:= \min_{\phi \in \mathfrak{m}} \operatorname{div}_E(\pi^*\phi) = b \\ \operatorname{div}_E(J\pi) &= a - 1. \end{aligned}$$

Here div_E denotes the order of vanishing along E, and $J\pi$ the Jacobian determinant of π. Note that the first equation shows that

$$\pi_* \operatorname{div}_E = b\,\nu_E. \tag{6.2}$$

This result allows us to give a local normal form for the contraction map; the following lemma is a key ingredient in the rest of the proof of Theorem 6.22.

Lemma 6.27. *Let $E \in \Gamma^*$ and $\pi \in \mathfrak{B}$ be as in the previous proposition, and pick any free point $p \in E$. Then one can find local coordinates (z,w) at p, local coordinates (x,y) at the origin, such that $E = \{z = 0\}$ and such that the contraction map is given by*

$$\pi(z,w) = (z^b, z^{a-b}w + z^b h(z)), \tag{6.3}$$

for some regular function h with $h(0) \neq 0$. Moreover, given any smooth curve V at p intersecting E transversely we may choose the coordinates such that $V = \{w = 0\}$.

We are now able to prove the relation between thinness and the Farey weight. Pick $\nu \in \mathcal{V}_{\mathrm{div}}$, choose $\pi \in \mathfrak{B}$, $E \subset \pi^{-1}(0)$ such that ν is equivalent to $\pi_* \operatorname{div}_E$. Let (a,b) be the Farey weight of E. We want to show $A(\nu) = a/b$. Pick a free point $p \in E$. By Proposition 6.26, $\pi_* \operatorname{div}_E = b\,\nu$. By Lemma 6.27 we can choose coordinates (z,w) at p such that $\pi(z,w) = (z^b, z^{a-b}w + z^b h(z))$, where $h(0) \neq 0$. Notice that $\nu(x) = b^{-1}\pi_* \operatorname{div}_E(x) = 1$. Recall from Chapter 4, that we wrote $k = \mathbf{C}(x)$, and let $\hat{k}$ be the field of Puiseux series over $\mathbf{C}$ that is the algebraic closure of k. Then the valuation ν extends to a valuation on $k[y]$ of the form $\hat{\nu} = \operatorname{val}[\hat{\phi}; \hat{\beta}]$ for a Puiseux series $\hat{\phi} \in \hat{k}$, and where $1 + \hat{\beta} \in \mathbf{Q}_+$ is the thinness of ν. See Theorem 4.17. On the other hand, $\hat{\beta}$ is also the maximum of $\nu(y - \hat{\psi})$ when $\hat{\psi}$ ranges over all Puiseux series. But

$$\hat{\nu}(y - \hat{\psi}) = b^{-1}(\pi_* \operatorname{div}_E)(y - \hat{\psi}) = b^{-1} \operatorname{div}_E\left(wz^{a-b} + z^b h(z) - \hat{\psi}(z^b)\right) \leq \frac{a-b}{b},$$

with equality for suitably chosen $\hat{\psi}$. Hence $\hat{\beta} = a/b - 1$ and $A(\nu) = 1 + \hat{\beta} = a/b$.

Proof (Proposition 6.26). First note that when π is the blowup of the origin, the proposition is clear as we have $E = \pi^{-1}(0)$, $\mathrm{div}_E \, \pi^*\mathfrak{m} = 1 = b(E)$, and $\mathrm{div}_E(J\pi) = 1 = a(E) - 1$. We then proceed by induction as follows. Pick $\pi \in \mathfrak{B}$, and suppose the proposition has been proved for all exceptional components $E \subset \pi^{-1}(0)$. We shall then prove it for the exceptional component obtained by blowing up an arbitrary point $p \in \pi^{-1}(0)$.

Fix coordinates (z,w) centered at p such that the exceptional divisor is given by $E = \{z = 0\}$ if p is free, and $E = \{z = 0\}$, $E' = \{w = 0\}$ if p is satellite. We let (a,b), and (a',b') be the Farey weights of E and E' respectively. The ideal $\pi^*\mathfrak{m}$ in the ring of regular functions at p is generated by the pull-back of a generic smooth element $\phi \in \mathfrak{m}$. It is hence principal. By the inductive step, it is generated by z^b when p is free , and $z^b w^{b'}$ when p is satellite. Similarly, the critical set of π coincides with $\pi^{-1}(0)$, so that the determinant of the Jacobian of π equals $J\pi = z^{a-1} \times$ units when p is free, and $z^{a-1}w^{a'-1} \times$ units when p is satellite.

Write $\tilde{\pi}$ for the blowup of p, and $F = \tilde{\pi}^{-1}(p)$. To compute the order of vanishing of $(\pi\circ\tilde{\pi})^*\mathfrak{m}$, and $J(\pi\circ\tilde{\pi})$ along F, we note that at a free point $p_1 \in F$, we may choose coordinates (z_1,w_1) such that $\tilde{\pi}(z_1,w_1) = (z_1, z_1(1+w_1))$. Whence $(\pi\circ\tilde{\pi})^*\mathfrak{m} = \tilde{\pi}^*\pi^*\mathfrak{m}$ is a principal ideal generated by

$$\begin{cases} \tilde{\pi}^* z^b = z_1^b & \text{when } p \text{ is free;} \\ \tilde{\pi}^* z^b w^{b'} = z_1^{b+b'} & \text{when } p \text{ is satellite.} \end{cases}$$

On the other hand, by the chain rule formula,

$$J\pi\circ\tilde{\pi} = J\pi\circ\tilde{\pi} \;\times J\tilde{\pi} =$$
$$\begin{cases} z_1^{a-1} \times \text{units} \times z_1 = z_1^a \times \text{units} & \text{when } p \text{ is free;} \\ z_1^{a-1+a'-1} \times \text{units} \times z_1 = z_1^{a+a'-1} \times \text{units} & \text{when } p \text{ is satellite.} \end{cases}$$

The Farey parameter (a_F, b_F) of F equals $(a+1,b)$ when p is free, and $(a+a', b+b')$ when p is satellite. Thus $\mathrm{div}_F(\tilde{\pi}^*\pi^*\mathfrak{m}) = b_F$, and $\mathrm{div}_F \, J(\pi\circ\tilde{\pi}) = a_F - 1$. This concludes the proof. □

Proof (Lemma 6.27). Pick local coordinates (x,y) at the origin such that the strict transforms of $\{x = 0\}$ and $\{y = 0\}$ do not pass through p. Also pick local coordinates (z,w) at p such that $E = \{z = 0\}$ and $V = \{w = 0\}$. By Proposition 6.26, π^*x and π^*y vanish to order b along E. After multiplying z by a unit, we thus have $\pi(z,w) = (z^b, z^b\xi(z,w))$ for some unit ξ. Similarly, by Proposition 6.26, the Jacobian determinant of π is given by $z^{a-1}\rho(z,w)$, where $\rho(0,0) \neq 0$. Thus $b^{-1}z^{b-1}z^b\frac{\partial\xi}{\partial w} = z^{a-1}\rho(z,w)$, which gives $\xi = h(z) + dwz^{a-2b}\xi_1(z,w)$, where $\xi_1 + w\frac{\partial\xi_1}{\partial w} = \rho(z,w)$. Since $\xi(0,0) \neq 0$ and $\rho(0,0) \neq 0$ we have $h(0) \neq 0$ and $\xi_1(0,0) \neq 0$. After multiplying w by $d\xi_1(z,w)$ we arrive at (6.3). □

6.5.3 Step 3: Φ and Φ^{-1} Are Order Preserving

This is the key (and hardest) part of the argument. We start by stating

Lemma 6.28. *Pick $\pi \in \mathfrak{B}$ and $p \in \pi^{-1}(0)$. Let $F \in \Gamma^*$ be the exceptional divisor of the blowup of p. Then:*

- *when p is free, it belongs to a unique component $E \subset \pi^{-1}(0)$, and we have $\nu_F > \nu_E$;*
- *when p is satellite, it lies at the intersection of two components E and E' and $\nu_F \in [\nu_E, \nu_{E'}]$.*

In order to prove that both Φ and its inverse are order preserving, it suffices to show that the map $\Phi : \Gamma^*_\pi \to \mathcal{V}_{\mathrm{div}}$ is a *tree embedding* for any $\pi \in \mathfrak{B}$, in the sense that $\Phi(E) < \Phi(E')$ iff $E < E'$. We then proceed by induction on the number of blowups necessary to decompose π, i.e. the cardinality of Γ^*_π. When this number equals one, π is the blowup of the origin, and Γ^*_π is reduced to one point, so that the statement is obvious.

Now suppose $\Phi : \Gamma^*_\pi \to \mathcal{V}_{\mathrm{div}}$ is a tree embedding and pick $p \in \pi^{-1}(0)$. Let $\tilde{\pi}$ denote the blowup of p, and set $F := \tilde{\pi}^{-1}(p)$. When p is a satellite point, it is the intersection of two divisors E and E'. The graph $\Gamma^*_{\pi\circ\tilde{\pi}}$ is obtained from Γ^*_π by adding one vertex in the segment $[E, E']$. By Lemma 6.28, $\nu_F \in [\nu_E, \nu_{E'}]$. This segment does not contain any other valuation $\nu_{E''}$ for any exceptional components $E'' \in \Gamma^*_\pi$ by the inductive assumption. Thus $\Phi : \Gamma^*_{\pi\circ\tilde{\pi}} \to \mathcal{V}_{\mathrm{div}}$ is also a tree embedding in this case.

When p is free, it belongs to a unique component $E \in \Gamma^*_\pi$. The graph $\Gamma^*_{\pi\circ\tilde{\pi}}$ is obtained from Γ^*_π by attaching a new vertex F to E. By Lemma 6.28, $\nu_F > \nu_E$. This shows that Φ is order preserving. To conclude the induction step, we also need to show that ν_F does not define the same tangent vector as $\nu_{E'}$ at ν_E for some other $E' \in \Gamma^*_\pi$. Equivalently, we have to show that $\nu_F \wedge \nu_{E'} \leq \nu_E$ for all $E' \in \Gamma^*_\pi$. We may assume $\nu_{E'} \geq \nu_E$. As $\Phi : \Gamma^*_\pi \to \mathcal{V}_{\mathrm{div}}$ is a tree embedding, it is also sufficient to consider the case when E' intersects E. Pick an irreducible curve $C = \{\phi = 0\}$, $\phi \in \mathfrak{m}$, at the origin, whose strict transform C' by π is smooth and contains $p' := E \cap E'$. Let b and b' be the generic multiplicities of E and E', respectively; and let F and F' be the exceptional divisors of the blowups at p and p'. Introduce μ_p and $\mu_{p'}$, the multiplicity valuations at p and p', respectively. From Proposition 6.26 we have $\pi_* \operatorname{div}_E = b\,\nu_E$ and $\pi_* \operatorname{div}_{E'} = b'\,\nu_{E'}$, so that $\pi_*\mu_p = b\,\nu_F$ and $\pi_*\mu_{p'} = (b+b')\nu_{F'}$. Finally, let ϕ' be a defining function for C': this is a regular function vanishing at p'. We then have:

$$\nu_F(\phi) = b^{-1}(\pi_*\mu_p)(\phi) = b^{-1}\mu_p(\pi^*\phi) = b^{-1}\operatorname{div}_E(\pi^*\phi) = \nu_E(\phi).$$

On the other hand,

$$\begin{aligned}\nu_{F'}(\phi) &= (b+b')^{-1}(\pi_*\mu_{p'})(\phi) = (b+b')^{-1}\mu_{p'}(\pi^*\phi) = \\ &= (b+b')^{-1}\left(\operatorname{div}_E(\pi^*\phi) + \operatorname{div}_{E'}(\pi^*\phi) + \mu_{p'}(\phi')\right) > \\ &\qquad > (b+b')^{-1}\left(b\nu_E(\phi) + b'\nu_{E'}(\phi)\right) > \nu_E(\phi).\end{aligned}$$

We conclude that $\nu_F(\phi) = \nu_E(\phi) < \nu_{F'}(\phi)$. But by Lemma 6.28, $\nu_{F'} \in [\nu_E, \nu_{E'}]$, so that $\nu_{F'}$ defines the same tangent vector as $\nu_{E'}$ at ν_E. Thus $\nu_F \wedge \nu_{F'} = \nu_F \wedge \nu_{E'} = \nu_E$, which completes the proof of Step 3.

Let us now turn to the proof of Lemma 6.28. We rely on the following well-known result, which may be viewed as the second basic ingredient in the proof of the isomorphism, the first one being the normal form in Lemma 6.27.

Lemma 6.29. *Fix a smooth curve D at the origin and consider a finite sequence $(p_j)_0^n$ of infinitely near points above the origin such that $p_1, \dots, p_{n_0}$ are all free and $p_{n_0+1}, \dots, p_n$ are all satellite. Here $0 \le n_0 \le n$. Let π be the composition of blowups at the points $p_0, \dots, p_{n-1}$. Then there exist coordinates (x, y) at the origin and (z, w) at p_n such that $D = \{x = 0\}$, $\{zw = 0\} \subset \pi^{-1}(D)$ and such that π is a monomial map in these coordinates.*

Here, "free/satellite point" is to be understood as "regular/singular point on the total transform of D". We postpone the proof of this lemma to the end of this section, and conclude the proof of Lemma 6.28.

Proof (Lemma 6.28). When p is free, it belongs to a unique component E. By Lemma 6.27, there exist coordinates (z, w) at p, (x, y) at the origin, such that $\pi(z, w) = (z^b, z^{a-b}w + z^b h(z))$, for some regular function h such that $h(0) \neq 0$. Here $E = \{z = 0\}$, and the form of π implies $\pi_* \operatorname{div}_z = b\,\nu_E$. If μ_p denotes the multiplicity valuation at p (i.e. the monomial valuation with weight $(1, 1)$ on (z, w)), then we also have $\pi_* \mu_p = b\,\nu_F$. As $\mu_p > \operatorname{div}_z$, we conclude that $\nu_F > \nu_E$.

Now assume that p is satellite. Let us first reduce to the case when π is a composition of blowups at infinitely near points. Consider the set of all $\pi' \in \mathfrak{B}$ such that $\pi = \varpi \circ \pi'$ for some composition of blowups ϖ such that ϖ is a local biholomorphism at p. This set admits a minimum thanks to Lemma 6.1. It is not difficult to check that this minimum is necessarily a composition of blowups at infinitely near points.

We may thus suppose that $\pi = \tilde{\pi}_0 \circ \dots \tilde{\pi}_{n-1}$ is a composition of point blowups at infinitely near points $p_0, \dots, p_{n-1}$ and that $p = p_n$ is a satellite point, say $p = E \cap E'$. Define the numbers n_g and $\bar{n}_{g+1}$ as in (6.12). Thus p_j is free for $n_g < j \le \bar{n}_{g+1}$ and satellite for $\bar{n}_{g+1} < j \le n$. Define $p' = p_{n_g+1}$, $\varpi = \tilde{\pi}_{n_g+1} \circ \dots \circ \tilde{\pi}_{n-1}$ and $\varpi' = \tilde{\pi}_0 \circ \dots \circ \tilde{\pi}_{n_g}$. Thus $\pi = \varpi' \circ \varpi$.

We may apply Lemma 6.29 to ϖ, with the origin being the point p' and the curve D the exceptional divisor of ϖ' at p'. This gives coordinates (z, w) at p and (z', w') at p' in which ϖ is a monomial map, say $\varpi(z, w) = (z^\alpha w^\beta, z^\gamma w^\delta)$ for some $\alpha, \beta, \gamma, \delta$. Moreover, the exceptional divisor of ϖ' is given by $\{z' = 0\}$ at p' and the exceptional divisor of π by $\{zw = 0\}$ at p.

After permuting coordinates we may assume $E = \{z = 0\}$ and $E' = \{w = 0\}$ at p. Moreover, F is the exceptional divisor of the blowup at p. Denote by μ_p the multiplicity valuation at p. Then ν_E, $\nu_{E'}$ and ν_F are proportional to the pushforwards by π of div_z, div_w and μ_p, respectively.

Since ϖ is monomial, the three valuations $\varpi_* \operatorname{div}_z$, $\varpi_* \operatorname{div}_w$, and $\varpi_* \mu_p$ are monomial in the coordinates (z', w'). at p'. Their values on (z', w') are given by (α, γ), (β, δ) and $(\alpha+\beta, \gamma+\delta)$, respectively. Write $\mu_{w',t}$ for the monomial valuation sending z' to 1 and w' to $t > 0$. This is an element of the relative valuative tree $\mathcal{V}_{z'}$ studied in Section 3.9. The three valuations $\varpi_* \operatorname{div}_z$, $\varpi_* \operatorname{div}_w$ and $\varpi_* \mu_p$ are then equivalent to $\mu_{w',\gamma/\alpha}$, $\mu_{w',\delta/\beta}$ and $\mu_{w',(\gamma+\delta)/(\alpha+\beta)}$, respectively. It is clear that $\mu_{w',(\gamma+\delta)/(\alpha+\beta)}$ belongs to the segment $[\mu_{w',\gamma/\alpha}, \mu_{w',\delta/\beta}]$, hence $\varpi'_* \mu_{w',(\gamma+\delta)/(\alpha+\beta)} \in [\varpi'_* \mu_{w',\gamma/\alpha}, \varpi'_* \mu_{w',\delta/\beta}]$.

On the other hand, the point p' is a free point, lying on the component $E_g = E_{n_g}$, so the pullback under ϖ' of a smooth generic function is given by $(z')^{b_g} \times$ units, where b_g is the generic multiplicity of E_g. See Proposition 6.26. Thus

$$\tilde{\pi}_* \mu_{w',(\gamma+\delta)/(\alpha+\beta)}(\mathfrak{m}) = \tilde{\pi}_* \mu_{w',\gamma/\alpha}(\mathfrak{m}) = \tilde{\pi}_* \mu_{w',\delta/\beta}(\mathfrak{m}) = b_g.$$

This gives

$$\nu_F = b_g^{-1} \tilde{\pi}_* \mu_{w',(\gamma+\delta)/(\alpha+\beta)}, \quad \nu_E = b_g^{-1} \tilde{\pi}_* \nu_{w',\gamma/\alpha} \quad \text{and} \quad \nu_{E'} = b_g^{-1} \tilde{\pi}_* \nu_{w',\delta/\beta}.$$

We conclude that $\nu_F \in [\nu_E, \nu_{E'}]$. □

Proof (Lemma 6.29). We denote by $\tilde{\pi}_j$ the blowup at p_j, $E_j := \tilde{\pi}_j^{-1}(p_j)$ (as well as its strict transform by further blowups). We shall construct coordinates (x, y) at the origin and (z_j, w_j) at each p_{j+1} such that $\tilde{\pi}_j$ is a monomial map for all $j < n$. This clearly implies the lemma.

First suppose $n_0 = n$ i.e. all points are free. The existence of the coordinates above is the proved in exactly the same way as in Proposition 6.12; the minor adjustments necessary are left to the reader.

Next suppose $n_0 = 0$, i.e. $p_1, \dots, p_n$ are all satellite (in the sense that they are singular points on the total transform of D). Fix arbitrary coordinates (x, y) at the origin such that $D = \{x = 0\}$. Our assumptions imply that p_1 is the intersection point of E_0 and the strict transform of D (also denoted by D). We may hence choose coordinates (z_0, w_0) at p_1 such that $\{z_0 = 0\} = E_0$, $\{w_0 = 0\} = D$ and $\tilde{\pi}_0(z_0, w_0) = (z_0, z_0 w_0)$. As p_2 is a satellite point, it is necessarily one of the two intersection points $E_1 \cap D$ (case 1) or $E_1 \cap E_0$ (case 2). We may now choose coordinates (z_1, w_1) at p_2 such that $\{z_1 = 0\} = E_1$ and $\{w_1 = 0\} = D$ in case 1, $= E_0$ in case 2; and $\tilde{\pi}_1(z_1, w_1) = (z_1, z_1 w_1)$ in case 1, or $= (z_1 w_1, z_1)$ in case 2. Inductively we obtain coordinates (z_j, w_j) at p_{j+1} such that $\{z_j = 0\} = E_j$, $\{w_j = 0\}$ represents the other exceptional component containing p_{j+1}, and $\tilde{\pi}_j(z_j, w_j) = (z_j, z_j w_j)$ or $= (z_j w_j, z_j)$. Thus $\tilde{\pi}_j$ is monomial for $0 \le j < n$.

Finally when $0 < n_0 < n$ we may compose the two constructions above. This completes the proof. □

6.5.4 Step 4: Φ Preserves Multiplicity

At this stage we may conclude that $\Phi : \Gamma^* \to \mathcal{V}_{\mathrm{div}}$ extends to an isomorphism $\Phi : \Gamma \to \mathcal{V}$ of rooted, nonmetric **R**-trees. Moreover, the Farey parameter defines a parameterization of Γ and $\Phi : (\Gamma, A) \to (\mathcal{V}, A)$ is an isomorphism of parameterized trees. Notice that the proofs of Corollaries 6.24 and 6.25 go through. Thus Φ maps any irreducible curve C, viewed as an end in Γ, to the corresponding curve valuation ν_C.

Lemma 6.30. *If C is any irreducible curve, then the multiplicity $m_\Gamma(C)$ of C as an element of Γ is the same as its multiplicity $m(C)$ as a curve.*

Now pick any $E \in \Gamma^o$. By the definition of the multiplicity in $\mathcal{V}$ and by Corollary 6.21 we have

$$\begin{aligned} m(\Phi(E)) = \min\{m(C)\ ;\ \nu_C > \Phi(E)\} &= \min\{m(C)\ ;\ C > E\} \\ &= \min\{m_\Gamma(C)\ ;\ C > E\} = m(E). \end{aligned}$$

This immediately implies that $m \circ \Phi = m$ on all of Γ, completing Step 4 and thus the whole proof of Theorem 6.22.

Proof (Lemma 6.30). Let $(p_j)_0^\infty$ be the sequence of infinitely near points associated to C. This is of Type 4 by Proposition 6.12. Pick n minimal such that p_j is free for $j > n$, let π be the composition of blowups at $p_0, \dots, p_n$ and denote by E the exceptional divisor obtained by blowing up p_n.

Let C' be the strict transform of C under π. Then C' intersects $\pi^{-1}(0)$ only at p_{n+1}. As the points p_j are free for all $j > n$, C' must be smooth and $m_\Gamma(C) = b$, where (a, b) is the Farey weight of E. See Corollary 6.20.

Lemma 6.27 provides us with local coordinates (z, w) at p_{n+1} and (x, y) at the origin, such that $E = \{z = 0\}$, $C' = \{w = 0\}$ and such that $\pi(z, w) = (z^b, z^{a-b}w + z^b h(z))$, where h is a regular function with $h(0) \neq 0$. A parameterization of C is then given by $t \mapsto \pi(t, 0) = (t^b, t^b h(t))$, showing that $m(C) = b = m_\Gamma(C)$. □

6.6 Applications

The fact that $\Phi : \Gamma \to \mathcal{V}$ is a tree isomorphism leads to interpretations in more geometrical terms of several constructions in the valuative tree. We shall give examples of this principle in subsequent sections. From the proof that Φ is order preserving (Step 3 of the proof), we also extract a monomialization procedure for arbitrary quasimonomial valuations. This is explained in Section 6.6.4.

6.6.1 Curvettes

Fix a divisorial valuation ν, and an irreducible curve C. We say that C defines a *curvette* for ν if there exist $\pi \in \mathfrak{B}$ and $E \in \Gamma_\pi^*$, such that $\nu = \nu_E$, and the strict transform of C is smooth and intersects E transversely at a free point. Before stating the proposition characterizing curvettes inside $\mathcal{V}$, let us introduce some terminology. For a fixed divisorial valuation ν, a tangent vector $\vec{v}$ at ν is *generic* if it is not represented by $\nu_{\mathfrak{m}}$, and its multiplicity is the generic multiplicity of ν. By Proposition 3.39, any divisorial valuation admits at most two tangent vectors which are not generic.

Proposition 6.31. *Pick $E \in \Gamma^*$ and let $\nu = \nu_E$ be the associated divisorial valuation. Then if C is an irreducible curve at the origin, the following assertions are equivalent:*

(i) $\nu_C > \nu$, $m(C) = b(\nu)$, the generic multiplicity of ν, and ν_C represents a generic tangent vector at ν;
(ii) $C > E$, $m(C) = b(E)$, the generic multiplicity of E, and C represents a generic tangent vector at E;
(iii) C is a curvette for ν.

It was observed by Spivakovsky, that a divisorial valuation acts by intersection with a curvette: see [Sp, Theorem 7.2]. In fact for any $\psi \in \mathfrak{m}$, $\nu(\psi) = \nu_C(\psi)$ as soon as C is a curvette for ν and does not define the same tangent vector at ν as the curve valuation associated to any irreducible factor of ψ.

Proof. The equivalence of (i) and (ii) is a direct consequence of the isomorphism between Γ and $\mathcal{V}$. Hence we need only show that (ii) and (iii) are equivalent. Notice that this is a statement purely inside Γ. Also recall that the multiplicity of an irreducible curve C coincides with its multiplicity in the universal dual graph Γ.

First suppose C is a curvette for ν. Pick $\pi \in \mathfrak{B}$, and an exceptional component $E \in \Gamma^*_\pi$ such that the strict transform of C by π intersects E at a free point p. By Corollary 6.18, we infer that $C > E$ and that C represents a generic tangent vector at E. Further, Corollary 6.20 gives $m(C) = b(E)$. Thus (ii) holds.

Conversely, suppose C satisfies (ii). Consider the sequence of infinitely near points $(p_j)_0^n$ associated to E, and let $\pi \in \mathfrak{B}$ be the composition of blowups at the points $p_0, \dots, p_n$. Then $E \in \Gamma^*_\pi$ is obtained by blowing up the last point p_n. The strict transform C' of C by π intersects E at a point p. As C defines a generic tangent vector at E, p is free by Corollary 6.18. As C' is smooth and transverse to E at p, $m(C) = b(E)$ by Corollary 6.20. Thus C is a curvette for $\nu = \nu_E$, which completes the proof. □

6.6.2 Centers of Valuations and Partitions of $\mathcal{V}$

Consider a valuation $\nu \in \mathcal{V}$ and a proper birational morphism $\pi \in \mathfrak{B}$, say $\pi : X \to (\mathbf{C}^2, 0)$. In Section 1.7 we defined the center of ν on X. Here we shall compute the center in terms of tree data, using the fundamental isomorphism between Γ and $\mathcal{V}$.

Recall more precisely that the center of ν on X is either an irreducible component E of $\pi^{-1}(0)$ (i.e. an element of Γ^*_π) or a (closed) point p on $\pi^{-1}(0)$. In the first case, ν is the divisorial valuation ν_E associated to E. In the second case, $\nu = \pi_* \mu$, where μ is a centered valuation on the ring R_p of formal power series at p, i.e. there exist local coordinates (z, w) at p such that $\mu(z), \mu(w) > 0$.

Let us analyze the second case in more detail. Consider the sequence $(q_j)_0^\infty$ of infinitely near points associated to μ. Thus $q_0 = p$ and $0 \leq n \leq \infty$. Let $\tilde{\pi}_j$ be the blowup at q_j, $F_j = \tilde{\pi}_j^{-1}(q_j)$ and set $\varpi_j = \tilde{\pi}_0 \circ \cdots \circ \tilde{\pi}_j$. For each j we may apply Corollary 6.4 to ϖ_j. This shows that F_j represents the tangent vector $\vec{v}_p$ defined by p, at E. In particular, the segment $[F, F_j]$ contains F for any $F \in \Gamma_\pi^*$.

When ν is divisorial, $n < \infty$ and $\nu = \nu_{F_n}$. As Φ is an isomorphism of rooted, nonmetric trees, we conclude that $[\nu_F, \nu]$ contains ν_E for any $F \in \Gamma_\pi^*$. When ν is nondivisorial, Corollary 6.24 implies that $\nu_{F_j} \to \nu$ as $n \to \infty$. Again we conclude that $[\nu_F, \nu]$ contains ν_E for any $F \in \Gamma_\pi^*$.

We may summarize our result as follows.

Proposition 6.32. *Pick $\pi \in \mathfrak{B}$ and $\nu \in \mathcal{V}$. Let $\mathcal{E}$ be the set consisting of divisorial valuations ν_E with $E \subset \pi^{-1}(0)$ such that $[\nu_E, \nu]$ contains no other divisorial valuations ν_F with $F \subset \pi^{-1}(0)$. It consists of one or two valuations.*

(i) When $\mathcal{E} = \{\nu_E\}$, either $\nu = \nu_E$ and the center of ν in $\pi^{-1}(0)$ equals E; or $\nu \neq \nu_E$ and the center of ν in $\pi^{-1}(0)$ is the (free) point on E associated by Proposition 6.3 to the tangent vector $\vec{v}$ at ν_E represented by ν.

(ii) When $\mathcal{E} = \{\nu_E, \nu_{E'}\}$, the center of ν is equal to the (satellite) point $E \cap E'$. This point is on E (E') the point associated to the tangent vector at ν_E ($\nu_{E'}$) represented by ν.

Remark 6.33. By applying this result to a curve valuation $\nu = \nu_C$ we obtain a description of the intersection point of the strict transform of C with the exceptional divisor $\pi^{-1}(0)$. (As follows from the proof, this could have been achieved without passing to the valuative tree.)

Given $\pi \in \mathfrak{B}$ we obtain a partition of $\mathcal{V}$ into sets of valuations having the same center on $\pi^{-1}(0)$. For any exceptional component $E \in \Gamma_\pi^*$, $\nu_E \in \mathcal{V}$ is the unique valuation in $\mathcal{V}$ having E as center. As follows from Proposition 6.32, the set $U(p) \subset \mathcal{V}$ of valuations whose center is a given closed point $p \in \pi^{-1}(0)$ can be computed as follows. See Figure 6.9.

Corollary 6.34. *Pick $\pi \in \mathfrak{B}$ and a closed point $p \in \pi^{-1}(0)$. Let $U(p) \subset \mathcal{V}$ be the set of valuations with center p on $\pi^{-1}(0)$. This is a weak open subset of $\mathcal{V}$ of the following type.*

(i) If p is a free point, belonging to a unique exceptional component $E \in \Gamma_\pi^*$, then $U(p) = U(\vec{v})$, where $\vec{v}$ is the tangent vector at ν_E associated to p as in Proposition 6.3.

(ii) If p is a satellite point, say $\{p\} = E \cap F$ for distinct exceptional components $E, F \in \Gamma_\pi^*$, then $U(p) = U(\vec{v}) \cap U(\vec{w})$, where $\vec{v}$ ($\vec{w}$) is the tangent vector at ν_E (ν_F) represented by ν_F (ν_E).

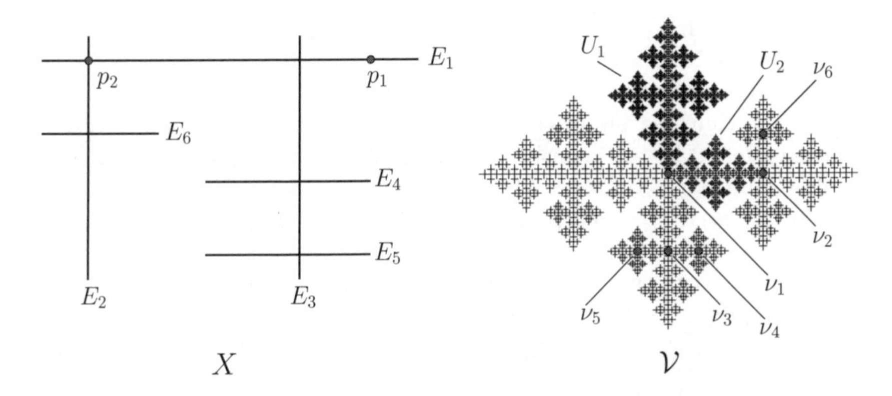

Fig. 6.9. Partition of the valuative tree. To the left is the exceptional divisor $\pi^{-1}(0)$ of a composition $\pi : X \to (\mathbf{C}^2, 0)$ of blowups. To the right is the valuative tree, with the dual graph Γ^*_π embedded. Associated to the points $p_1, p_2 \in \pi^{-1}(0)$ are weak open subsets $U_1, U_2 \subset \mathcal{V}$ consisting of valuations whose centers on X are p_1 and p_2, respectively. See Corollary 6.34. For $i = 1, \cdots, 6$, ν_i denotes the divisorial valuation associated to E_i.

In Figure 6.9 it is also possible to interpret the segment joining the two divisorial valuations ν_1 and ν_2.

Proposition 6.35. *Pick $\pi \in \mathfrak{B}$ and a closed point $p \in \pi^{-1}(0)$, lying at the intersection of two exceptional divisors E and F. Choose local coordinates (z, w) at p such that $E = \{z = 0\}$ and $F = \{w = 0\}$. Let $\mu_{s,t}$ be the monomial valuation in the coordinates (z, w) with weights s on z and t on w; and denote by ν_E, ν_F the divisorial valuations associated to E and F respectively.*

Then π_* induces a homeomorphism from the set $\{\mu_{s,t};\ s, t \geq 0,\ sb_E + tb_F = 1\}$ onto the segment $[\nu_E, \nu_F] \subset \mathcal{V}$.

Proof. The pull-back by π of the maximal ideal $\mathfrak{m}$ is locally principal and generated by $z^{b_E} w^{b_F}$. We infer that $\pi_*\mu_{s,t}(\mathfrak{m}) = sb_E + tb_F$, and thus π_* maps $\{\mu_{s,t};\ s, t \geq 0,\ sb_E + tb_F = 1\}$ into $\mathcal{V}$. It is clearly injective and continuous. By definition $\mu_{1,0} = \mathrm{div}_z$, $\mu_{0,1} = \mathrm{div}_w$, so $\pi_*\mu_{b_E^{-1},0} = \nu_E$ and $\pi_*\mu_{0,b_F^{-1}} = \nu_F$. As $\mathcal{V}$ is a tree, the only continuous and injective map from a real segment into $\mathcal{V}$ joining ν_E to ν_F is $[\nu_E, \nu_F]$. This concludes the proof. □

6.6.3 Potpourri on Divisorial Valuations

In this section we prove four results on divisorial valuations.

First, we describe which divisorial valuations are obtained by blowing up a free point p on some exceptional component.

Proposition 6.36. *Let ν be a divisorial valuation with associated sequence $(p_0, \dots, p_n)$ of infinitely near points. Then $m(\nu) = b(\nu)$ iff p_n is free.*

Proof. If p_n is free, then by Lemma 6.16 the multiplicity of the exceptional divisor $F \in \Gamma^*$ associated to the blowup at p_n satisfies $m(F) = b(F)$. As Φ preserves multiplicity, $m(\nu) = b(\nu)$.

For the converse, suppose p_n is satellite, say the intersection of two components E and E' of generic multiplicity b and b' respectively. Then the generic multiplicity of F is $b + b'$, whereas its multiplicity is bounded by $\max\{m(E), m(E')\} \leq \max\{b, b'\}$. Hence $b(F) > m(F)$, and $b(\nu) > m(\nu)$. □

Second, let us summarize what happens in general when blowing up a point p on an exceptional component.

Proposition 6.37. *Fix $\pi \in \mathfrak{B}$, pick a point $p \in \pi^{-1}(0)$, and define ν_F to be the divisorial valuation associated to the blowup at p.*

(i) If p is a free point, i.e. p belongs to a unique exceptional component E of π, then:
  (a) $\nu_F > \nu_E$ and ν_F does not define the same tangent vector at ν_E as any $\nu_{E'}$ for $E' \in \Gamma^*_\pi \setminus \{E\}$;
  (b) the multiplicity $m(\nu_F)$ of ν_F is equal to its generic multiplicity $b(\nu_F)$, and both coincide with $b(\nu_E)$; moreover, the multiplicity is constant, equal to $b(\nu_E)$, on the segment $]\nu_E, \nu_F]$;
  (c) $A(\nu_F) = A(\nu_E) + b(\nu_E)^{-1}$.

(ii) If p is a satellite point, i.e. p is the intersection point of two exceptional components E and E', then:
  (a) $\nu_{E'} > \nu_F > \nu_E$ or $\nu_E > \nu_F > \nu'_E$;
  (b) the multiplicity is constant, equal to $\max\{m(\nu_E), m(\nu'_E)\}$ on the segment $]\nu_E, \nu_{E'}[$; moreover, the generic multiplicity of $\nu(F)$ is given by $b(\nu_F) = b(\nu_E) + b(\nu_{E'})$;
  (c) $A(\nu_F) = (a(\nu_E) + a(\nu_{E'}))/(b(\nu_E) + b(\nu_{E'}))$, where $a(\nu_E) - 1$ and $a(\nu_{E'}) - 1$ are the orders of vanishing of the Jacobian of π along E and E', respectively.

Proof. By the fundamental isomorphism $\Phi : \Gamma \to \mathcal{V}$ it suffices to prove the corresponding statements in the universal dual graph. Most of them are then straightforward consequences of the combinatorial definition of the partial ordering and Farey weights. Specifically, assertions (a) and (c) in both (i) and (ii) are immediate; (see also Proposition 6.26) and assertion (b) in (i) follows from Lemma 6.16.

Let us prove the first statement in assertion (b) in (ii) for completeness and as we shall use it below. For this, we show in general that if $\pi \in \mathfrak{B}$ and E, E' are adjacent elements in Γ^*_π (i.e. E and E' intersect), then either $E < E'$ or $E' < E$, and the multiplicity is constant on the segment $]E, E'[$.

For this, we may assume that π is minimal such that $E, E' \in \Gamma^*_\pi$. The proof now goes by induction on the cardinality of Γ^*_π. Without loss of generality, E' is obtained by blowing up a point on E. If this point is free, then $E' > E$ and the multiplicity is constant equal to b on $]E, E'[$, where (a, b) is the Farey weight of E. If the point is not free, it is the intersection point between E and another irreducible components E''. By induction we may assume $E'' > E$ and that the multiplicity is constant on $]E, E''[$. As $E' \in]E, E''[$, this completes the proof. □

Third, let us define two divisorial valuations ν, ν' to be *adjacent* if there exists a composition of blowups $\pi \in \mathfrak{B}$ such that ν and ν' are proportional to the pushforward of the order of vanishing along two irreducible components E, E' of $\pi^{-1}(0)$ with nonempty intersection. In other words, $\nu = \nu_E$, $\nu' = \nu_{E'}$ with E, E' being adjacent vertices in some Γ_π.

We wish to characterize when two divisorial valuations are adjacent, purely in terms of quantities on the valuative tree $\mathcal{V}$. It follows from Proposition 6.37 that if ν and ν' are adjacent divisorial valuations, then either $\nu < \nu'$ or $\nu' < \nu$. Further, the multiplicity is constant, equal to $m = \max\{m(\nu), m(\nu')\}$, on the segment $]\nu, \nu'[$. By lower semicontinuity we then conclude that the multiplicity is constant on $]\nu, \nu']$ if $\nu < \nu'$ and $]\nu', \nu]$ if $\nu' < \nu$.

Proposition 6.38. *Let ν and ν' be divisorial valuations with generic multiplicities b and b', respectively. Assume that $\nu < \nu'$ and that the multiplicity is constant, equal to $m = m(\nu')$ on the segment $]\nu, \nu']$. Then*

$$A(\nu') - A(\nu) \geq \frac{m}{b\,b'}, \tag{6.4}$$

with equality iff ν and ν' are adjacent.

As an important consequence, we may recover the parameterization by *skewness* from the Farey weights.

Corollary 6.39. *Let ν and ν' be divisorial valuations with generic multiplicities b and b', respectively. Assume that $\nu < \nu'$. Then*

$$\alpha(\nu') - \alpha(\nu) \geq \frac{1}{b\,b'}, \tag{6.5}$$

with equality iff ν and ν' are adjacent. In the latter case, the multiplicity is constant, equal to $a'b - ab'$ on the segment $]\nu, \nu']$. Here (a, b) and (a', b') denote the Farey weights of (the elements of Γ^* associated to) ν and ν', respectively.

Proof (Corollary 6.39). First suppose that the multiplicity is constant, equal to m on the segment $]\nu, \nu']$. Then (6.4) immediately gives

$$\alpha(\nu') - \alpha(\nu) = \frac{A(\nu') - A(\nu)}{m} \geq \frac{1}{bb'}, \tag{6.6}$$

with equality iff ν and ν' are adjacent.

In general let $\nu = \nu_0 < \nu_1 < \cdots < \nu_n = \nu'$ be the divisorial valuations such that the multiplicity is constant, equal to m_i on $[\nu_{i-1}, \nu_i[$ for $1 \leq i \leq n$, and $m_1 < m_2 < \cdots < m_n$. Then the generic multiplicities $b_i = b(\nu_i)$ satisfy $b_1 < b_2 < \cdots < b_n$. From the first part of the proof we obtain

$$\alpha(\nu') - \alpha(\nu) = \sum_1^n (\alpha(\nu_i) - \alpha(\nu_{i-1})) \geq \sum_1^n \frac{1}{b_i b_{i-1}} \geq \frac{1}{b_0 b_n} = \frac{1}{bb'}.$$

Here equality can only hold if $n = 1$, i.e. the multiplicity is constant on $]\nu, \nu']$. In this case, the formula $m = a'b - ab'$ is an immediate consequence of (6.6). □

Proof (Proposition 6.38). Let us first prove (6.4). By Proposition 3.39, the multiplicity $m = m(\nu')$ divides $b' = b(\nu')$. As $\nu' > \nu$, the tangent vector $\vec{v}$ represented by ν' at ν is not represented by $\nu_{\mathfrak{m}}$. By assumption $m(\vec{v}) = m$. Again by Proposition 3.39, either $m(\vec{v}) = m(\nu)$ or $m(\vec{v}) = b(\nu)$. In both cases, m divides $b(\nu)$. Write $\nu = \nu_E$ and $\nu' = \nu_{E'}$, for $E, E' \in \Gamma^*$ and let (a, b), (a', b') be the Farey weights of E and E', respectively. Note that $b = b(\nu)$, $b' = b(\nu')$, hence m divides $a'b - ab'$. Then (6.4) holds since

$$A(\nu') - A(\nu) = \frac{a'}{b'} - \frac{a}{b} = \frac{m}{bb'} \frac{a'b - ab'}{m} \geq \frac{m}{bb'}.$$

Next we show that for any $\pi \in \mathfrak{B}$, and any adjacent vertices in Γ^*_π, equality in (6.4) holds. We proceed by induction on the number of blowups in π, writing $\pi = \pi' \circ \tilde{\pi}$ with $\tilde{\pi}$ being the blowup at a point p. We let E be the exceptional divisor of $\tilde{\pi}$. Using the induction step, we only need to prove (6.4) for E and a vertex adjacent to E. When p is free, lying on a divisor F with Farey weight (a, b), the Farey weight of E is $(a + 1, b)$ by definition, and the multiplicity in $]E, F[$ is constant (this is clear in Γ^*), equal to $m := b = b(E) = b(F)$. Whence $A(E) - A(F) = \frac{(a+1)b - ba}{b^2} = \frac{m}{b^2}$. This proves the induction step in this case, as E is adjacent to a unique vertex F. When p is satellite, intersection of two divisors F_1, F_2 with Farey weights (a_1, b_1), (a_2, b_2), E belongs to the segment $[F_1, F_2]$ by construction and has Farey weight $(a_1 + a_2, b_1 + b_2)$ by definition. The multiplicity on the segment $]F_1, F_2[$ is constant (again look in Γ^*), equal to $m = |a_1 b_2 - a_2 b_1|$ by the induction step. This immediately implies (6.4) for both pairs E, F_1 and E, F_2. As E is adjacent to either F_1 or F_2 this completes the induction step.

Finally suppose $E < E'$, that the multiplicity is constant equal to m on $]E, E'[$, and that $|a'b - ba'| = m$ where (a, b) and (a', b') are the Farey weights of E and E', respectively. We want to prove that E and E' are adjacent. Let $\pi \in \mathfrak{B}$ be minimal (for the order relation $\trianglerighteq$, see Lemma 6.1) such that $E, E' \in \Gamma^*_\pi$. Clearly E and E' are adjacent iff they are adjacent vertices in this dual graph Γ_π. We can write $\pi = \pi' \circ \tilde{\pi}$, where $\pi' \in \mathfrak{B}$, $\tilde{\pi}$ is the blowup at $q \in (\pi')^{-1}(0)$ and $\tilde{\pi}^{-1}(q) \in \{E, E'\}$.

First suppose q is a free point lying on a single component F. Then $\tilde{\pi}^{-1}(q) = E'$ or else $E \not< E'$ by Proposition 6.37. The Farey weight of F equals $(a' - 1, b')$. If $F \neq E$, then $F \in]E, E'[$ so that the multiplicity of F is equal to m. The preceding argument shows $\frac{m}{bb'} \leq \frac{a'-1}{b'} - \frac{a}{b} = \frac{m}{bb'} - \frac{1}{b'}$, a contradiction. Thus $F = E$ and E' is adjacent to E. When q is satellite, it is the intersection point of $F_1 < F_2$, whose Farey weights are (a_1, b_1) and

(a_2, b_2), respectively. If $E \neq F_1, F_2$, then $E < F_1 < E' < F_2$, thus F_1 and F_2 have multiplicity m. By what precedes,

$$A(E') - A(E) = A(E') - A(F_1) + A(F_1) - A(E) \geq \frac{m}{b'b_1} + \frac{m}{b_1 b} = \frac{m}{bb'} \frac{b+b'}{b_1}.$$

But $b' = b_1 + b_2 > b_1$, thus $A(E') - A(E) > \frac{m}{bb'}$. This contradiction concludes the proof. □

Finally we prove the following useful result:

Proposition 6.40. *Suppose ν is a divisorial valuation, and pick $\nu' > \nu$. Suppose the multiplicity function jumps (is discontinuous) at ν along the segment $[\nu_{\mathrm{m}}, \nu'[$. Then there exists a composition of blowups $\pi : X \to (\mathbf{C}^2, 0)$ such that ν is the divisorial valuation associated to some exceptional divisor $E \subset \pi^{-1}(0)$ and the center of ν' on X is a* free *point on E.*

Proof. By Proposition 3.39, the tangent vector at ν represented by ν' has multiplicity $b(\nu)$. By Corollary 6.18, there exists a composition of blowups $\pi : X \to (\mathbf{C}^2, 0)$ and an exceptional component $E \subset \pi^{-1}(0)$ such that this tangent vector corresponds to a free point p on E. Now by Corollary 6.34, the center of ν' on X is necessarily p. □

6.6.4 Monomialization

Let ν be a quasimonomial valuation. By Theorem 6.22, $\Phi^{-1}(\nu)$ is either a branch point or a regular point in Γ, and by Theorem 6.10 the sequence of infinitely near points $\bar{p} = (p_j)$ associated to ν is either of Type 0 or 2. Let n be the smallest integer such that for some $k \geq 0$, $p_{n+1}, \cdots, p_{n+k-1}$ are all free, and p_{n+k+l} are satellite for all $l \geq 0$. Thus $n = n_g$ in the notation of Section 6.2. Let $\pi \in \mathfrak{B}$ be the composition of the point blowups at $p_0, \cdots, p_n$. By Lemma 6.29, the valuation determined by $p_{n+1}, \ldots, p_{n+l}$ is monomial in suitable coordinates at $p = p_{n+1}$, for all $l \geq 1$. When $\bar{p}$ is of Type 2, the valuation associated to the infinite sequence $(p_j)_{n+1}^{\infty}$ is also monomial by continuity. Therefore the valuation ν is equivalent to $\pi_* \mu$ for some monomial valuation μ at p. We have thus proved

Proposition 6.41. *Let $\nu \in \mathcal{V}_{\mathrm{qm}}$ be a quasimonomial valuation which is not monomial. Then there exists a proper birational morphism $\pi \in \mathfrak{B}$, a smooth point $p \in \pi^{-1}(0)$ and a monomial valuation μ centered at p, such that $\pi_* \mu$ is proportional to ν.*

Note that in [ELS] it is proved that, in any dimension, an Abhyankar valuation (i.e. giving equality in (1.1)) of rank one can be made monomial by a proper birational morphism. In our setting, the Abhyankar valuations of rank one are exactly the quasimonomial ones.

Proposition 6.41 above is, however, more precise than what can be extracted from [ELS]. First, the monomial valuation is centered at a *point*—something that cannot be guaranteed in higher dimensions. Second, the birational morphism is explicitly constructed. In fact, it can be detected in terms of data in the valuative tree as follows:

Corollary 6.42. *The divisorial valuation associated to the sequence $p_0, \dots, p_n$ of infinitely near points above is exactly the last element ν_g in the approximating sequence of ν.*

Proof. This is a consequence of the fact that the isomorphism $\Phi : \Gamma \to \mathcal{V}$ preserves the partial ordering and multiplicity. More precisely, we described in Section 6.2 the dual graph associated to the sequence $\bar{p} = (p_j)$ of infinitely near points of Type 0 and Type 2. See Figures 6.2 and 6.6. In particular, this analysis shows that $E_g < \gamma(\bar{p})$, where $E_g \in \Gamma^*$ is the exceptional divisor of the blowup at p_n, $n = n_g$. In Section 6.3.2 we described the restriction of the multiplicity function to this dual graph. See Figure 6.8. In particular, E_g can be characterized as the maximum element in the segment $[E_0, \gamma(\bar{p})]$ having multiplicity strictly smaller than $m(\gamma(\bar{p}))$. On the other hand, by the definition of the approximating sequence, ν_g is the maximum element in the segment $[\nu_{\mathfrak{m}}, \nu]$ having multiplicity strictly smaller than $m(\nu)$. As Φ preserves the partial ordering and multiplicity, and as $\Phi(\gamma(\bar{p})) = \nu$, this shows that $\Phi(E_g) = \nu_g$. □

6.7 The Dual Graph of the Minimal Desingularization

A *desingularization* of a (reduced, formal) curve C is a composition of point blowups $\pi \in \mathfrak{B}$ such that the total transform $\pi^{-1}(C)$ has normal crossings. By Lemma 6.1, the set of all desingularization maps admits a minimal element, the *minimal desingularization*, which we denote by π_C.

Our aim in this section is two-fold. First we describe the embedding of the dual graph Γ_C of π_C inside the universal dual graph. In particular, when C is irreducible, we shall see that the branch points of Γ_C correspond to the approximating sequence of C. Then we explain how to recover Γ_C from the following finite set of data: the Farey parameters of all elements in the approximating sequences of the irreducible components of C, and the intersection multiplicities between these components. It is well-known since the work of Zariski that these data determine exactly the topology of the embedding of C. In the literature it also is referred to as the *equisingularity type* of C. In Appendix D, we shall discuss a way of encoding these data in a tree called the Eggers tree.

An algorithm describing the dual graph Γ_C in terms of the equisingularity type of C is already described in [Ga, Section 1.4.3-1.4.5] using decompositions into continued fractions. In our algorithm this decomposition is included in the recursive computation of the "Farey weights" of the exceptional divisors. The algorithm of [Ga] was implemented in a computer by the Spanish researchers A. and J. Castellanos. Unfortunately, to our present knowledge these works have not been published yet, and there seems to be no other precise reference concerning this problem. We hope that our approach will lead the interested reader to read the excellent work of E. G. Barroso.

The dual graph may be viewed equivalently as a simplicial tree (i.e. a collection of vertices and edges) Γ_C, or as an **N**-tree (i.e. a finite poset) Γ_C^*. Then Γ_C^* is the set of vertices in Γ_C^*; see Section 3.1.7.

6.7.1 The Embedding of Γ_C^* in Γ^*

First suppose C is irreducible. Then C defines an end in Γ. Recall that this end is defined in terms of the sequence $\bar{p} = (p_j)_0^\infty$ of infinitely near points associated to Γ: as $\bar{p}$ is of Type 4, the points p_j are free for large j, and the components $E'_j \in \Gamma^*$ increase to C as $j \to \infty$. Define indexes n_i, $0 \leq i \leq g$ and $\bar{n}_i$, $1 \leq i \leq g$ as in (6.1), i.e. p_j is free for $n_i < j \leq \bar{n}_{i+1}$ and satellite for $\bar{n}_i < j \leq n_i$. In particular, p_j is free for $j > n_g$.

For j large enough, the strict transform of C is smooth and transverse at p_{j+1}. We denote this curve by C_{j+1}. Suppose moreover that $j > n_g$. Then the contraction of the exceptional divisor containing p_{j+1} maps it onto a free point, namely p_j. Thus C_j is still smooth and transverse at p_j. On the other hand, when $j = n_g$, the contraction map sends C_{j+1} to a curve passing through p_{n_g} which is a satellite point. Whence the total transform of C by the blowups at $p_0, \dots, p_{n_g-1}$ does not have normal crossing singularities. This yields

Proposition 6.43. *The minimal desingularization π_C of C is given by the composition of blowups at the sequence of infinitely near points $p_0, \dots, p_{n_g}$.*

Write E'_j for the exceptional divisor obtained by blowing up p_j, and set $E_i = E'_{n_i}$, $\bar{E}_i = E'_{\bar{n}_i}$. Thus $E_1, \dots, E_g$ is the approximating sequence of C and the dual graph Γ_C^* looks as in Figure 6.10.

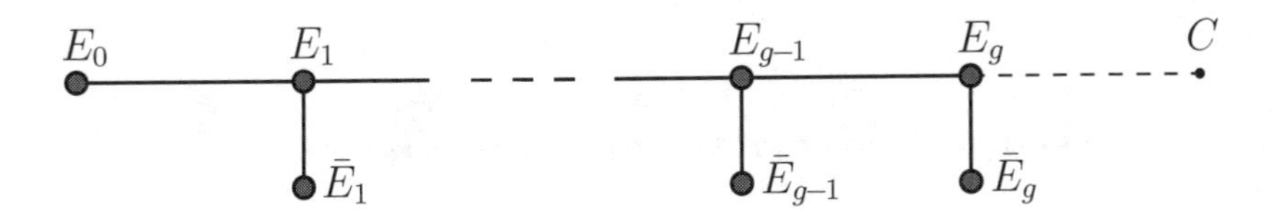

Fig. 6.10. The dual graph of the minimal desingularization of an irreducible curve. Not all points are marked.

Remark 6.44. This yields a geometric interpretation of the approximating sequence of C (and hence of ν_C). Let $\hat{\Gamma}_C$ be the simplicial graph (or tree) whose vertices are irreducible components of the total transform $\pi^{-1}(C)$ and where two vertices are joined by an edge iff they intersect. Then $E_1, \dots, E_g$ are exactly the branch points in $\hat{\Gamma}_C$. Moreover, E_g is the unique vertex with an edge joining the strict transform of C.

Now suppose C is reducible, with irreducible components C_j. The minimal desingularization π_C of C clearly dominates the minimal desingularizations π_{C_j}, hence also their join, which we denote by π'_C. The strict transform C'_j of each C_j by π'_C is smooth and intersects the exceptional divisor $(\pi')^{-1}(0)$ in a free point. However, it may happen that C'_j and C'_k intersect for $j \neq k$. It is then clear that π_C is obtained from π'_C by blowing up these possible intersection points until the strict transforms of C_j and C_k no longer intersect for $j \neq k$. Notice that this involves only free blowups, and that the number of blow ups needed to separate these components is given by the order of contact of C'_j with C'_k.

This allows us to describe the dual graph Γ^*_C as follows. Consider the dual graph of π'_C. This is simply the union of the dual graphs $\Gamma^*_j := \Gamma^*_{C_j}$. For each j, define the exceptional component $E_{j,g_j} \in \Gamma^*_j$ as above. If $j \neq k$ and $C'_j \cap C'_k \neq \emptyset$, then we must have $E_{j,g_j} = E_{k,g_k} =: E_{jk}$ and C'_j and C'_k define the same tangent vector $\vec{v}_{jk}$ at E_{jk}. Thus the exceptional component $C_j \wedge C_k \in \Gamma^*$ also represents $\vec{v}_{jk}$. This component is obtained by blowing up a sequence of infinitely near (free) points, above $C'_j \cap C'_k$. From these observations we conclude

Proposition 6.45. *The dual graph Γ^*_C of the minimal desingularization π_C is the union of $\bigcup_j \Gamma^*_{C_j}$ and all exceptional components $E \in \Gamma^*$ such that $E_{jk} < E \leq C_j \wedge C_k$ and $b(E) = b(E_{jk})$ for some $j \neq k$.*

In the discussion of the Eggers tree in Appendix D we shall need the following consequence of the proposition

Corollary 6.46. *All branch points in Γ^*_C are dominated by some C_j.*

In Figure 6.11 we illustrate the dual graph Γ^*_C, where C has two tangential cusps as irreducible components.

6.7.2 Construction of Γ_C from the Equisingularity Type of C

As before, we decompose C into irreducible components C_j, and view them as points in Γ. We let E_{ji} for $i = 1, \dots, g_j$ be the approximating sequence of C_j. Our aim is to present an algorithm which has as input all Farey parameters $A(j,i) := A(E_{ji})$ and $A(C_j \wedge C_{j'})$, and as output the dual graph of the minimal desingularization of C.

Note that we preferred working with the Farey parameter A of $C_j \wedge C_{j'}$ instead of the skewness α of the corresponding valuation, because the Farey

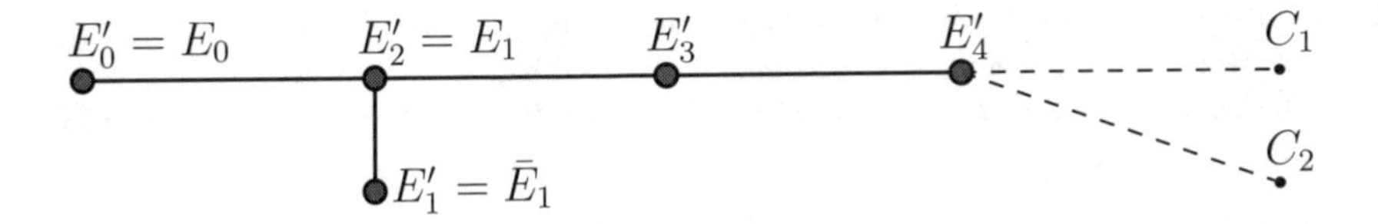

Fig. 6.11. The dual graph of the minimal desingularization of the curve whose irreducible components C_1 and C_2 are parameterized by $t \mapsto (t^2, t^3)$ and $t \mapsto (t^2(1+t^2), t^3(1+t^2)^2)$, respectively.

parameter appears more naturally in our algorithm. However, the intersection multiplicity between C_j and $C_{j'}$ is more easily related to α than to A, see (3.16). But note that the knowledge of the Farey parameters of the approximating sequence is essentially equivalent to the knowledge of its skewness. Indeed, we may use results of Section 3.7 to compute the multiplicities, and (3.8) to compute the Farey parameters (i.e. the thinness thanks to Theorem 6.22), see also Table D.1.

For convenience we denote by J the set of irreducible components of C. At each step our data consist of several elements: a finite simplicial graph $\mathcal{T}$; a function w defined on the set of vertices of $\mathcal{T}$ with values in $(\mathbf{N}^*)^2$; to each vertex $E \in \mathcal{T}$ (resp. each edge e of $\mathcal{T}$) a subset $J(E), J(e) \subset J$ such that the collection $\{J(E), J(e)\}$ form a partition of J; finally to each $j \in J$ an integer $I(j) \in \{1, \cdots, g_j\}$.

Before explaining the algorithm, let us interpret these data in geometric terms. Each step corresponds to the blow up of the finitely many intersection points of the strict transform of C with the exceptional divisor. The graph $\mathcal{T}$ is the dual graph of the exceptional divisor at the current step. In particular, a vertex in $\mathcal{T}$ corresponds to a divisor $E \in \Gamma^*$. If E is a vertex of $\mathcal{T}$, then $w(E)$ is exactly the Farey weight of E. The subset $J(E)$ is the set of branches of C which intersects the divisor E at a free point. An edge e of $\mathcal{T}$ corresponds to the intersection point of two divisors: $J(e)$ is the set of branches of C which intersects the exceptional divisor at this point. Finally a branch C_j intersects one or two irreducible components of the exceptional divisor which have the same multiplicity m. The number $I(j)$ is then the unique integer such that $m(E_{j,I(j)}) = m$.

When $w = (a, b) \in (\mathbf{N}^*)^2$, we shall write $A(w) = a/b$. This is consistent with our previous notation: when w is the Farey weight of $E \in \Gamma^*$, then $A(w)$ is the Farey parameter of E.

▲ *Initiation*: $\mathcal{T}$ consists of one point E_0, $w(E_0) = (2, 1)$, $J(E_0) = J$, and $I(j) = 1$ for any $j \in J$.

▲ *Loop*: we modify $\mathcal{T}$ by adding a certain number of vertices/edges to it, leaving the weight w of the previous vertices unchanged. Modifications of the graph are done exactly at the vertices and edges for which $J(E)$ or $J(e)$ is non-empty.

-*Modification at vertices*: to each vertex E of $\mathcal{T}$ for which $J(E) \neq \emptyset$ apply the following procedure.

- Step 1. Define a partition of $J(E)$ using the equivalence relation $j \sim j' \in J(E)$ iff $A(C_j \wedge C_{j'}) > A(E)$.
- Step 2. For each element of this partition $J' \subset J(E)$, add a new vertex F attached to $\mathcal{T}$ by one edge f at E. Set $w(F) = w(E) + (1,0)$.
  (a) Define $J(F)$ to be the elements $j \in J'$ for which $A(F) < A(j, I(j))$. For any $j \in J'$, leave $I(j)$ unchanged.
  (b) Define $J(f)$ to be the elements $j \in J'$ for which $A(F) > A(j, I(j))$. For any $j \in J'$, leave $I(j)$ unchanged.
- Step 3. Set $J(E) = \emptyset$.

-*Modification at edges*: to each edge e of $\mathcal{T}$ for which $J(e) \neq \emptyset$ apply the following procedure. We adopt the following notation. The boundary of e consists of two vertices E_1 and E_2, and we will assume $A(E_1) < A(E_2)$.

- Step 1. Replace the edge e by a vertex F and two edges, e_1 joining F to E_1, and e_2 joining F to E_2. Define $w(F) = w(E_1) + w(E_2)$.
- Step 2. Define $J(F)$ to be the collection of $j \in J(e)$ for which $A(F) = A(j, I(j))$. For any $j \in J(F)$, replace $I(j)$ by $I(j) + 1$.
- Step 3. Define $J(e_1) := \{j \in J(e),\ A(F) > A(j, I(j))\}$, and $J(e_2) := \{j \in J(e),\ A(F) < A(j, I(j))\}$. For any elements in $J(e_1) \cup J(e_2)$, leave $I(j)$ unchanged.

▲ *End*: the algorithm stops when

(i) $J(e)$ is empty for all edges of $\mathcal{T}$;
(ii) for any couple $j \neq j' \in J(E)$, $A(C_j \wedge C_{j'}) = A(E)$;
(iii) for any j, $I(j) = g_j$.

Let us explain a bit what is going on. We start with the dual graph of the blowup at the origin. Suppose we have made the loop finitely many times. This produces a finite graph $\mathcal{T}$ which corresponds to a certain modification $\pi \in \mathfrak{B}$. We let C' be the strict transform of C by π.

The set of vertices with $J(E) \neq \emptyset$ is the set of exceptional component E that C' intersects at a free point. The step "modification at a vertex" corresponds to the blowup at all these free points. Pick any component E that C' intersects at a free point. We first subdivide $J(E)$ into subsets consisting of branches intersecting E at the same point (Step 1). Then we add to $\mathcal{T}$ a vertex for each of the intersection points, and compute the Farey weight of the exceptional components which are created (Step 2). After these blowups some branches will intersect the exceptional divisor at a free point (Case (a)) or at a satellite point (Case (b)). In any case, being created by a blowup at a free point, the new divisor cannot belong to the approximating sequence of a branch of C. Thus $I(j)$ is left unchanged in this case.

The set of edges with $J(e) \neq \emptyset$ is the set of singular points of the exceptional set that C' contains. Choose one of these points p. This is a satellite point, which is the intersection of two components E_1 and E_2. Blowing up p induces an elementary modification of $\mathcal{T}$ which consists of adding a vertex F in between E_1 and E_2, and the Farey weight of the new exceptional divisor is the sum of the Farey weights of E_1 and E_2 (Step 1 of "modification at edges"). Then several cases appear for the strict transform of a given branch C_j of C containing p. First it may intersect F at a free point. This happens precisely when F lies in the approximating sequence of C_j i.e. when $A(F) = A(j, I(j))$. We thus have "superseded" the $I(j)$-th element of the approximating sequence of C_j, and thus we add 1 to $I(j)$ (Step 2). Or second, it still intersects the exceptional divisor at a satellite point, $F \cap E_1$ when $A(F) > A(j, I(j))$, or $F \cap E_2$ when $A(F) < A(j, I(j))$. In these two cases, F can not be in the approximating sequence of C_j. Whence $I(j)$ is left unchanged (Step 3).

Finally, we have achieved the desingularization of C when three conditions are satisfied. First, the strict transforms of all branches of C intersect the exceptional divisor at a free point (this is the condition $J(e) = \emptyset$ for all edges). Second, no two branches intersect the exceptional divisor at the same point (when C_j and C_{kj} intersect E, this happens precisely when $C_j \wedge C_{j'} = E$). Finally, the strict transform of each branch C_j is smooth and transverse to the exceptional divisor (this happens precisely when the last element of the approximating sequence of each branch has "appeared" in the process of blowup, i.e. $I(j) = g_j$).

We leave to the reader to check that this shows that the algorithm stops after finitely many steps and produces Γ_C.

6.8 The Relative Tree Structure

We described in Section 3.9 the relative valuative $\mathcal{V}_x$ i.e. the union of div_x and the set of centered valuations normalized by $\nu(x) = 1$, where $x \in \mathfrak{m}$ is an element determining a smooth formal curve $\{x = 0\}$ at the origin. We explain here how to recover $\mathcal{V}_x$ geometrically from the dual graphs of compositions of blowups. More precisely we will construct a *relative universal dual graph* Γ_x and show, similarly to Theorem 6.22, that $\mathcal{V}_x$ is isomorphic to Γ_x.

We already saw in Chapter 4 that $\mathcal{V}_x$ was useful for the approach to the valuative tree through Puiseux series. In Section 6.8.4 we shall describe another situation where $\mathcal{V}_x$ appears naturally.

6.8.1 The Relative Dual Graph

Let us first construct the relative universal dual graph as a nonmetric tree. Fix $x \in \mathfrak{m}$ with $m(x) = 1$ and denote the curve $\{x = 0\}$ by E_x.

Consider the set $\mathfrak{B}$ of proper birational morphisms above the origin as before. To each $\pi \in \mathfrak{B}$ we associate a *relative dual graph* $\Gamma_{x,\pi}$. This is a finite graph whose set of vertices $\Gamma^*_{x,\pi}$ is in bijection with the union of (the strict transform of) E_x and the set of all irreducible components of $\pi^{-1}(0)$. Two vertices are joined by an edge iff they intersect. Note that $\Gamma_{x,\pi}$ is always a simplicial tree, that $\Gamma^*_{x,\pi} = \Gamma^*_\pi \cup \{E_x\}$, and that E_x is joined by exactly one other vertex.

We endow $\Gamma^*_{x,\pi}$ with the natural partial ordering in which E_x is the unique minimal element. As before, the set of all posets $\Gamma^*_{x,\pi}$ defines an injective system and we denote the limit by Γ^*_x. It is naturally a nonmetric **Q**-tree rooted in E_x. All of its points are branch points, except E_x which is an end. In fact, Γ^*_x is naturally isomorphic to the **Q**-tree $\Gamma^* \cup \{E_x\}$ rooted in E_x. We construct a canonical nonmetric **R**-tree Γ^o_x from Γ^*_x using Proposition 3.12 and also consider its completion Γ_x. Then Γ^o_x is naturally isomorphic to the **R**-tree $\Gamma^o \cup \{E_x\}$ rooted in E_x, and Γ_x is isomorphic to Γ, also rooted in E_x.

As in the case of Γ (see Section 6.2), the points in Γ_x can be uniquely encoded by sequences of infinitely near points of Types 0,1,2 and 4. The only difference is that E_x is encoded by the empty sequence, and that the unique sequence of Type 4 that encodes the end in Γ corresponding to E_x, now encodes the unique tangent vector at E_x in Γ_x.

Identify Γ_x with Γ as sets but write $\leq_x$ and $\leq$ for their partial orderings. Also denote the associated infimum operators by $\wedge_x$ and $\wedge$. As before, let E_0 be the exceptional divisor obtained by blowing up the origin (which lies on E_x). Then we have

Lemma 6.47. *If I is a segment in Γ, then the partial orderings $\leq$ and $\leq_x$ on I coincide iff the intersection of I with the segment $[E_x, E_0]$ contains at most one point. On the other hand, on $[E_x, E_0]$ we have $E' \leq_x E''$ iff $E' \geq E''$.*

6.8.2 Weights, Parameterization and Multiplicities

Next we define *relative Farey weights* (a_x, b_x). These are defined using elementary modifications just like their nonrelative counterparts, with the following exception: E_x is set to have relative Farey weight $(1,1)$. As E_0 is obtained by a free blowup of E_x, it has relative Farey weight $(2,1)$, which is also its nonrelative Farey weight. This easily gives that the relative and nonrelative Farey weights agree on the tree $\{E \geq_x E_0\}$. In general we have the following relation:

Lemma 6.48. *The two Farey weights of $E \in \Gamma^*$ are related by*

$$(a_x, b_x) = \left(a, b\left(\frac{c}{d} - 1\right)\right), \tag{6.7}$$

where (c,d) is the nonrelative Farey weight of $E \wedge_x E_0$.

Proof. We have to prove that (6.7) holds for any $\pi \in \mathfrak{B}$, and any $E \in \Gamma^*_\pi$. The proof goes by induction on the cardinality of Γ^*_π and is left to the reader. □

The *relative Farey parameter* of $E \in \Gamma_x^*$ is defined by $A_x(E) = a_x/b_x$, where (a_x, b_x) is the relative Farey weight of E. It follows from Lemma 6.48 that

$$A_x(E) = A(E)/(A(E \wedge_x E_0) - 1). \tag{6.8}$$

This formula easily implies that A_x defines a parameterization of Γ_x. Moreover, $A_x = A$ on the subtree $E \geq_x E_0$.

We can also use the relative Farey weights to define a *relative multiplicity* function m_x on Γ_x^o. Namely, we set

$$m_x(E) = \min\{b_x(F) \ ; \ F \in \Gamma_x^*, F \geq_x E\}. \tag{6.9}$$

This multiplicity function naturally extends to the completion Γ_x. Lemma 6.47 and Lemma 6.48 easily imply

Lemma 6.49. *For any $E \in \Gamma$ we have $m_x(E) = 1$ if $E \in [E_x, E_0]$ and $m_x(E) = m(E)(A(E \wedge_x E_0) - 1)$ otherwise.*

6.8.3 The Isomorphism

We can now define a relative version of the fundamental isomorphism $\Phi : \Gamma \to \mathcal{V}$ in Theorem 6.22. Let us define a map Φ_x from Γ_x^* into $\mathcal{V}_x$ by setting $\Phi_x(E_x) := \mathrm{div}_x$ and $\nu_{x,E} = b_x^{-1}\pi_* \mathrm{div}_E(\phi)$ for $E \in \Gamma_x^* \setminus \{E_x\}$. In view of Lemma 6.48 it is clear that

$$\Phi_x(E) = \Phi(E)\,(A(E \wedge_x E_0) - 1) \tag{6.10}$$

Using Theorem 6.22 we have $A(E \wedge_x E_0) = A(\nu_E \wedge_x \nu_{\mathfrak{m}}) = 1 + \nu_E(x)$. Thus (6.10) implies $\nu_{x,E} = \nu_E/\nu_E(x)$, i.e. $\nu_{x,E}$ is the centered valuation on R equivalent to ν_E and normalized by $\nu_{x,E}(x) = 1$.

Theorem 6.50. *The map $\Phi_x : \Gamma_x^* \to \mathcal{V}_x$ extends uniquely to an isomorphism of parameterized trees $\Phi_x : (\Gamma_x, A_x) \to (\mathcal{V}_x, A_x)$ Here A_x denotes the relative Farey parameter on Γ_x and relative thinness on $\mathcal{V}_x$. Further, Φ_x preserves relative multiplicity.*

Proof. Let $\mathcal{V}_{\mathrm{div},x}$ denote the set of divisorial valuations in $\mathcal{V}_x$, with the convention that $\mathrm{div}_x \in \mathcal{V}_{\mathrm{div},x}$. Clearly Φ_x maps Γ_x^* into $\mathcal{V}_{\mathrm{div}x}$. It suffices to prove that $\Phi_x : \Gamma_x^* \to \mathcal{V}_{\mathrm{div},x}$ is an isomorphism of rooted, nonmetric $\mathbf{Q}$-trees, and that Φ_x preserves the parameterization and multiplicity.

As in Section 3.9 we denote by $N : \mathcal{V} \to \mathcal{V}_x$ the map sending ν_x to div_x and $\nu \neq \nu_x$ to $\nu/\nu(x)$. Then N restricts to an isomorphism of nonrooted, nonmetric $\mathbf{Q}$-trees $\mathcal{V}_{\mathrm{div}} \cup \{\nu_x\} \to \mathcal{V}_{\mathrm{div},x}$.

Above we identified Γ_x^* and $\Gamma^* \cup \{E_x\}$ as sets. Let us for the purpose of this proof think of this identification as an isomorphism $N : \Gamma^* \cup \{E_x\} \to \Gamma_x^*$ of nonrooted, nonmetric $\mathbf{Q}$-trees. We saw above that $A_x \circ N = A$.

It is then clear from the construction that $N \circ \Phi = \Phi_x \circ N$ on $\Gamma^* \cup \{E_x\}$. Since by Theorem 6.22, Φ is an isomorphism of nonrooted, nonmetric $\mathbf{Q}$-trees, so is Φ_x. As Φ_x maps E_x to div_x, it is an isomorphism of rooted, nonmetric **Q**-trees. Since $A_x \circ N \circ \Phi = A$ and $A_x \circ N = A$ on $\Gamma^* \cup \{E_x\}$ we get $A_x \circ \Phi_x = A_x$ on Γ_x^*, so that Φ_x preserves the parameterization. The same reasoning, in conjunction with Lemma 6.49 and Proposition 3.65, shows that Φ_x preserves multiplicity. This completes the proof. □

6.8.4 The Contraction Map at a Free Point

Let us end this section by exhibiting a situation where the relative point of view appears naturally.

Pick $\pi \in \mathfrak{B}$, an exceptional component $E \in \Gamma_\pi^*$, and a free point $p \in E$. Let (a,b) be the Farey weight of E. Pick a regular function z at p such that $E = \{z = 0\}$, and let Γ_z and $\mathcal{V}_z$ be the corresponding relative universal dual graph and valuative tree, respectively.

The point p defines a tangent vector in Γ at E. Write U_Γ for the corresponding (weak) open set, i.e. the set of elements in $\Gamma \setminus \{E\}$ representing this tangent vector. The closure of U_Γ is given by $\overline{U_\Gamma} = U_\Gamma \cup \{E\}$ (see Lemma 7.3 below). Notice that $\overline{U_\Gamma}$ is naturally a tree rooted in E.

Let $\nu_E \in \mathcal{V}$ be the divisorial valuation defined by E. The point p also defines a tangent vector at ν_E in $\mathcal{V}$. Write $U_\mathcal{V}$ for the corresponding (weak) open set, i.e. the set of valuations $\nu \in \mathcal{V} \setminus \{\nu_E\}$ representing the same tangent vector at ν_E as the divisorial valuation obtained by blowing up p. Again, $\overline{U_\mathcal{V}} = U_\mathcal{V} \cup \{\nu_E\}$ and $\overline{U_\mathcal{V}}$ is naturally a tree rooted in ν_E.

Define a map $\varpi_\Gamma : \Gamma_z \to \Gamma$ in terms of sequences of infinitely near points. Let $(p_j)_0^n$ be the (finite) sequence of infinitely near points associated to E. Then ϖ_Γ maps a sequence $(q_k)_0^l$, $0 \le l \le \infty$, not of Type 3, with $q_0 = p$, to the concatenated sequence $p_0, \dots, p_n, q_0, \dots, q_l$. In particular, the empty sequence is mapped to $(p_0)_0^n$. Notice that the concatenated sequence is never of Type 3.

We also define a map $\varpi_\mathcal{V} : \mathcal{V}_z \to \mathcal{V}$ by $\varpi_\mathcal{V}(\mu) = b^{-1}\pi_*\mu$. As $\mu(z) = 1$, it follows from Lemma 6.27 that $(\pi_*\mu)(\mathfrak{m}) = b$, so that $\varpi_\mathcal{V}(\mu) \in \mathcal{V}$.

Theorem 6.51. *The following properties hold:*

(i) the map ϖ_Γ gives an isomorphism $\varpi_\Gamma : \Gamma_z \to \overline{U_\Gamma}$ of rooted, nonmetric trees. Moreover, for any $F \in \Gamma_z$ we have:

$$m(\varpi_\Gamma(F)) = b\, m_z(F) \tag{6.11}$$

$$A(\varpi_\Gamma(F)) = A(E) + b^{-1}(A_z(F) - 1) \tag{6.12}$$

(ii) the map $\varpi_\mathcal{V}$ gives an isomorphism $\varpi_\mathcal{V} : \mathcal{V}_z \to \overline{U_\mathcal{V}}$ of rooted, nonmetric trees. Moreover, for any $\mu \in \mathcal{V}_z$ we have:

$$m(\varpi_\mathcal{V}(\mu)) = b\, m_z(\mu) \tag{6.13}$$

$$A(\varpi_\mathcal{V}(\mu)) = A(\nu_E) + b^{-1}(A_z(\mu) - 1) \tag{6.14}$$

$$\alpha(\varpi_\mathcal{V}(\mu)) = \alpha(\nu_E) + b^{-2}\alpha_z(\mu) \tag{6.15}$$

(iii) the isomorphisms in (i) and (ii) respect the identification of the universal dual graph and the valuative tree: if $\Phi : \Gamma \to \mathcal{V}$ and $\Phi_z : \Gamma_z \to \mathcal{V}_z$ are as in Theorem 6.22 and Theorem 6.50, then $\varpi_{\mathcal{V}} \circ \Phi_z = \Phi \circ \varpi_\Gamma$.

The theorem shows that the valuative tree (and the universal dual graph) has a self-similar, or fractal, structure. This is illustrated in Figure 6.12.

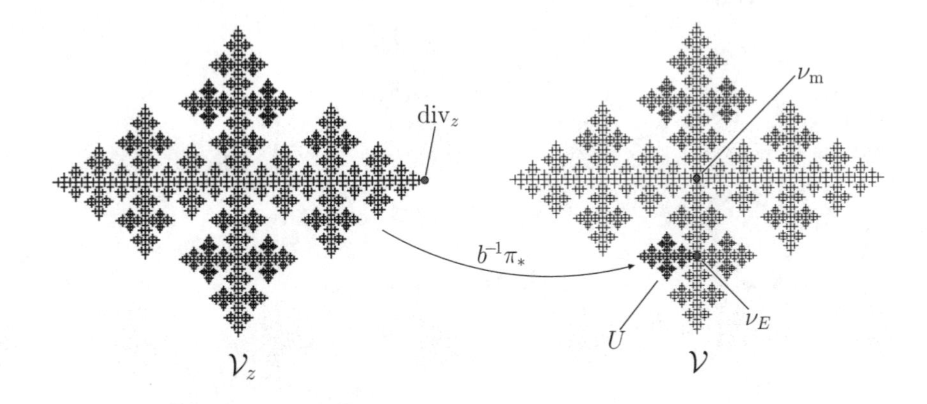

Fig. 6.12. The contraction map π at a free point p on an exceptional component E induces an isomorphism between the relative valuative tree $\mathcal{V}_z$ and a subset $\overline{U}$ of the valuative tree $\mathcal{V}$. See Theorem 6.51

Proof. It is immediate from the definition of ϖ_Γ in terms of concatenations of sequences of infinitely near points, from the definition of the partial orderings on Γ_z^* and Γ^*, and from Corollary 6.13, that ϖ_Γ gives an isomorphism of rooted, nonmetric **Q**-trees between Γ_z^* and $\overline{U_\Gamma} \cap \Gamma^*$. Thus it also gives an isomorphism between the corresponding rooted, nonmetric **R**-trees. To prove (6.11) and (6.12) we have to control the Farey weights. The relative Farey weight of E_z is $(1,1)$. Write (a,b) for the Farey weight of E. A simple induction shows that if F has relative Farey weight (c,d), then $\varpi_\Gamma(F)$ has relative Farey weight $(c+(a-1), bd)$. This easily implies both (6.11) and (6.12) and completes the proof of (i).

Both (ii) and (iii) now follow from Theorem 6.22 and Theorem 6.50 together with Corollary 6.24 and its corresponding relative version. Notice that (6.15) is a consequence of (6.14) and (6.13).

We can also give a direct proof of (ii), not using the universal dual graph. As we already noticed, Lemma 6.27 implies that $\varpi_{\mathcal{V}}$ is a well-defined map from $\mathcal{V}_z$ into $\mathcal{V}$. It is clearly order-preserving and (weakly) continuous. It is also injective as π is a birational map and hence induces an isomorphism between the fraction fields of the rings of formal power series at p and at the origin. If C is an irreducible curve at the origin, then it follows from

Proposition 6.32 that $\nu_C \in U_{\mathcal{V}}$ iff the strict transform of C contains p. This immediately implies that $\varpi_{\mathcal{V}}$ restricts to a bijection between curve valuations in $\mathcal{V}_z$ and curve valuations in $U_{\mathcal{V}}$. By continuity, this shows that $\varpi_{\mathcal{V}}$ is a surjective map of $\mathcal{V}_z$ onto $\overline{U_{\mathcal{V}}}$. Equations (6.13)-(6.15) can also be proved directly, using Lemma 6.27. The details are left to the reader. □

7

Tree Measures

One of the main motivations behind the present monograph is the fact that the valuative tree $\mathcal{V}$ provides an efficient means of encoding singularities. We shall see in the next chapter that this encoding can be nicely given in terms of measures on $\mathcal{V}$. These measures are obtained indirectly by passing through functions defined on the tree $\mathcal{V}_{\mathrm{qm}}$ of quasimonomial valuations, a procedure analogous to the identification of positive measures on the real line with suitably normalized concave functions.

Here we develop a general methodology for identifying certain classes of functions on $\mathcal{V}_{\mathrm{qm}}$ with measures on $\mathcal{V}$. In fact, the analysis is purely tree-theoretic and does not depend on the fact that the elements of the valuative tree are valuations. We shall therefore work on a general complete, rooted nonmetric tree, equipped with an increasing parameterization taking value 1 at the root. The main example we have in mind is $\mathcal{V}$ with the skewness parameterization.

An outline of the theory, as well as a more precise comparison to the situation on the real line, is given in Section 7.1. We refer to that section for the organization of the remainder of the chapter.

While the main results in this chapter are of fundamental importance for applications, the techniques are quite different from those of the rest of the monograph. Hence many of the details can be skipped on a first reading. A similar study appears in [BR].

7.1 Outline

As the analysis that we are about to undertake is somewhat delicate, we start in this section by giving an outline of what we are trying to accomplish.

7.1.1 The Unbranched Case

It is a standard result that finite positive measures on $\mathbf{R}$ can be identified with either decreasing functions or increasing, concave functions (up to certain normalizations). More precisely we have

Proposition 7.1. *The following objects are naturally in 1-1 correspondence with each other:*

(i) finite positive measures ρ on $\mathbf{R}$;
(ii) left continuous decreasing functions $f : \mathbf{R} \to \mathbf{R}$ with $f(\infty) = 0$;
(iii) increasing concave functions $g : \mathbf{R} \to \mathbf{R}$ with $g(0) = 0$ and $g'(\infty) = 0$.

This correspondence goes as follows:

- *to a measure ρ is associated the function $f(x) = \rho\,[x, \infty)$;*
- *to a function f is associated the function $g(x) = \int_0^x f(y)\,dy$;*
- *to a function g is associated the measure $\rho = -d^2g/dx^2$.*

We shall extend this correspondence in two ways. First, we shall work with complex measures in (i). The corresponding functions in (ii) are then known as (normalized) functions of bounded variation, and the functions in the analogue of (iii) are characterized by having a complex measure as second derivative (in the sense of distributions); again subject to a normalization.

The more difficult extension consists of replacing $\mathbf{R}$ with a general, complete, parameterized tree. This presents two challenges. First, as opposed to a complete tree, the real line has neither a maximal nor a minimal element. Essentially, this means that the normalizations of the functions corresponding to the classes (ii) and (iii) above have to be done with great care. Second, and more importantly, a tree typically has branch points. This forces us to redefine notions such as "increasing", "concave" and "bounded variation". Notice that in general there is no bound on the amount of branching that a tree can exhibit. Already in the case of the valuative tree, the tangent space at a divisorial valuation has the cardinality of the continuum. Luckily, we shall not have to worry about the amount of branching, and in any case it turns out that the measures and functions that we shall consider all "live" on completions of countable unions of finite subtrees.

7.1.2 The General Case

We now outline the analysis that we will undertake in the remainder of the chapter. Our objective is to provide the reader with an intuitive picture rather than stating all the precise definitions.

Consider a complete, parameterizable, rooted nonmetric tree $\mathcal{T}$. Let $\mathcal{T}^o$ denote the set of non-ends in $\mathcal{T}$. By convention, the root τ_0 of $\mathcal{T}$ belongs to $\mathcal{T}^o$. Also fix an increasing parameterization α of $\mathcal{T}$ with $\alpha(\tau_0) = 1$. The choice of parameterization is important for some, but not all, of the constructions below.

We will work exclusively with the *weak topology* on $\mathcal{T}$. Any parameterizable complete tree is weakly compact. Denote by $\mathcal{M}$ the set of *complex Borel measures* on $\mathcal{T}$. We think of the elements of $\mathcal{M}$ as complex valued functions on the Borel σ-algebra generated by the weak open sets, but $\mathcal{M}$ can also be viewed as the dual of $C(\mathcal{T})$. The latter point of view provides $\mathcal{M}$ with a norm, turning it into a Banach space.

Our main objective is to identify $\mathcal{M}$ with two classes of functions on $\mathcal{T}^o$. This goes as follows. To any complex Borel measure $\rho \in \mathcal{M}$ we associate two complex-valued functions f_ρ and g_ρ on $\mathcal{T}^o$. These are defined by

$$f_\rho(\tau) = \rho\{\sigma \geq \tau\} \tag{7.1}$$

$$g_\rho(\tau) = \int_{\mathcal{T}} \alpha(\sigma \wedge \tau)\, d\rho(\sigma). \tag{7.2}$$

We let $\mathcal{N}$ and $\mathcal{P}$ be the set of all functions of the form f_ρ and g_ρ, respectively.

The key point is now that the spaces $\mathcal{N}$ and $\mathcal{P}$ can be characterized intrinsically, without any reference to measures. We shall refrain from giving the precise definitions here, but the functions in $\mathcal{N}$ are *left continuous* and of *bounded variation.* The *total variation* $TV()$ provides $\mathcal{N}$ with a norm. As for the set $\mathcal{P}$, its elements are called *complex tree potentials.* They can be characterized by the fact that they have a well-defined *left derivative* at every point and this left-derivative defines a function in $\mathcal{N}$. This defines a natural isomorphism $\delta : \mathcal{P} \to \mathcal{N}$, which enables us to define a norm on $\mathcal{P}$ by $\|g\| = TV(\delta g)$.

The mappings $\rho \mapsto f_\rho$ and $\rho \mapsto g_\rho$ defined in (7.1)-(7.2) can then be thought of as Banach space isomorphisms $\mathcal{M} \to \mathcal{N}$ and $\mathcal{M} \to \mathcal{P}$. We denote their inverses by $d : \mathcal{N} \to \mathcal{M}$ and $\Delta : \mathcal{P} \to \mathcal{M}$. The mapping Δ is of particular interest: we call it the *Laplace operator* as it naturally generalizes the Laplace operator both on the real line and on a simplicial tree.

Inside the three isometric Banach spaces $\mathcal{M}$, $\mathcal{N}$ and $\mathcal{P}$ are located three natural (positive) convex cones $\mathcal{M}^+$, $\mathcal{N}^+$ and $\mathcal{P}^+$. The cone $\mathcal{M}^+$ consists of all (finite) *positive Borel measures* on $\mathcal{T}$, and $\mathcal{N}^+$ of all *left-continuous, nonnegative, strongly decreasing* functions on $\mathcal{T}^o$. As before, these properties have to be interpreted in the tree sense. The cone $\mathcal{P}^+$ is of particular interest. It consists of all *positive tree potentials*, and can be characterized intrinsically, without passing to $\mathcal{N}$. The positive tree potentials should be thought of as nonnegative functions that are concave in a (strong) way that reflects the branching of the tree. The isomorphisms $d : \mathcal{N} \to \mathcal{M}$ and $\Delta : \mathcal{M} \to \mathcal{P}$ then restrict to bijections $d : \mathcal{N}^+ \to \mathcal{M}^+$ and $\Delta : \mathcal{M}^+ \to \mathcal{P}^+$.

In addition to the topologies induced from the norms, the cones above also carry *weak topologies*, of great importance for applications. In the case of $\mathcal{M}^+$, the weak topology is given by the usual weak (or vague) topology on measures. On $\mathcal{N}^+$, it is given by pointwise convergence outside a countable subset of $\mathcal{T}^o \setminus \{\tau_0\}$. On the cone $\mathcal{P}^+$ of positive tree potentials, it is given by pointwise convergence on $\mathcal{T}^o$.

The bijections $d : \mathcal{N}^+ \to \mathcal{M}^+$ and $\Delta : \mathcal{M}^+ \to \mathcal{P}^+$ then become homeomorphisms in the weak topologies. As the subset of probability measures in $\mathcal{M}^+$ is weakly compact, the cone $\mathcal{P}^+$ possesses strong compactness properties as well, reminiscent of those of concave functions on the real line or superharmonic functions on the unit disk. The analogy to the latter two spaces is reinforced by the fact that the infimum of any family of positive tree potentials remains a positive tree potential.

We also consider *inner products* on the cones above. On positive measures, the inner product is defined as the bilinear extension of an inner product on Dirac masses, defined as follows:

$$\tau \cdot \tau' := \alpha(\tau \wedge \tau'), \quad \tau, \tau' \in \mathcal{T} \tag{7.3}$$

On the cone $\mathcal{N}^+$, it is given by

$$\langle f, f' \rangle := f(\tau_0) f'(\tau_0) + \int_{\mathcal{T}^o} f f' \, d\lambda, \quad f, f' \in \mathcal{N}^+, \tag{7.4}$$

where λ denotes the one-dimensional *Hausdorff measure* on $\mathcal{T}^o$, induced by the parameterization α. Finally, the inner product is defined on the cone $\mathcal{P}^+$ of positive tree potentials by declaring $\langle g, g' \rangle = \langle \delta g, \delta g' \rangle$. These inner products are preserved under the isomorphisms of $\mathcal{M}^+$, $\mathcal{N}^+$ and $\mathcal{P}^+$.

Integrability problems prevent us from extending these inner products to the full Banach spaces $\mathcal{M}$, $\mathcal{N}$ and $\mathcal{P}$, but as we show, they are well defined on suitable subspaces $\mathcal{M}_0$, $\mathcal{N}_0$ and $\mathcal{P}_0$.

Finally, let us comment on the role of the parameterization α. It does not appear in the definition of the space $\mathcal{M}$ of complex Borel measures, nor in the space $\mathcal{N}$. It does, however, appear in the definition of complex tree potentials, i.e. the space $\mathcal{P}$.

Moreover, the parameterization is needed to define the inner products in all three cases (the Hausdorff measure depends on the parameterization).

Throughout the chapter we shall assume that the parameterization is increasing and takes value 1 at the root τ_0. This is motivated by the applications in Chapter 8 where we work with the valuative tree parameterized by skewness. Suitable adaptations should be made when working with other types of parameterizations.

7.1.3 Organization

Let us end this section by giving a guide to the organization of the remainder of the chapter. See the introduction to each section for further details.

The weak topology on a tree, which was introduced in Section 3.1.4, is analyzed in more detail in Section 7.2. In particular we characterize connected open sets and prove that a complete, parameterizable, nonmetric tree is weakly compact.

We analyze (weak) Borel measures in Section 7.3. Among other things we show that every Borel measure is Radon (i.e. regular), supported on the completion of a countable union of finite subtrees, and determined by its values on appropriate generalizations of half-open intervals on the real line.

In Section 7.4 we turn to functions of bounded variation. They are defined and analyzed in much the same way as on the real line (see e.g. [Fo]), but considerable care has to be taken because of the branching in the tree. Here we define the space $\mathcal{N}$, which is the first space of functions to be identified with complex Borel measures. This identification is proved in Section 7.5.

Then, in Section 7.6, we turn to complex tree potentials. These constitute the second—and for applications most important—class of functions to be identified with complex Borel measures. They are constructed by integrating functions in $\mathcal{N}$ along segments. The space $\mathcal{P}$ of complex tree potentials is by definition isomorphic to $\mathcal{N}$, hence to $\mathcal{M}$. The isomorphism between $\mathcal{P}$ and $\mathcal{M}$ defines the Laplacian of a complex tree potential and is spelled out in Section 7.7.

Atomic measures play an important role in the two applications in Chapter 8. The corresponding functions in $\mathcal{N}$ and $\mathcal{P}$ are easy to characterize; this is done in Section 7.8.

Another important class of measures is given by the positive cone $\mathcal{M}^+ \subset \mathcal{M}$ of positive Borel measures. In Section 7.9 we describe its preimages $\mathcal{N}^+ \subset \mathcal{N}$ and $\mathcal{P}^+ \subset \mathcal{P}$ under the isomorphism of $\mathcal{N}$ and $\mathcal{P}$ with $\mathcal{M}$. Moreover, we show that the real-valued elements of $\mathcal{M}$, $\mathcal{N}$ and $\mathcal{P}$ admit canonical Jordan decompositions into differences of elements in $\mathcal{M}^+$, $\mathcal{N}^+$ and $\mathcal{P}^+$, respectively.

The cone $\mathcal{P}^+$ is very important in applications and consists of the positive tree potentials. As opposed to general complex tree potentials, these can be easily characterized directly as functions on $\mathcal{T}^o$,

On the cones $\mathcal{M}^+$, $\mathcal{N}^+$ and $\mathcal{P}^+$ it is natural to consider weak topologies, in addition to the restriction of the norm topologies. The weak topologies are defined and analyzed in Section 7.10. We show that the three cones are homeomorphic in the weak topology, a fact that yields compactness properties in $\mathcal{P}^+$.

In Section 7.11 we show that the three cones behave well when passing to and from a complete subtree $\mathcal{S} \subset \mathcal{T}$.

Finally, in Section 7.12, we analyze inner products, first on the cones $\mathcal{M}^+$, $\mathcal{N}^+$ and $\mathcal{P}^+$, then on suitable complex subspaces of $\mathcal{M}$, $\mathcal{N}$ and $\mathcal{P}$. We shall use them in the next chapter.

The exact assumptions on the tree will vary somewhat from section to section, but are stated in the respective introductions.

7.2 More on the Weak Topology

Before studying measures and functions on trees, we review the weak topology defined in Chapter 3 in more detail.

In this section, $\mathcal{T}$ denotes a nonmetric tree. Except if we mention it otherwise, $\mathcal{T}$ is supposed nonrooted.

7.2.1 Definition

Recall the definition of the weak topology on $\mathcal{T}$. If $\vec{v} \in T\tau$ is a tangent vector at a point $\tau \in \mathcal{T}$, we define

$$U(\vec{v}) = \{\sigma \in \mathcal{T} \setminus \{\tau\} \; ; \; \sigma \text{ represents } \vec{v}\}. \tag{7.5}$$

The weak topology is generated by the sets $U(\vec{v})$ in the sense that the open sets are arbitrary unions of finite intersections of sets of the form $U(\vec{v})$. See Figure 3.2.

7.2.2 Basic properties

First we state and prove some basic results.

Lemma 7.2. *The weak topology is Hausdorff.*

Proof. Suppose τ, τ' are distinct points in $\mathcal{T}$. Pick a point $\tau'' \in]\tau, \tau'[$ and let $\vec{v}$ and $\vec{v}'$ be the tangent vectors at τ'' represented by τ and τ', respectively. Then $U(\vec{v})$ and $U(\vec{v}')$ are disjoint open neighborhoods of τ and τ'. □

Lemma 7.3. *If $\vec{v} \in T\tau$, then the weak closure of $U(\vec{v})$ equals $U(\vec{v}) \cup \{\tau\}$.*

Proof. Any open set of the form $\bigcap_1^n U(\vec{v}_i)$ containing τ must intersect $U(\vec{v})$. Thus $\tau \in \overline{U(\vec{v})}$. On the other hand, if $\sigma \notin U(\vec{v}) \cup \{\tau\}$, let $\vec{w}$ be the tangent vector at τ represented by σ. Then $\sigma \in U(\vec{w})$ and $U(\vec{v}) \cap U(\vec{w}) = \emptyset$ so $\sigma \notin \overline{U(\vec{v})}$.
□

Lemma 7.4. *If $\mathcal{T}$ has no branch points, then $\mathcal{T}$ is homeomorphic to a real interval.*

Proof. Suppose for simplicity that $\mathcal{T}$ is complete. Then $\mathcal{T}$ is isomorphic as a nonmetric tree to the real interval $I = [0, 1]$. Moreover, the sets $U(\vec{v})$ correspond to intervals of the form $(x, 1]$ or $[0, x)$ with $0 \leq x \leq 1$. Such intervals generate the topology on I so $\mathcal{T}$ and I are homeomorphic. □

Proposition 7.5. *If $(\tau_k)_1^\infty$ is a sequence of points in $\mathcal{T}$, then $\tau_k \to \tau \in \mathcal{T}$ iff for all subsequence $(k_j)_1^\infty$, the segments $]\tau, \tau_{k_j}]$ have empty intersection. In particular, if the points τ_k all represent distinct tangent vectors at τ, then $\tau_k \to \tau$.*

Proof. If there is a subsequence (k_j) such that the segments $]\tau, \tau_{k_j}]$ have a point $\sigma \in \mathcal{T}$ in common, let $\vec{v}$ be the tangent vector at σ represented by τ. Then $U(\vec{v})$ is an open neighborhood of τ not containing any of the points τ_{k_j}, so $\tau_{k_j} \not\to \tau$, implying $\tau_k \not\to \tau$.

Conversely, suppose that there is no such subsequence. It suffices to show that for any open set of the form $U(\vec{v})$ containing τ we have $\tau_k \in U(\vec{v})$ for large k. Here $\vec{v}$ is a tangent vector at some point $\sigma \neq \tau$ and represented by τ. But this is clear, since otherwise we could find infinitely many k with $\sigma \in [\tau, \tau_k]$. The proof is complete. □

7.2.3 Subtrees

As we show next, the weak topology is well behaved with respect to subtrees.

Lemma 7.6. *If $\mathcal{S}$ is a subtree of $\mathcal{T}$, then the weak topology on $\mathcal{S}$ coincides with the topology on $\mathcal{S}$ induced from the weak topology on $\mathcal{T}$. In other words, the inclusion map $\imath : \mathcal{S} \to \mathcal{T}$ is an embedding.*

Proof. The weak topology on $\mathcal{S}$ is generated by subsets of $\mathcal{S}$ of the form $U_{\mathcal{S}}(\vec{v})$ defined as in (7.5), but where $\vec{v}$ is a tangent vector in $\mathcal{S}$. The induced topology is generated by sets of the form $U(\vec{v}) \cap \mathcal{S}$, where $\vec{v}$ ranges over tangent vectors in $\mathcal{T}$.

On the one hand, every tangent vector in $\mathcal{S}$ also defines a tangent vector in $\mathcal{T}$ and $U_{\mathcal{S}}(\vec{v}) = U(\vec{v}) \cap \mathcal{S}$. On the other hand, if $\vec{v}$ is a tangent vector at some point $\tau \in \mathcal{T}$ such that $U(\vec{v}) \cap \mathcal{S} \neq \emptyset$, then either $\tau \in \mathcal{S}$ and $\vec{v}$ is a tangent vector in $\mathcal{S}$; or $\tau \notin \mathcal{S}$ and $U(\vec{v}) \cap \mathcal{S} = \mathcal{S}$. □

Lemma 7.7. *The relative closure of a subtree $\mathcal{S} \subset \mathcal{T}$ consists of $\mathcal{S}$ and all points in $\mathcal{T}$ that are ends in $\mathcal{S}$. In particular, if $\mathcal{T}$ is complete, then the closure of $\mathcal{S}$ in $\mathcal{T}$ is equal to the completion of $\mathcal{S}$.*

Proof. Consider a point $\tau \in \mathcal{T} \setminus \mathcal{S}$, and let $\vec{w}$ be the unique tangent vector $\vec{w}$ at τ such that $\mathcal{S} \subset U(\vec{w})$. If τ is not an end in $\mathcal{S}$, then we can find a point $\tau' \in U(\vec{w})$ such that $U(\vec{v}) \cap \mathcal{S} = \emptyset$, where $\vec{v}$ is the tangent vector at τ' represented by τ. Thus τ does not belong to the closure of $\mathcal{S}$ in $\mathcal{T}$. On the other hand, if τ is an end in $\mathcal{S}$, then it is easy to see that if $\vec{v}$ is a tangent vector in $\mathcal{T}$ and $\tau \in U(\vec{v})$, then $U(\vec{v}) \cap \mathcal{S} \neq \emptyset$. This concludes the proof. □

If $\mathcal{S}$ is a complete subtree of $\mathcal{T}$, then we can define a natural mapping $p_{\mathcal{S}} : \mathcal{T} \to \mathcal{S}$ as follows. Pick a root $\tau_0 \in \mathcal{S}$ of $\mathcal{T}$ and set $p_{\mathcal{S}}(\tau) = \max([\tau_0, \tau] \cap \mathcal{S})$; this does not depend on the choice of τ_0. Clearly $p_{\mathcal{S}} = \mathrm{id}$ on $\mathcal{S}$.

Lemma 7.8. *The mapping $p_{\mathcal{S}} : \mathcal{T} \to \mathcal{S}$ is continuous, hence defines a retraction of $\mathcal{T}$ onto $\mathcal{S}$.*

Proof. Any tangent vector $\vec{v}$ in $\mathcal{S}$ defines both an open set $U_{\mathcal{S}}(\vec{v})$ in $\mathcal{S}$ and an open set $U_{\mathcal{T}}(\vec{v})$ in $\mathcal{T}$. Clearly $U_{\mathcal{T}}(\vec{v}) = p_{\mathcal{S}}^{-1}(U_{\mathcal{S}}(\vec{v}))$. Thus $p_{\mathcal{S}}$ is continuous. □

Corollary 7.9. *If $(\mathcal{T}, \leq)$ is a rooted, nonmetric tree, then the map $\sigma \mapsto \tau \wedge \sigma$ is (weakly) continuous for any $\tau \in \mathcal{T}$.*

Proof. Apply Lemma 7.8 to $\mathcal{S} = [\tau_0, \tau]$, where τ_0 is the root of $\mathcal{T}$. □

7.2.4 Connectedness

On the real line, the connected open sets are very easy to describe, simply being the open intervals. On a general tree, the situation is more complicated, yet understandable. The following result will play an important role later in the chapter.

Proposition 7.10. *On a nonmetric tree $\mathcal{T}$, any connected open subset $U \subsetneq \mathcal{T}$ is a countable increasing union of (open) sets of the form*

$$\bigcap_{i=1}^{n} U(\vec{v}_i), \ \vec{v}_i \text{ tangent vectors.} \tag{7.6}$$

When $\mathcal{T}$ is rooted at $\tau_0 \notin U$, one can write the previous equation in the form

$$U(\vec{v}) \setminus \bigcup_{i=1}^{n} \{\sigma \geq \tau_i\}; \tag{7.7}$$

where $\vec{v}$ is a tangent vector, not represented by τ_0, at some $\tau \in U$, and $\tau_i \in U(\vec{v})$ for all i.

Remark 7.11. By definition of the weak topology, *any* open set is a union of sets of the form $\bigcap_{i=1}^{n} U(\vec{v}_i)$. The point of Proposition 7.10 is that the union can be taken to be increasing and countable.

Remark 7.12. Note that any open set of the form $U(\vec{v})$ is also a countable increasing union of sets of the form $\{\sigma \geq \tau\}$ when $\vec{v}$ is not represented by the root. Proposition 7.10 can hence be rephrased by saying that any connected open set is a countable increasing union of sets of the form

$$\{\sigma \geq \tau, \, \sigma \not\geq \tau_i, \, 1 \leq i \leq n\},$$

where τ, τ_i are in $\mathcal{T}$. These sets play a role similar to that of half-open intervals on the real line and will be explored systematically in Section 7.3.5.

Proof (Proposition 7.10). As $U \neq \mathcal{T}$, we may pick an element $\tau_0 \in \mathcal{T} \setminus U$, and choose it as the root of $\mathcal{T}$. Define $\tau := \inf U$. As U is open, $U \cap [\tau_0, \tau]$ is an open set in the segment $[\tau_0, \tau]$. It is hence empty, and thus $\tau \notin U$. For simplicity of notation, let us change the root if necessary, so that $\tau = \tau_0$.

Thus U is a subset of $\mathcal{T} \setminus \{\tau_0\}$, which is a disjoint union of weak open sets $U(\vec{v})$ where $\vec{v}$ ranges over all tangent at τ_0. As U is connected, there exists a unique tangent vector $\vec{v}$ such that $U \subset U(\vec{v})$. As U is open, $U(\vec{v}) \setminus U$ is closed in $U(\vec{v})$,

Let F be the set of all minimal elements in $U(\vec{v}) \setminus U$. More precisely, $\tau \in F$ iff $\tau \in U(\vec{v}) \setminus U$ and $]\tau_0, \tau[\subset U$. We claim that

$$U = U(\vec{v}) \setminus \bigcup_{\tau \in F} \{\sigma \geq \tau\}. \tag{7.8}$$

Indeed, on the one hand, for any $\tau \in F$ we may cover U by the disjoint open sets $U \cap \{\sigma > \tau\}$ and $U \cap U(\vec{v}_0)$, where $\vec{v}_0$ is the tangent vector at τ represented by τ_0. As U is connected, we conclude that $U \subset U(\vec{v}) \setminus \bigcup_{\tau \in F}\{\sigma \geq \tau\}$. On the other hand, if $\sigma \in U(\vec{v}) \setminus U$, then set $\tau = \inf\,]\tau_0, \sigma] \setminus U$. As U is open and nonempty we have $\tau \notin U$. Thus $\tau \in F$ and we obtain $U \supset U(\vec{v}) \setminus \bigcup_{\tau \in F}\{\sigma \geq \tau\}$. Hence (7.8) holds.

Notice that $\inf F = \inf(U(\vec{v}) \setminus U)$. Let us prove that U can be written in the form (7.7) under the assumption $\inf F > \tau_0$. We shall explain later how to handle the other case. Set $\tau_1 = \inf F$. Then either $\tau_1 \in F$ in which case $U = U(\vec{v}) \setminus \{\sigma \geq \tau_1\}$; or $F \subset \{\sigma > \tau_1\}$. In the latter case, we define V to be the set of tangent vectors at τ_1 which are represented by some element in F. This set is finite, since otherwise we could find an infinite sequence of elements in $F \subset U(\vec{v}) \setminus U$ representing distinct tangent vectors at τ_1. By Proposition 7.5, and the fact that $U(\vec{v}) \setminus U$ is closed this would imply that $\tau_1 \notin U$, a contradiction. For any $\vec{w} \in V$, we define $\tau(\vec{w}) := \inf(F \cap U(\vec{w}))$. Since $U(\vec{v}) \setminus U$ is closed in $U(\vec{v})$ and $\tau_1 \in U(\vec{v})$, we have $\tau(\vec{w}) > \tau_1$. Set $F_1 := \bigcup_{\vec{w} \in V} \tau(\vec{w})$.

Inductively we construct finite sets F_k with the following properties:

(i) any element in F_{k+1} dominates some element in F_k;
(ii) any element in F_k is dominated by some element in F;
(iii) any element in F dominates some element in F_k;
(iv) any increasing sequence $(\tau_k)_1^\infty$ with $\tau_k \in F_k$ for all k converges to an element in F.

The set F_1 was already defined. The construction of F_{k+1} in terms of F_k is done as follows. Fix $\tau \in F_k$, and suppose $\tau \notin F$. Let $V(\tau)$ be the set of tangent vectors at τ represented by at least one point in F. By the same argument as before, $V(\tau)$ is a finite set, and we let

$$F_{k+1} := (F_k \cap F) \cup \bigcup_{\tau \in F_k \setminus F} \; \bigcup_{\vec{w} \in V(\tau)} \inf(F \cap U(\vec{w})).$$

Clearly (i)-(iii) hold. As for (iv), pick an increasing sequence $\tau_{k+1} \geq \tau_k \in F_k$. By construction, if $\tau_k = \tau_{k+1}$, then $\tau_k \in F$. We may hence suppose $\tau_{k+1} > \tau_k$ for all k. As $\mathcal{T}$ is complete, the sequence converges to a point $\tau \in \mathcal{T}$. We need to prove $\tau \in F$. By construction, any τ_k is dominated by some $\sigma_k \in F$ which does not represent the same tangent vector as τ at τ_k. This implies that $\sigma_k \to \tau$ by Proposition 7.5, hence $\tau \in U(\vec{v}) \setminus U$. But τ_k belongs to U for all k and the sequence increases to τ, so $\tau \in F$ and (iv) holds.

Define $U_k := U(\vec{v}) \setminus \bigcup_{\tau \in F_k}\{\sigma \geq \tau\}$. This is a set of the form (7.7). From (i) it is clear that $U_{k+1} \supset U_k$. By (ii) and (7.8), $U_k \subset U$. We leave it to the reader to check that (iii) and (iv) imply $\bigcup U_k = U$. Hence we are done in the case $\inf F > \tau_0$.

When $\inf F = \tau_0$, we fix a strictly decreasing sequence $(\tau_j)_1^\infty$ in U converging to τ_0, and apply the preceding result to $U_j := U(\vec{v}_j)$, where $\vec{v}_j$ is the tangent vector at τ_j represented by τ_{j+1}. This is a connected open set with $\inf(F \cap U_j) > \inf U_j = \tau_j$.

The proof is complete. □

7.2.5 Compactness

We end this section by proving

Proposition 7.13. *Any parameterizable, complete, nonmetric tree $\mathcal{T}$ is weakly compact.*

Proof. We will embed $\mathcal{T}$ as a closed subspace of a product space. Pick a root τ_0 of $\mathcal{T}$ and let $\leq$ be the partial ordering rooted at τ_0. Also pick a parameterization $\alpha : \mathcal{T} \to [0,1]$. Let E be the set of functions $\chi : \mathcal{T} \to [0,1]$ endowed with the product topology. This is a compact space, and convergence in this topology is given by pointwise convergence. Define a mapping $\jmath : \mathcal{T} \to E$ by

$$\jmath(\tau)(\sigma) = \alpha(\sigma \wedge \tau).$$

It is clear that $\jmath$ is injective: if $\tau \neq \tau'$, then without loss of generality $\tau \wedge \tau' < \tau = \tau \wedge \tau$, and so $\jmath(\tau)(\tau) \neq \jmath(\tau)(\tau')$. Let us show that $\jmath(\mathcal{T})$ is closed in E. For this, notice that if $\tau \in \mathcal{T}$, then $\chi := \jmath(\tau) : \mathcal{T} \to [0,1]$ has the following two properties:

(i) for any end σ in $\mathcal{T}$, the restriction of χ to $[\tau_0, \sigma]$ is of the form

$$\chi(\sigma') = \max\{\alpha(\sigma'), s\}$$

for some $s \in [0, \alpha(\sigma)]$ (we have $s = \alpha(\sigma \wedge \tau)$);

(ii) $\chi(\sigma \wedge \sigma') = \inf\{\chi(\sigma), \chi(\sigma')\}$ for any $\sigma, \sigma' \in \mathcal{T}$.

Conversely, we claim that any function $\chi : \mathcal{T} \to [0,1]$ satisfying (i) and (ii) is of the form $\chi = \jmath(\tau)$ for some $\tau \in \mathcal{T}$. Indeed, let $\mathcal{S} = \mathcal{S}_\chi \subset \mathcal{T}$ be the set of points σ such that χ is strictly increasing on the segment $[\tau_0, \sigma]$. If χ is nonconstant, then $\mathcal{S}$ is nonempty by (i) and is a totally ordered set by (ii). Since $\mathcal{T}$ is complete, $\mathcal{S}$ has a maximal element τ. It is then easy to check from (i) and (ii) that $\chi = \jmath(\tau)$. If χ is constant, then $\chi = \jmath(\tau_0)$. This proves the claim.

It is then clear that in E both conditions (i) and (ii) define a closed set in the product topology, hence $\jmath(\mathcal{T})$ is closed.

Thus $\jmath$ gives a bijection of $\mathcal{T}$ onto the closed subset $\jmath(\mathcal{T})$ of E. We claim that $\jmath$ is a homeomorphism onto its image. As $\mathcal{T}$ is Hausdorff and E compact, it suffices to show that $\jmath$ is an open map onto its image. By the definition of the weak topology it suffices to show that $\jmath(U(\vec{v}))$ is relatively open in $\jmath(\mathcal{T})$ for every tangent vector $\vec{v}$ at any point τ in $\mathcal{T}$. If $\vec{v}$ is represented by τ_0, then

$$U(\vec{v}) = \{\sigma \in \mathcal{T} \; ; \; \alpha(\sigma \wedge \tau) < \alpha(\tau)\} = \{\sigma \in \mathcal{T} \; ; \; \jmath(\sigma)(\tau) < \alpha(\tau)\}$$

and if $\vec{v}$ is represented by some $\tau' > \tau$, then

$$U(\vec{v}) = \{\sigma \in \mathcal{T} \; ; \; \alpha(\sigma \wedge \tau') > \alpha(\sigma \wedge \tau)\} = \{\sigma \in \mathcal{T} \; ; \; \jmath(\sigma)(\tau) < \jmath(\sigma)(\tau')\}.$$

In both cases it follows, after unwinding definitions, that $\jmath(U(\vec{v}))$ is the intersection of an open set in E and $\jmath(\mathcal{T})$, hence is relatively open in $\jmath(\mathcal{T})$. This concludes the proof. □

7.3 Borel Measures

In this section we let $\mathcal{T}$ be a complete, nonmetric tree. We also assume that $\mathcal{T}$ is weakly compact: as we have seen, this is the case if $\mathcal{T}$ is parameterizable. We shall study complex Borel measures on $\mathcal{T}$.

Since a tree can be a quite "large" space in the sense that there is no a priori bound on the amount of branching, we shall be very detailed in our analysis. In particular we shall (temporarily) distinguish between Borel measures and Radon measures. Our general reference is [Fo].

7.3.1 Basic Properties

Let $\mathcal{B}$ be the Borel σ-algebra on $\mathcal{T}$, i.e. the smallest σ-algebra containing all (weakly) open sets in $\mathcal{T}$. A *complex Borel measure* is a function $\rho : \mathcal{B} \to \mathbf{C}$ satisfying $\rho(\emptyset) = 0$ and $\rho(\bigcup_1^\infty E_j) = \sum_1^\infty \rho(E_j)$ whenever $(E_j)_1^\infty$ is a sequence of disjoint Borel sets; *real* (or *signed*) Borel measures are defined in the same way, with $\mathbf{C}$ replaced by $\mathbf{R}$.[1] A real Borel measure ρ is *positive* if $\rho(E) \geq 0$ for all $E \in \mathcal{B}$. Obviously a function $\rho : \mathcal{B} \to \mathbf{C}$ is a complex Borel measure iff $\operatorname{Re}\rho$ and $\operatorname{Im}\rho$ are real Borel measures. Less trivial is the fact that a real Borel measure ρ has a unique *Jordan decomposition* $\rho = \rho_+ - \rho_-$, where $\rho_\pm$ are mutually singular positive measures: there exist disjoint Borel sets $E_\pm$ with $\rho_+(E_-) = \rho_-(E_+) = 0$ and $\rho_+(\mathcal{T} \setminus E_+) = \rho_-(\mathcal{T} \setminus E_-) = 0$. The measures ρ_+ and ρ_- are called the *positive and negative parts* of ρ. These decompositions allow us to reduce many questions about complex measures to positive measures.

7.3.2 Radon Measures

A positive measure ρ is a *Radon measure*[2] if

$$\rho(E) = \inf\{ \rho(U) \; ; \; U \supset E, U \text{ open} \}$$

for any Borel set E and

[1] We do not allow sets of measure $\pm\infty$.

[2] Some authors would call Radon measures *regular Borel measures*.

$$\rho(U) = \sup\{\, \rho(K) \; ; \; K \subset U, K \text{ compact } \}$$

for any open set U; the latter equality then holds when U is replaced by a general Borel set E [Fo, Proposition 7.5]. A real Borel measure is said to be Radon iff its positive and negative parts are Radon, and a complex measure is Radon iff its real and imaginary parts are.

To any complex Borel measure ρ is associated its *total variation measure* $|\rho|$. This is a positive measure which can be defined by $|\rho|(E) = \sup \sum_1^n |\rho(E_i)|$, where the supremum is taken over collections of finitely many disjoint Borel sets $E_1, \dots, E_n$ such that $E = \cup E_i$. We call $\|\rho\| := |\rho|(\mathcal{T}) \in \mathbf{R}_+$ the *total variation* of ρ. If ρ is a real measure, then $|\rho| = \rho_+ + \rho_-$, where $\rho = \rho_+ - \rho_-$ is the Jordan decomposition of ρ.

7.3.3 Spaces of Measures

Let $C(\mathcal{T})$ be the space of continuous functions on $\mathcal{T}$ with the topology of uniform convergence. By the Riesz representation theorem [Fo, Theorem 7.17], complex Radon measures can be identified with continuous linear functionals on $C(\mathcal{T})$: to ρ is associated the operator $\varphi \mapsto \int \varphi \, d\rho$. (In fact this operator is well-defined for any Borel measure—the point is that there is a canonical way of associating a unique Radon measure to each continuous linear functional.) The total variation of ρ is equal to its norm as a linear functional on $C(\mathcal{T})$.

On many topological spaces every finite complex Borel measure is Radon. This is the case for $\mathbf{R}^n$ and, more generally, any locally compact Hausdorff space in which every open set is a countable union of compact subsets [Fo, Theorem 7.8]. While a tree may not fall into the latter category (if, for instance, the tangent space at some point is uncountable) we still have

Proposition 7.14. *Every complex Borel measure on a weakly compact nonmetric tree is a Radon measure.*

The proof is given in Section 7.3.6 below. We shall consequently omit the adjective "Radon" in what follows, and we shall identify Borel measures with continuous linear functionals on $C(\mathcal{T})$.

We let $\mathcal{M}$ be the set of complex Borel measures on $\mathcal{T}$. Let us recall a few facts about $\mathcal{M}$; they follow from the fact that $\mathcal{T}$ is a compact Hausdorff space (and that every Borel measure is Radon), but do not use the tree structure of $\mathcal{T}$.

There are two important topologies on $\mathcal{M}$. First, we have the *strong topology* induced by the norm $\|\rho\| := \sup\{|\int_{\mathcal{T}} \varphi \, d\rho| \; ; \; \sup_{\mathcal{T}} |\varphi| = 1\}$. Under this norm, $\mathcal{M}$ is a Banach space. We shall also consider the *weak topology*, defined in terms of convergence by $\rho_k \to \rho$ iff $\int \varphi \, d\rho_k \to \int \varphi \, d\rho$ for all $\varphi \in C(\mathcal{T})$. This topology is Hausdorff. The unit ball $\{\rho \in \mathcal{M} \; ; \; \|\rho\| \le 1\}$ is weakly compact.

We denote by $\mathcal{M}^+$ the set of positive Borel measures on $\mathcal{V}$. This is a closed subset of $\mathcal{M}$ in both the weak and strong topologies. The subset of

$\mathcal{M}^+$ consisting of probability measures (i.e. positive measure of total mass one) is weakly compact.

Atomic measures will play an important role in our study. To alleviate notation we shall identify $\tau \in \mathcal{T}$ with the corresponding point mass $\delta_\tau \in \mathcal{M}^+$. Indeed, the mapping $\tau \mapsto \delta_\tau$ gives an embedding of $\mathcal{T}$ as a weakly closed subset of $\mathcal{M}^+$ (see [Bo, III, §1, n.9, p.59] for instance).

7.3.4 The Support of a Measure

A general tree can exhibit quite wild branching, but as the following result asserts, Borel measures live on reasonably well-behaved subsets.

Lemma 7.15. *The support of any complex Borel measure ρ on $\mathcal{T}$ is contained in the completion of a countable union of finite subtrees. Moreover, the weak topology on the support is metrizable.*

Proof. We may suppose that ρ is a positive measure of mass 1. Consider the decreasing function $f : \mathcal{T} \to [0, 1]$ defined by $f(\tau) = \rho\{\sigma \geq \tau\}$. The support of ρ is included in the completion of the tree $\mathcal{S} := \{f > 0\}$ (see Lemma 7.7). Let $\mathcal{S}_n = \{f \geq n^{-1}\}$, $n \geq 1$. By construction the tree $\mathcal{S}_n$ has at most $n + 1$ ends, hence at most $n + 2$ branch points, so $\mathcal{S}_n$ is finite.

According to the Urysohn Metrization Theorem (see [Fo, p.139]), any normal, second countable space is metrizable. Now the support $\operatorname{supp} \rho$ is closed, hence compact, and therefore normal: see [Fo, Proposition 4.25]. To conclude we need to show that it is also second countable, i.e. that it admits a countable basis for its topology. To do so, write $\operatorname{supp} \rho$ as the completion of an increasing union of finite trees $\mathcal{S}_n$ as before. Take a countable dense set F in $\bigcup \mathcal{S}_n$. Then the collection of all open sets $\{\sigma > \tau\}$ over $\tau \in F$, and $\{\sigma > \tau,\ \sigma \not\geq \tau'\}$ over $\tau, \tau' \in F$, form a countable basis. □

Remark 7.16. In general, a Borel measure is not necessarily supported on a countable union of finite trees. An example can be constructed as follows.

Consider the Cantor set $X := \{0, 1\}^{\mathbf{N}}$ endowed with the (ultra-)metric $d(x, x') = \max 2^{-i}|x_i - x'_i|$ where $x = (x_i)_{i \geq 0}$, $x' = (x'_i)_{i \geq 0}$. Following Section 3.1.6, define $\mathcal{T}$ to be the quotient of $X \times [0, 1]$ by the equivalence relation $(x, t) \sim (x', t')$ iff $t = t' < d(x, x')$.

On the Cantor set X there exists a natural probability measure ρ_X giving measure 2^{-n} to any cylinder $\{x\ ;\ x_1, \dots, x_n \text{ are fixed}\}$. The pushforward of ρ_X by the inclusion map $X \ni x \mapsto (x, 1) \in \mathcal{T}$ is a probability measure on $\mathcal{T}$ which is not supported on a countable union of finite trees.

7.3.5 A Generating Algebra

Next we generalize the fact that on the real line, complex Borel measures are determined by their values on half-open intervals.

Define collections $\mathcal{E}$ and $\mathcal{A}$ of subsets of $\mathcal{T}$ as follows. First, $\mathcal{E}$ consists of $\mathcal{T}$ and all sets of the form $\{\sigma \in \mathcal{T} ; \sigma \geq \tau, \sigma \not\geq \tau_i, 1 \leq i \leq n\}$ where $\tau, \tau_i \in \mathcal{T}$. It is clear that the complement of an element in $\mathcal{E}$ is a finite disjoint union of elements in $\mathcal{E}$. Hence $\mathcal{E}$ is an *elementary family* in the sense of [Fo, p. 22]. Second, $\mathcal{A}$ is the set of finite disjoint unions of elements in $\mathcal{E}$. Then $\mathcal{A}$ is an *algebra* in the sense that $\mathcal{A}$ is closed under complements and finite unions and intersections (see [Fo, Proposition 1.7]).

Still following [Fo], we define a *premeasure* (on $\mathcal{A}$) to be a function $\rho : \mathcal{A} \to [0, \infty[$ such that $\rho(\emptyset) = 0$ and $\rho(\bigcup_1^\infty E_j) = \sum_1^\infty \rho(E_j)$ whenever $(E_j)_1^\infty$ is a sequence of disjoint sets in $\mathcal{A}$ such that $\bigcup_1^\infty E_j \in \mathcal{A}$.

Lemma 7.17. *Any premeasure on $\mathcal{A}$ has a unique extension to a (positive) Borel measure on $\mathcal{B}$.*

Proof. This follows from Theorem 1.14 in [Fo]. □

Recall from Remark 7.12 that any connected open sets is a countable increasing union of elements in $\mathcal{E}$. Another reason why the elementary family $\mathcal{E}$ is important lies in the following fact. Any set $E \in \mathcal{E}$ is a countable intersection of sets of the form $U(\vec{v}) \setminus \bigcup_1^n U(\vec{v}_i)$, where the $\vec{v}_i$'s are tangent vectors, and $U(\vec{v}_i)$'s are disjoint subsets of $U(\vec{v})$. Conversely, any set $U(\vec{v}) \setminus \bigcup_1^n U(\vec{v}_i)$ as above, is a countable increasing union of elements in $\mathcal{E}$.

From this and Lemma 7.17, one deduces

Lemma 7.18. *Let ρ, ρ' be any two Borel measures on $\mathcal{T}$. If $\rho(U(\vec{v})) = \rho'(U(\vec{v}))$ for any tangent vector $\vec{v}$, then $\rho = \rho'$.*

As a trivial consequence we have

Corollary 7.19. *Any complex Borel measure on $\mathcal{T}$ is uniquely determined by its values on open subsets of the form $\bigcap_{i=1}^n U(\vec{v}_i)$.*

7.3.6 Every Complex Borel Measure is Radon

Finally we prove that every complex Borel measure is Radon:

Proof (Proposition 7.14). It suffices to show that every positive Borel measure ρ on $\mathcal{T}$ is Radon. Following the proof of Theorem 7.8 in [Fo] we can associate a positive Radon measure ρ' to ρ as follows:

$$\rho'(U) := \sup\left\{\int_{\mathcal{T}} \varphi \, d\rho \; ; \; \varphi \in C(\mathcal{T}), \, 0 \leq \varphi \leq 1, \, \operatorname{supp} \varphi \subset U\right\},$$

for U open and $\rho'(E) := \inf\{\rho'(U); \, E \subset U, U \text{ open}\}$ for any Borel set E. We claim that $\rho' = \rho$. Consider a tangent vector $\vec{v}$ at a point $\tau \in \mathcal{T}$. First assume that $\vec{v}$ is not represented by the root τ_0. Pick a sequence $(\tau_k)_1^\infty$ in $\mathcal{T}$, decreasing to τ, and such that τ_k represents $\vec{v}$ for all k. Choose nondecreasing continuous functions $\chi_k : [\tau, \tau_k] \to [0, 1]$ such that $\chi_k(\tau) = 0$ and $\chi_k(\tau_k) = 1$.

Then define $\varphi_k : \mathcal{T} \to [0,1]$ by $\varphi_k(\sigma) := \chi_k(\sigma \wedge \tau_k)$ if $\sigma \geq \tau$, and $\varphi_k(\sigma) = 0$ otherwise. Using Corollary 7.9, we see that φ_k is weakly continuous, Moreover, $0 \leq \varphi_k \leq 1$, $\operatorname{supp} \varphi_k \subset U$, and φ_k increases pointwise to the characteristic function of $U(\vec{v})$ as $k \to \infty$. Monotone convergence gives

$$\rho(U(\vec{v})) = \lim_{k\to\infty} \int_{\mathcal{T}} \varphi_k \, d\rho = \rho'(U(\vec{v})).$$

A similar argument shows that $\rho(U(\vec{v})) = \rho'(U(\vec{v}))$ also when $\vec{v}$ is represented by τ_0. In view of Lemma 7.18 we conclude that $\rho = \rho'$. Hence ρ is a Radon measure. □

7.4 Functions of Bounded Variation

Our objective now is to define a space $\mathcal{N}$ of functions that can be naturally identified with the space $\mathcal{M}$ of complex Borel measures. The elements in $\mathcal{N}$ are functions of bounded variation. These functions are defined and studied in a way similarly to the classical case of the real line, but some details turn out to be trickier because of the branching in the tree.

Unless otherwise stated, $(\mathcal{T}, \leq)$ will denote a rooted, nonmetric tree (not necessarily complete) and τ_0 its root. Recall that, by convention, the ends of $\mathcal{T}$ are the maximal elements of $\mathcal{T}$. In particular, the root τ_0 is not an end.

7.4.1 Definitions

Consider a function $f : \mathcal{T} \to \mathbf{C}$. Let us define what it means for f to be of bounded variation, taking into account the (possible) branching in $\mathcal{T}$. If F is a finite subset of $\mathcal{T}$, let $F_{\max}$ be the set of maximal elements of F and write $\sigma \succ \tau$ when $\sigma, \tau \in F$ and σ is an immediate successor of τ in F, i.e. $\sigma > \tau$ and there is no $\sigma' \in F$ with $\tau < \sigma' < \sigma$. Given a function $f : \mathcal{T} \to \mathbf{C}$ and a finite set F, define the *variation of f on F* by

$$V(f;F) := \sum_{\tau \in F \setminus F_{\max}} \left| f(\tau) - \sum_{\sigma \succ \tau} f(\sigma) \right| ; \tag{7.9}$$

if $F_{\max} = F$ then we declare $V(f;F) = 0$.

Remark 7.20. In contrast to the unbranched case, it can happen $V(f;F) > V(f;F')$ when $F \subset F'$. Indeed, let $\mathcal{T}$ be a rooted, nonmetric tree having exactly two ends $\tau_\pm$. Let $F = \{\tau_0, \tau_+\}$ and $F' = \{\tau_0, \tau_+, \tau_-\}$. Define $f : \mathcal{T} \to \mathbf{C}$ by $f = 0$ on $\mathcal{T} \setminus \{\tau_+, \tau_-\}$ and $f(\tau_\pm) = \pm 1$. Then $V(f;F') = 0$ but $V(f;F) = 1$.

However, we have

Lemma 7.21. *Let $(\mathcal{T}, \leq)$ be a nonmetric tree and let $F \subset F'$ be finite subsets of $\mathcal{T}$ such that all the maximal points in F' belong to F. Then $V(f; F) \leq V(f; F')$.*

Proof. By induction it suffices to consider the case when $F' = F \cup \{\tau'\}$, where τ' is not a maximal element in F'. First assume τ' is not a minimal element of F'. Then τ' has an immediate predecessor $\tau \in F$. Let E' (E) be the set of immediate successors σ of τ in F such that $\tau' \not< \sigma$ $(\tau' < \sigma)$. Then we have

$$V(f;F')-V(f;F) = \left|f(\tau) - f(\tau') - \sum_{E'} f\right| + \left|f(\tau') - \sum_{E} f\right| - \left|f(\tau) - \sum_{E \cup E'} f\right|$$

which is nonnegative by the triangle inequality. The case when τ' is a minimal element of F' is similar, but easier, and left to the reader. □

Definition 7.22. *A function $f : \mathcal{T} \to \mathbf{C}$ is of* bounded variation *if*

$$TV(f) := \sup_F V(f; F) < \infty. \tag{7.10}$$

We call $TV(f)$ the total variation *of f (on $\mathcal{T}$).*

Clearly a function $f : \mathcal{T} \to \mathbf{C}$ is of bounded variation iff $\mathrm{Re}\, f$ and $\mathrm{Im}\, f$ are. Notice that a constant function is *not* necessarily of bounded variation. In fact, the constant function 1 is of bounded variation iff $\mathcal{T}$ is finite.

If $f : \mathcal{T} \to \mathbf{C}$ has bounded variation, then so has the restriction of f to any subtree $\mathcal{S}$ and $TV(f|_{\mathcal{S}}) \leq TV(f)$.

If $\mathcal{T}$ is totally ordered (i.e. has no branch points), then we have $TV(f) = \sup \sum_1^{n-1} |f(\tau_{i+1}) - f(\tau_i)|$, where the supremum is taken over all finite, strictly increasing sequences $(\tau_i)_1^n$ in $\mathcal{T}$. In particular, if $\mathcal{T}$ is a real interval, then the definition above coincides with the classical definition.

7.4.2 Decomposition

On the real line, any bounded decreasing function is of bounded variation, and it is a standard result [Fo, Theorem 3.27] that any real-valued function of bounded variation is in fact a difference of two nonnegative, decreasing functions.[3]

In order for these results to generalize we have to interpret "bounded" and "decreasing" in a way that reflects the branching. Let us declare a finite subset E of $\mathcal{T}$ to be *without relations* if $\sigma \not< \tau$ whenever $\sigma, \tau \in E$.

Definition 7.23. *We say that*

[3]Folland [Fo] works with *increasing* functions but the theory is completely analogous.

(i) a function $f : \mathcal{T} \to \mathbf{C}$ is strongly bounded *if* $\sup_E |\sum_E f| < \infty$, *where the supremum is taken over finite sets E without relations;*
(ii) a function $f : \mathcal{T} \to \mathbf{R}$ is strongly decreasing *if $f(\tau) \geq \sum_E f$ whenever E is a finite set without relations and $\tau < \sigma$ for all $\sigma \in E$.*

Remark 7.24. The condition in (i) is equivalent to $\sup_E \sum_E |f| < \infty$: see [Ru, Lemma 6.3]. Also, any nonnegative strongly decreasing function is clearly strongly bounded.

Lemma 7.25. *The following properties hold:*

(i) any function $f : \mathcal{T} \to \mathbf{C}$ of bounded variation is strongly bounded;
(ii) any strongly bounded and strongly decreasing function $f : \mathcal{T} \to \mathbf{R}$ is of bounded variation.

The proof is given in Section 7.4.6 below.

Corollary 7.26. *The support of a function of bounded variation (i.e. the locus where it is nonzero) is contained in the completion of a countable union of finite subtrees.*

Proof. If f is has bounded variation, Lemma 7.25 and Remark 7.24 give us $C > 0$ such that $\sum_E |f| < C$ for any finite set E without relations. Thus the smallest subtree of $\mathcal{T}$ containing the set $\{|f| > n^{-1}\}$ has at most Cn ends. □

Next we introduce an auxiliary function that plays the same role as the total variation measure associated to a complex Borel measure.

Proposition 7.27. *Suppose $\mathcal{T}$ is complete, and consider a function $f : \mathcal{T} \to \mathbf{R}$ of bounded variation that vanishes on $\mathcal{T} \setminus \mathcal{T}^o$. Define the* total variation function $T_f : \mathcal{T} \to [0, \infty[$ *by*

$$T_f(\tau) = TV(f|_{\{\sigma \geq \tau\}}). \tag{7.11}$$

Then the three functions T_f, $T_f + f$ and $T_f - f$ are nonnegative, strongly decreasing functions on $\mathcal{T}$ and vanish identically on $\mathcal{T} \setminus \mathcal{T}^o$.

As $f = \frac{1}{2}(T_f + f) - \frac{1}{2}(T_f - f)$, we obtain the following decomposition result: any function of bounded variation is the difference of two functions in the convex cone of nonnegative, strongly decreasing functions on $\mathcal{T}$, vanishing on $\mathcal{T} \setminus \mathcal{T}^o$.

Proof (Proposition 7.27). Let us start with the function $T_f + f$. It is clearly nonnegative and nonincreasing. Our conventions imply that $T_f + f = 0$ on $\mathcal{T} \setminus \mathcal{T}^o$. Let us show that $T_f + f$ is strongly decreasing. Consider any finite set $E \subset \mathcal{T}$ without relations and $\tau \in \mathcal{T}$ such that $\tau < \sigma$ for every $\sigma \in E$. Let $\varepsilon > 0$ and for every $\sigma \in E$ pick a finite subset $F_\sigma \subset \{\sigma' \geq \sigma\}$ such that $T_f(\sigma) \leq V(f; F_\sigma) + \varepsilon/n$, where n is the number of elements of E. Set $F = \{\tau\} \cup \bigcup_{\sigma \in E} F_\sigma$. Then we have

$$T_f(\tau) + f(\tau) - \sum_E (T_f + f) \geq V(f; F) - \sum_{\sigma \in E} V(f; F_\sigma) - \varepsilon = -\varepsilon.$$

Letting $\varepsilon \to 0$ we conclude that the function $T_f + f$ is strongly decreasing.

As $T_{-f} = T_f$, this also shows that $T_f - f$ is strongly decreasing and non-negative. Finally $2T_f = (T_f + f) + (T_f - f)$, thus T_f is also strongly decreasing and non-negative. This completes the proof. □

7.4.3 Limits and Continuity

If $\vec{v}$ is a tangent vector at some point $\tau \in \mathcal{T}$ then we say that a sequence $(\tau_k)_1^\infty$ *converges to* τ *along* $\vec{v}$ if τ_k represent $\vec{v}$ for all k and $]\tau, \tau_k]$ form a decreasing sequence of segments with empty intersection.

If f is a complex-valued function on $\mathcal{T}$ and $\vec{v}$ is a tangent vector then we say that f has a *limit along* $\vec{v}$ if there exists a number $a \in \mathbf{C}$ such that $f(\tau_k) \to a$ for all sequences $(\tau_k)_1^\infty$ converging to τ along $\vec{v}$; we then write $f(\vec{v}) = a$. If $\tau \neq \tau_0$, then we define $f(\tau-) = f(\vec{v})$, where $\vec{v}$ is the tangent vector at τ represented by τ_0 (assuming $f(\vec{v})$ exists). For convenience we also define $f(\tau_0-) = f(\tau_0)$. Let us say that a function $f : \mathcal{T} \to \mathbf{C}$ is *left continuous* at $\tau \neq \tau_0$ if $f(\tau-) = f(\tau)$. A function f is *left continuous* if it is left continuous at every point $\tau \neq \tau_0$.

Lemma 7.28. *If $f : \mathcal{T} \to \mathbf{C}$ has bounded variation, then $f(\vec{v})$ exists for every tangent vector $\vec{v}$.*

Proof. If $\tau \in \mathcal{T}$, then the restriction of f to the segment $[\tau_0, \tau]$ has bounded variation, allowing us to invoke the corresponding result on the real line, see [Fo, Theorem 3.27]. □

Next we wish to introduce a number measuring roughly the discontinuity at a point of a function of bounded variation, in a way that reflects the tree structure. In the case of the real line, this number equals the absolute value of the difference between the left and right limits at a point, and can be used to prove that a function of bounded variation has at most countably many discontinuities. Note that on a general tree, a function of bounded variation may have uncountably many points of discontinuity in the usual sense.

Proposition-Definition 7.29. *Suppose $f : \mathcal{T} \to \mathbf{C}$ has bounded variation. Then for each $\tau \in \mathcal{T}$, the series $\sum_{\vec{v} \in T\tau} f(\vec{v})$ is absolutely convergent. In particular, there are at most countably many tangent vectors $\vec{v}$ at τ for which $f(\vec{v}) \neq 0$.*

We can hence set

$$d(\tau; f) := f(\tau-) - \sum_{\vec{v}} f(\vec{v}) \tag{7.12}$$

where the sum is over all tangent vectors $\vec{v}$ at τ not represented by τ_0.

Recall that we defined $f(\tau_0-) = f(\tau_0)$.

Proof. Fix $\tau \in \mathcal{T}$ and pick an arbitrary finite set V of tangent vectors at τ, none represented by the root τ_0. Fix $\varepsilon > 0$. For each of these vectors, choose a representing element $\sigma > \tau$, sufficiently close to τ so that $|f(\vec{v}) - f(\sigma)| \leq \varepsilon$. We have $\sum |f(\vec{v})| \leq \sum |f(\sigma)| + \varepsilon \# V$. The collection of $\{\sigma\}$ defines a set without relation, and f is strongly bounded, thus $\sum |f(\vec{v})| \leq B + \varepsilon \# V$, where $B > 0$ is independent on the choices of tangent vectors (and on τ also). By letting ε tend to zero, we conclude $\sum |f(\vec{v})| \leq B$. This implies that $\sum_{T\tau} |f(\vec{v})|$ is absolutely convergent. □

Lemma 7.30. *Suppose $f : \mathcal{T} \to \mathbf{C}$ has bounded variation. Then, we have $d(\tau; f) = 0$ for all but countably many points τ, and the series $\sum_{\tau \in \mathcal{T}} d(\tau; f)$ is absolutely convergent; in fact $\sum_{\tau \in \mathcal{T}} |d(\tau; f)| \leq TV(f)$.*

Proof. Fix $\varepsilon > 0$ and consider any finite subset $Z \subset \mathcal{T}$ such that $d(\tau; f) \neq 0$ for all $\tau \in Z$. For each $\tau \in Z$, pick finitely many points $\tau_1, \ldots, \tau_n$ close to τ with the following properties: $\tau_1 \leq \tau$ and $\tau_1 < \tau$ unless $\tau = \tau_0$; $\tau_i > \tau$ for $2 \leq i \leq n$ and these τ_i's represent distinct tangent vectors at τ; $|f(\tau_1) - \sum_2^n f(\tau_i) - d(\tau; f)| < \varepsilon |d(\tau; f)|$. We may assume that the sets $\{\tau_1, \ldots, \tau_n\}$ are disjoint as τ varies over Z. Let F be their union: then $\sum_{\tau \in Z} |d(\tau; f)| < (1 - \varepsilon)^{-1} V(f; F) \leq (1 - \varepsilon)^{-1} TV(f)$. We conclude by letting $\varepsilon \to 0$. □

7.4.4 The Space $\mathcal{N}$

We are now in position to define the first space of functions that can be identified with complex Borel measures. Assume that the rooted, nonmetric tree $(\mathcal{T}, \leq)$ is *complete*. Let $\mathcal{T}^o$ be the set of nonmaximal elements of $\mathcal{T}$ (i.e. the points of $\mathcal{T}$ that are not ends).

Definition 7.31. *We let $\mathcal{N}$ be the set of functions $f : \mathcal{T} \to \mathbf{C}$ of bounded variation such that f is left continuous at any point in $\mathcal{T}^o$ and $f = 0$ on $\mathcal{T} \setminus \mathcal{T}^o$.*

Definition 7.32. *We let $\mathcal{N}^+$ be the set of nonnegative, strongly decreasing functions on $\mathcal{T}$, which are left continuous on $\mathcal{T}^o$ and vanish on $\mathcal{T} \setminus \mathcal{T}^o$.*

Notice that $\mathcal{N}^+ \subset \mathcal{N}$, thanks to Lemma 7.25, and that $\mathcal{N}^+$ is a convex cone. We shall prove the following crucial decomposition result (see Proposition 7.27).

Proposition 7.33. *For any $f \in \mathcal{N}$, the function $T_f(\tau) := TV(f|_{\{\sigma \geq \tau\}})$ is left continuous, and the functions T_f, $T_f + f$ and $T_f - f$ belong to $\mathcal{N}^+$.*

We can thus write any real-valued function in $\mathcal{N}$ as the difference of two functions in $\mathcal{N}^+$, $f = \frac{1}{2}(T_f + f) - \frac{1}{2}(T_f - f)$. This decomposition is called the Jordan decomposition *of f.*

We shall later see that this decomposition $f = f_+ - f_-$, $f_\pm \in \mathcal{N}^+$ is unique if one imposes $TV(f) = TV(f_+) + TV(f_-)$ (see Proposition 7.62).

We can equivalently view $\mathcal{N}$ as the space of left continuous functions of bounded variation *on the subtree* $\mathcal{T}^o$. In fact, this point of view is natural in many situations when $\mathcal{T}$ is the valuative tree $\mathcal{V}$ and $\mathcal{T}^o$ the subtree $\mathcal{V}_{\mathrm{qm}}$ of quasimonomial valuations. The equivalence follows from

Lemma 7.34. *Let $f : \mathcal{T}^o \to \mathbf{C}$ be a function of bounded variation. Extend f to $\mathcal{T}$ by declaring $f|_{\mathcal{T}\setminus\mathcal{T}^o} = 0$. Then $f : \mathcal{T} \to \mathbf{C}$ also has bounded variation.*

Proof. Pick any finite subset $F' \subset \mathcal{T}$. Set $F := F' \cap \mathcal{T}^o$ and let E be the set of points in F that are not maximal elements of F' but have no immediate successors in F. Then $V(f; F') = V(f; F) + \sum_E |f|$. Here the first term is bounded by $TV(f|_{\mathcal{T}^o})$ and the second term is uniformly bounded in view of Lemma 7.25 (i). □

In the sequel we shall often consider the functions in $\mathcal{N}$ as defined only on $\mathcal{T}^o$. However, it is important to notice that the norm $TV(f)$ of $f \in \mathcal{N}$ is defined as the total variation of f as a function on the complete tree $\mathcal{T}$.

Notice that the functions in $\mathcal{N}$ are not necessarily left continuous at the ends of $\mathcal{T}$ but they do have left limits there in view of Lemma 7.28.

Proposition 7.35. *$(\mathcal{N}, TV)$ is a normed, complex vector space.*

Proof. That $\mathcal{N}$ is a complex vector space is obvious, as are the facts that $TV(\lambda f) = \lambda TV(f)$ for $\lambda \in \mathbf{C}$, $f \in \mathcal{N}$, and $TV(f) = 0$ iff $f = 0$. Finally, if $f_1, f_2 \in \mathcal{N}$, then for any finite subset $F \subset \mathcal{T}$ we have $V(f_1 + f_2; F) \leq V(f_1; F) + V(f_2; F)$ by the triangle inequality. Hence $TV(f_1 + f_2) \leq TV(f_1) + TV(f_2)$. □

We shall later see that the norm $TV(\cdot)$ is complete, so that $\mathcal{N}$ is in fact a Banach space (see Corollary 7.41).

7.4.5 Finite Trees

The total variation $TV(f)$ of a function $f : \mathcal{T} \to \mathbf{C}$ captures both the variation of f along segments of $\mathcal{T}$ and the discontinuities of f at branch points. This assertion can be made precise as long as we deal with finite, complete trees and functions in the space $\mathcal{N}$:

Proposition 7.36. *Assume that $\mathcal{T}$ is complete and finite. Let $B \subset \mathcal{T}^o$ be any finite subset containing all the branch points of $\mathcal{T}$ and let $\mathcal{I}$ be the set of connected components of $\mathcal{T} \setminus B$. Then for any $f \in \mathcal{N}$ we have*

$$TV(f) = \sum_{\tau \in B} |d(\tau; f)| + \sum_{I \in \mathcal{I}} TV(f|_I). \tag{7.13}$$

Remark 7.37. It is important that f be zero on the ends of $\mathcal{T}$. Indeed, let $\mathcal{T}$ be a rooted, nonmetric tree having two ends $\tau_\pm$ and one branch point τ'. Define $f : \mathcal{T} \to \mathbf{R}$ by $f|_{[\tau_0,\tau']} = 0$, and $f|_{]\tau',\tau_\pm]} = \pm 1$. Then $TV(f) = 1$, but if $B = \{\tau'\}$, then the right hand side of (7.13) is zero.

In view of this remark, the proof of this proposition, which is given below, is trickier than might be expected.

7.4.6 Proofs

We end this section by supplying the proofs of various assertions above.

Proof (Lemma 7.25). First suppose $f : \mathcal{T} \to \mathbf{C}$ has bounded variation. Pick any finite subset E without relations and not containing the root τ_0, and set $F = E \cup \{\tau_0\}$. Then $|\sum_E f| \le V(f;F) + |f(\tau_0)| \le TV(f) + |f(\tau_0)| < \infty$ so f is strongly bounded, proving (i).

As for (ii), suppose $f : \mathcal{T} \to \mathbf{R}$ is strongly bounded and strongly decreasing. Set $C = \inf_E \sum_E f$, where E runs over all finite subsets without relations. Then $C > -\infty$ as f is strongly bounded. We claim that $TV(f) = f(\tau_0) - C$. To see this, first pick any $\varepsilon > 0$ and choose E finite without relations such that $\sum_E f \le C + \varepsilon$. As f is strongly decreasing, we may assume that $\tau_0 \notin E$. Set $F = E \cup \{\tau_0\}$. Then $TV(f) \ge V(f;F) \ge f(\tau_0) - C - \varepsilon$, hence $TV(f) \ge f(\tau_0) - C$ when $\varepsilon \to 0$. For the converse inequality, consider any finite subset $F \subset \mathcal{T}$. Let us show that $V(f;F) \le f(\tau_0) - C$. We may assume that $\tau_0 \in F$ as $V(f;F) \le V(f;F \cup \{\tau_0\})$. Let E be the set of maximal elements in F. Then E is without relations and

$$\sum_{\tau \in F \setminus F_{\max}} \left| f(\tau) - \sum_{F_\tau} f \right| = \sum_{\tau \in F \setminus F_{\max}} \left(f(\tau) - \sum_{F_\tau} f \right) = f(\tau_0) - \sum_E f$$
$$\le f(\tau_0) - C,$$

where F_τ stands for the set of immediate successor of τ in F. This completes the proof. □

Proof (Proposition 7.33). In view of Proposition 7.27, it suffices to show that the function T_f defined in (7.11) is left continuous.

For this, pick any $\tau \in \mathcal{T}$, $\tau \ne \tau_0$. Suppose $T_f(\tau-) > T_f(\tau)$ and fix $\varepsilon > 0$ with $0 < 3\varepsilon < T_f(\tau-) - T_f(\tau)$. Pick $\tau' < \tau$ such that $|f(\sigma) - f(\tau)| \le \varepsilon$ and $T_f(\sigma) \ge T_f(\tau) + 3\varepsilon$ for all $\sigma \in [\tau', \tau[$.

We shall construct a strictly increasing sequence $\tau_n \in [\tau', \tau[$ converging to τ, and finite sets $F_n \subset \{\sigma \,;\, \sigma > \tau_n, \sigma \not> \tau_{n+1}\}$, such that $V(f;F_n) \ge \varepsilon$ and $\tau_{n+1} \in F_n$. This implies $V(f;\bigcup_1^n F_i) \ge \sum_1^n V(f;F_i)$, thus $TV(f) \ge V(f;\bigcup_1^n F_i) \ge \sum_1^n V(f;F_i) \ge n\varepsilon$, which gives a contradiction as $n \to \infty$.

We define τ_n and F_n by induction. First set $\tau_1 := \tau'$. Given $\tau_n \in [\tau', \tau)$, let us show how to construct F_n. First pick a finite subset F_n' of $\{\sigma > \tau_n\}$

such that $V(f;F'_n) \geq T_f(\tau) + 2\varepsilon$. Pick any end $\bar{\tau}$ in $\mathcal{T}$ with $\bar{\tau} > \tau$. As f vanishes on $\mathcal{T} \setminus \mathcal{T}^o$, we have $f(\bar{\tau}) = 0$, so that we may assume that $\bar{\tau} \in F'_n$. By Lemma 7.21 we may then also assume that $\tau \in F'_n$. In addition, we may assume that F'_n contains a point τ'_n such that $\tau_n < \tau'_n < \tau$ and $F'_n \cap \{\sigma > \tau'_n\} = F'_n \cap \{\sigma \geq \tau\}$. Let $F_n = F'_n \setminus \{\sigma \geq \tau\}$ and $F''_n = F'_n \cap \{\sigma \geq \tau\}$. Then $V(f;F'_n) = V(f;F_n) + |f(\tau'_n) - f(\tau)| + V(f;F''_n)$. Since $V(f;F''_n) \leq T_f(\tau)$ and $|f(\tau'_n) - f(\tau)| < \varepsilon$, this gives $V(f;F_n) \geq \varepsilon$.

Finally we set $\tau_{n+1} := \tau'_n (= \max F_n \cap [\tau', \tau[)$. It is clear that $F_n \subset \{\sigma\, ;\, \sigma > \tau_n, \sigma \not> \tau_{n+1}\}$, and that τ_{n+1} belongs to F_n. This completes the inductive construction of τ_n and F_n and ends the proof. □

Finally we address the formula for the total variation on a finite tree.

Proof (Proposition 7.36). Fix $\varepsilon > 0$. For each open segment $I \in \mathcal{I}$ pick a finite set $F_I \subset I$ such that $V(f;F_I) \geq TV(f|_I) - \varepsilon$.

For a point $\tau \in B$, denote by F_τ the set of points $\sigma \geq \tau$ in $\bigcup_I F_I$ such that $[\tau, \sigma[\cap F_I$ is empty. This set consists of $\min F_I$ for the segments I for which $\min I = \tau_0$. As any segment I is totally ordered, one can add to F_I points arbitrarily closed to $\min I$. One can therefore assume $|d(\tau; f) - V(f; F_\tau \cup \{\tau\})| \leq \varepsilon$. Declare F to be the union of B and all the F_I's, and define $\underline{\tau} := \max\{\sigma \in F_\tau,\, \sigma < \tau\}$ for any $\tau \in B$. We then have

$$TV(f) \geq V(f;F) = \sum_{\tau \in B} V(f;F_\tau) + \sum_{I \in \mathcal{I}} V(f;F_I) + \sum_{\tau \in B} |f(\underline{\tau}) - f(\tau)| \geq$$
$$\geq \sum_{\tau \in B} |d(\tau;f)| + \sum_{I \in \mathcal{I}} TV(f|_I) - (2|B| + |\mathcal{I}|)\varepsilon.$$

Letting $\varepsilon \to 0$, we get $TV(f) \geq \sum_{\tau \in B} |d(\tau;f)| + \sum_{I \in \mathcal{I}} TV(f|_I)$.

Conversely, pick a finite set F such that $V(f;F) \geq TV(f) - \varepsilon$. As f vanishes at the ends of $\mathcal{T}$, we can suppose F contains all ends of $\mathcal{T}$. For each point $\tau \in B$, and each segment I containing τ in its boundary, pick a point $\tau_I \in I$ closed enough to τ such that $|f(\tau_I) - f(\vec{v}_I)| \leq \varepsilon$. Here $\vec{v}_I$ denotes the tangent vector represented by I at τ. When $\vec{v}_I$ is represented by the root, we let as above $\underline{\tau} := \tau_I$. As f is left continuous, one can assume $|f(\underline{\tau}) - f(\tau)| \leq \varepsilon$, for all $\tau \in B$. Otherwise, we denote by F_τ the union of all $\tau_I \geq \tau$.

By Lemma 7.21, we can add all points $\tau \in B$ and τ_I to F, this may only increase $V(f;F)$. Define $F_I := F \cap I$. Then

$$TV(f) \leq \varepsilon + V(f;F) = \varepsilon + \sum_{\tau \in B} V(f;F_\tau) + \sum_{I \in \mathcal{I}} V(f;F_I) + \sum_{\tau \in B} |f(\underline{\tau}) - f(\tau)| \leq$$
$$\leq \sum_{\tau \in B} |d(\tau;f)| + \sum_{I \in \mathcal{I}} TV(f|_I) + (1 + M)\varepsilon,$$

where M is the sum of the number of branches at all points in B. We conclude the proof by letting $\varepsilon \to 0$. □

7.5 Representation Theorem I

We are now ready to relate complex Borel measures and normalized functions of bounded variation.

Theorem 7.38. *Let $(\mathcal{T}, \leq)$ be a weakly compact, complete, rooted nonmetric tree.*[4] *Then for any complex measure, the function $I\rho = f_\rho : \mathcal{T}^o \to \mathbf{C}$ defined by*

$$f_\rho(\tau) := \rho\{\sigma \geq \tau\} \tag{7.14}$$

belongs to $\mathcal{N}$. Moreover, the map

$$I : (\mathcal{M}, \|\cdot\|) \to (\mathcal{N}, TV),$$

is an isometry and restricts to a bijection between the set of positive measures $\mathcal{M}^+$, and $\mathcal{N}^+$.

When $f \in \mathcal{N}$, we shall denote by df the unique complex Borel measure such that $I(df) = f$.

Remark 7.39. If $\mathcal{T} = [1, \infty]$, with the natural parameterization $\alpha(x) = x$, then $df = -df/dx$.

If $\vec{v}$ is a tangent vector not represented by τ_0, then $U(\vec{v})$ is a countable decreasing intersection of sets of the form $\{\sigma \geq \tau\}$; and if $\vec{v}$ is represented by τ_0, then $U(\vec{v})$ is the complement of a set $\{\sigma \geq \tau\}$. By regularity of Borel measures, we immediately obtain

Proposition 7.40. *Pick $f \in \mathcal{N}$, and $\rho \in \mathcal{M}$ such that $\rho = df$. Then*

(i) $\rho\{\sigma \geq \tau\} = f(\tau)$ for every $\tau \in \mathcal{T}^o$;
(ii) $\rho U(\vec{v}) = f(\vec{v})$ for every tangent vector $\vec{v}$ not represented by τ_0;
(iii) $\rho U(\vec{v}) = f(\tau_0) - f(\vec{v})$ for every tangent vector $\vec{v}$ represented by τ_0;
(iv) $\rho\{\tau\} = d(\tau; f)$ for every $\tau \in \mathcal{T}^o$.

Since $\mathcal{M}$ is complete in the norm $\|\cdot\|$ we also infer

Corollary 7.41. *$(\mathcal{N}, TV)$ is a Banach space.*

We split the proof of the theorem into three parts.

7.5.1 First Step

We show that $\rho \in \mathcal{M}$ (resp. in $\mathcal{M}^+$) implies $I\rho \in \mathcal{N}$ (resp. in $\mathcal{N}^+$) for any complex Borel measure; and $I : \mathcal{M} \to \mathcal{N}$ is injective.

Write $f = f_\rho$, i.e. $f(\tau) = \rho\{\sigma \geq \tau\}$ for any $\tau \in \mathcal{T}^o$. Let us show that f is left continuous and of bounded variation. After decomposing ρ into real and imaginary parts, and further into positive and negative parts, we may assume that ρ is a positive measure. We need to show that f is left continuous and

[4] For instance a complete, parameterizable, rooted nonmetric tree.

strongly decreasing. That f is left continuous is easy to prove: if τ_k increases to τ then $\rho\{\sigma \geq \tau_k\}$ decreases to $\rho\{\sigma \geq \tau\}$. To see that f is strongly decreasing, consider a finite subset E of $\mathcal{T}$ without relations and $\tau \in \mathcal{T}^o$ with $\tau < \sigma$ for all $\sigma \in E$. Then the subsets $\{\sigma' \geq \sigma\}_{\sigma \in E}$ are mutually disjoint and contained in $\{\sigma \geq \tau'\}$, implying that $f(\tau) \geq \sum_E f$, so that f is strongly decreasing. This shows that I sends $\mathcal{M}$, $\mathcal{M}^+$ to $\mathcal{N}$, $\mathcal{N}^+$ respectively.

To prove the injectivity of I, suppose $I\rho = I\rho'$ for some Borel measures ρ, ρ'. Then $\rho\{\sigma \geq \tau\} = \rho'\{\sigma \geq \tau\}$ for any $\tau \in \mathcal{T}$, thus $\rho(E) = \rho'(E)$ for any element of the elementary family $\mathcal{E}$ defined in Section 7.3.5. And $\rho = \rho'$ by Lemma 7.17. Thus I is injective.

7.5.2 Second Step: from Functions to Measures

Given $f \in \mathcal{N}^+$ we shall find a *positive* Borel measure $\rho \in \mathcal{M}$ such that $f(\tau) = \rho\{\sigma \geq \tau\}$ for any $\tau \in \mathcal{T}^o$. This shows that $I : \mathcal{M}^+ \to \mathcal{N}^+$ is surjective, hence bijective. Note that this implies $I : \mathcal{M} \to \mathcal{N}$ to be surjective too. Indeed, if f belongs to $\mathcal{N}$, we may write $f = \mathrm{Re}(f) + i\,\mathrm{Im}(f)$. In view of Proposition 7.33, $\mathrm{Re}(f)$ and $\mathrm{Im}(f)$ are both differences of elements in $\mathcal{N}^+$, thus lie in the image of I. Hence so does f.

So pick $f \in \mathcal{N}^+$. Recall that this means that f is nonnegative, left continuous, strongly decreasing, and vanishing on $\mathcal{T} \setminus \mathcal{T}^o$. To construct the positive measure ρ, we proceed as on the real line (see [Fo, Section 1.5]), with suitable adaptations to our setting, using the elementary family $\mathcal{E}$ and the algebra $\mathcal{A}$ from Section 7.3.5. First define ρ as a function on $\mathcal{E}$ by

$$\rho\{\sigma \geq \tau, \sigma \not\geq \tau_k\} = f(\tau) - \sum_k f(\tau_k).$$

As f is strongly decreasing, ρ is nonnegative. Recall that an element E of the algebra $\mathcal{A}$ is a finite disjoint union $\bigcup E_i$ of elements in $\mathcal{E}$. We can therefore try to extend ρ to $\mathcal{A}$ by declaring $\rho(E) = \sum \rho(E_i)$. A priori, this is not well-defined, as the decomposition of E into elements of $\mathcal{E}$ is not unique. However, as in the proof of Proposition 1.15 in [Fo], if $E = \bigcup E_i = \bigcup F_j$ are two decompositions, then it is easy to see that $\sum_i \rho(E_i) = \sum_{i,j} \rho(E_i \cap F_j) = \sum_j \rho(F_j)$.

Hence ρ is well defined on $\mathcal{A}$. We claim that it defines a premeasure on $\mathcal{A}$. Clearly ρ is finitely additive. It remains to show that if $(E_i)_1^\infty$ is a disjoint sequence of elements in $\mathcal{A}$ such that $E = \bigcup E_i \in \mathcal{A}$, then $\rho(E) = \sum_i \rho(E_i)$. By finite additivity we may assume that $E \in \mathcal{E}$, and we have

$$\rho(E) = \rho\left(\bigcup_1^n E_i\right) + \rho\left(E \setminus \bigcup_{n+1}^\infty E_i\right) \geq \rho\left(\bigcup_1^n E_i\right) = \sum_1^n \rho(E_i).$$

Letting $n \to \infty$ yields $\rho(E) \geq \sum_1^\infty \rho(E_i)$. For the converse inequality, fix $\varepsilon > 0$ and write $E = \{\sigma \geq \tau, \sigma \not\geq \tau_k\}$ and $E_i = \{\sigma \geq \tau_i, \sigma \not\geq \tau_{ij}\}$. For each k, one can pick $\tilde{\tau}_k \in]\tau, \tau_k[$ such that $0 \leq f(\tilde{\tau}_k) - f(\tau_k) < \varepsilon 2^{-k}$ as f is left continuous. The set $K := \{\sigma \geq \tau, \sigma \not> \tilde{\tau}_k\}$ is then a compact set included in E, and as f is decreasing

$$\rho(K) \geq \rho(E) - \sum (f(\tilde{\tau}_k) - f(\tau_k)) \geq \rho(E) - \varepsilon.$$

Similarly, for each i pick $\tilde{\tau}_i < \tau_i$ such that $0 \leq f(\tilde{\tau}_i) - f(\tau_i) < \varepsilon 2^{-i}$. Again this is possible, at least as long as τ_i is not the root τ_0. If $\tau_i = \tau_0$ for some (unique) i, then we set $\tilde{\tau}_i = \tau_i = \tau_0$. The set $V_i := \{\sigma > \tilde{\tau}_i, \sigma \not\geq \tau_{ij}\}$ is then an open set containing E_i, and

$$\rho(V_i) \leq \rho(E_i) + (f(\tilde{\tau}_i) - f(\tau_i)) \leq \rho(E_i) + \varepsilon 2^{-i}.$$

The compact set K can now be covered by finitely many of the open sets V_i, say by $V_1, \ldots, V_N$. This gives

$$\rho(E) \leq \rho(K) + \varepsilon \leq \sum_1^N \rho(V_i) + \varepsilon \leq \sum_1^\infty \rho(E_i) + 2\varepsilon.$$

Letting $\varepsilon \to 0$ we conclude that $\rho(E) \leq \sum \rho(E_i)$, implying that equality holds.

Thus ρ is a premeasure on $\mathcal{A}$. By Lemma 7.17 it extends uniquely to a positive Borel measure ρ which obviously satisfies $I\rho = f$.

7.5.3 Total Variation

To complete the proof of Theorem 7.38 we now show that $d : \mathcal{N} \to \mathcal{M}$ preserves the norm. Thus pick any $f \in \mathcal{N}$ and write $\rho = df \in \mathcal{M}$. We must show that the total variation of f equals the total mass of ρ.

Let $T_f : \mathcal{T}^o \to [0, \infty)$ be the function defined in (7.11), i.e. $T_f(\tau) = TV(f|_{\{\sigma \geq \tau\}})$. By Proposition 7.33, T_f belongs to $\mathcal{N}^+$. Therefore there exists a positive measure $\rho' := d(T_f)$ such that $T_f\{\tau\} = \rho'\{\sigma \geq \tau\}$ for all $\tau \in \mathcal{T}^o$. We shall prove that $\rho' = |\rho|$, which implies $TV(f) = \rho'(\mathcal{T}) = \rho(\mathcal{T}) = \|\rho\|$.

Consider $\tau \in \mathcal{T}^o$, fix $\varepsilon > 0$, and pick a finite set $F \subset \{\sigma \geq \tau\}$ such that $T_f(\tau) \leq V(f; F) + \varepsilon$. Following the notation in (7.9) we have

$$V(f;F) = \sum_{\tau \in F \setminus F_{\max}} \left| f(\tau) - \sum_{\sigma \succ \tau} f(\sigma) \right| = \sum_{\tau \in F \setminus F_{\max}} |\rho\{\sigma' \geq \tau, \sigma' \not\geq \sigma\}| \\ \leq |\rho|\{\sigma' \geq \tau\}.$$

After letting $\varepsilon \to 0$ we conclude that $\rho'\{\sigma' \geq \tau\} = T_f(\tau) \leq |\rho|\{\sigma' \geq \tau\}$. This implies that $\rho'(E) \leq |\rho(E)|$ for all $E \in \mathcal{E}$, so that $\rho' \leq |\rho|$ by Lemma 7.17.

Let us now show that $|\rho| \leq \rho'$. First consider a set E in the elementary family $\mathcal{E}$, i.e. $E = \{\sigma \geq \tau, \sigma \not\geq \tau_i\}$. It is then clear that

$$|\rho(E)| = |f(\tau) - \sum_i f(\tau_i)| \leq |T_f(\tau) - \sum_i T_f(\tau_i)| = \rho'(E).$$

Next consider a connected open set U. By Remark 7.12 there exists a countable increasing sequence $(E_i)_1^\infty$ such that $E_i \in \mathcal{E}$ and $\bigcup E_i = U$. Thus

$$|\rho(U)| = \lim |\rho(E_i)| \leq \limsup \rho'(E_i) = \rho'(U).$$

This easily implies that $|\rho(U)| \leq \rho'(U)$ for all (not necessarily connected) open sets U. Finally consider an arbitrary Borel set E. Since ρ and ρ' are Radon measures we may find a decreasing sequence $(U_j)_1^\infty$ of open neighborhoods of E such that $\lim \rho(U_j) = \rho(E)$ and $\lim \rho'(U_j) = \rho'(E)$. This gives $|\rho(E)| \leq \rho'(E)$. By the definition of the total variation measure, we conclude that $|\rho| \leq \rho'$. This ends the proof of Theorem 7.38.

For further reference we note that we proved

Proposition 7.42. *For any function $f \in \mathcal{N}$, the total variation measure of df can be computed as follows:*

$$|df| = d\,(T_f)\,. \tag{7.15}$$

7.6 Complex Tree Potentials

We now turn to the second type of functions that will be identified with complex Borel measures. They are slightly more complicated to define, and their relationship to measures is less direct, but they are the functions that appear most naturally in applications. Their analogues on the real line are (normalized) antiderivatives of functions of bounded variation, and we shall define them accordingly also in the tree setting.

In this section we work with a complete, rooted, nonmetric tree $(\mathcal{T}, \leq)$ and a fixed increasing parameterization $\alpha : \mathcal{T} \to [1, \infty]$ of $\mathcal{T}$. While the parameterization was not important when studying functions of bounded variation, here we must fix a choice. We require that α be increasing and $\alpha(\tau_0) = 1$, where τ_0 is the root of $\mathcal{T}$. As before, denote by $\mathcal{T}^o$ the ends of $\mathcal{T}$, i.e. the set of nonmaximal points in $\mathcal{T}$.

7.6.1 Definition

Recall the Banach space $\mathcal{N}$ defined in Section 7.4.4: its elements are complex-valued, left continuous functions on $\mathcal{T}^o$ of bounded variation. The norm $TV(f)$ of $f \in \mathcal{N}$ is the total variation of the extension of f to $\mathcal{T}$ obtained by setting $f = 0$ on $\mathcal{T} \setminus \mathcal{T}^o$.

As in Section 3.6 we need to integrate with respect to the parameterization α. Fix $\sigma_0 < \sigma_1 \in \mathcal{T}$ and write $\alpha_i = \alpha(\sigma_i)$ for $i = 0, 1$. For $t \in [\alpha_0, \alpha_1]$, define σ_t to be the unique element in $[\sigma_0, \sigma_1]$ such that $\alpha(\sigma_t) = t$. For any measurable function f on $\mathcal{T}$, set $\int_{\sigma_0}^{\sigma_1} f(\sigma)\, d\alpha(\sigma) := \int_{\alpha_0}^{\alpha_1} f(\sigma_t)\, dt$.

Definition 7.43. *A function $g : \mathcal{T}^o \to \mathbf{C}$ is a* complex tree potential *if there exists a function $f : \mathcal{T}^o \to \mathbf{C}$ such that $f \in \mathcal{N}$ and*

$$g(\tau) = f(\tau_0) + \int_{\tau_0}^{\tau} f(\sigma)\, d\alpha(\sigma) \tag{7.16}$$

for any $\tau \in \mathcal{T}^o$. We shall write $f = \delta g$, and $g = If$, when f, g are related by (7.16). *We denote by $\mathcal{P}$ the set of all complex tree potentials.*

Notice that f appears both in the constant term and in the integrand. Also note that $f = \delta g$ is uniquely determined by g:

$$f(\tau_0) = g(\tau_0) \quad \text{and} \quad f(\tau) = \frac{dg}{d\alpha}(\tau) := \lim_{\sigma \to \tau-} \frac{g(\tau) - g(\sigma)}{\alpha(\tau) - \alpha(\sigma)} \quad \text{for } \tau \neq \tau_0. \tag{7.17}$$

Definition 7.44. *We refer to $\frac{dg}{d\alpha}$ as the* left derivative *of g.*

It is clear that $\mathcal{P}$ is a complex vector space as $\mathcal{N}$ is. Moreover, $I : \mathcal{N} \to \mathcal{P}$ is a bijection. If $g \in \mathcal{P}$, we may thus define a norm on $\mathcal{P}$ by $\|g\| := TV(\delta g)$. It is clear from the definitions and Corollary 7.41 that

Proposition 7.45. *The set $(\mathcal{P}, \|\cdot\|)$ is a Banach space and the mappings $I : \mathcal{N} \to \mathcal{P}$ and $\delta : \mathcal{P} \to \mathcal{N}$ defined by* (7.16) *and* (7.17) *are isometric isomorphisms.*

Define the *support* of a complex tree potential g to be the smallest subtree $\mathcal{S}$ of $\mathcal{T}$ such that g is constant on any segment in $\mathcal{T}$ disjoint from $\mathcal{S}$. We obtain from Corollary 7.26:

Corollary 7.46. *The support of any complex tree potential is contained in the completion of a countable union of finite subtrees.*

Remark 7.47. The definition of a complex tree potential is a bit indirect. It is possible to give an equivalent, more direct definition by saying that $g \in \mathcal{P}$ iff:

(i) the restriction of g to any segment $[\tau_0, \tau]$ admits a left-derivative at all points;
(ii) the function f given by (7.17) is left continuous and has bounded variation.

7.6.2 Directional Derivatives

Consider a function $g : \mathcal{T}^o \to \mathbf{C}$. and pick a tangent vector $\vec{v}$ at a point $\tau \in \mathcal{T}^o$. We define the *derivative* of g along $\vec{v}$ (when it exists) by

$$D_{\vec{v}} g = \lim_{k \to \infty} \frac{g(\tau_k) - g(\tau)}{|\alpha(\tau_k) - \alpha(\tau)|}$$

for any sequence $(\tau_k)_1^\infty$ converging to τ along $\vec{v}$. Note that if $\vec{v}$ is represented by τ_0, then $D_{\vec{v}} g = -\frac{dg}{d\alpha}$, where $\frac{dg}{d\alpha}$ is the left derivative as in (7.17). It is then clear from Lemma 7.28 that

Lemma 7.48. *If $g \in \mathcal{P}$ is a complex tree potential, then $D_{\vec{v}}g$ exists for every tangent vector $\vec{v}$; in fact $D_{\vec{v}}g = (\delta g)(\vec{v})$.*

Next we define the analogue of the quantity $d(\tau; f)$ for $f \in \mathcal{N}$ defined in (7.12). Namely, if g is a complex tree potential, then we set

$$\Delta(\tau_0; g) = g(\tau_0) - \sum_{\vec{v} \in T\tau_0} D_{\vec{v}}g \quad \text{and} \quad \Delta(\tau; g) = - \sum_{\vec{v} \in T\tau} D_{\vec{v}}g \text{ for } \tau \neq \tau_0. \tag{7.18}$$

Note that this makes sense as the series $\sum_{\vec{v} \in T\tau} D_{\vec{v}}g$ for any $\tau \in \mathcal{T}^o$ is absolutely convergent thanks to Proposition 7.29. Lemma 7.30 immediately implies

Lemma 7.49. *Let $g : \mathcal{T}^o \to \mathbf{C}$ be a complex tree potential. Then $\Delta(\tau; f) = 0$ for all but countably many points τ, and the series $\sum_{\tau \in \mathcal{T}} \Delta(\tau; g)$ is absolutely convergent; in fact $\sum_{\tau \in \mathcal{T}} |\Delta(\tau; g)| \leq \|g\|$.*

7.7 Representation Theorem II

We can now state the relation between complex tree potentials and complex Borel measures.

Theorem 7.50. *Let $(\mathcal{T}, \leq)$ be a complete, rooted nonmetric tree equipped with an increasing parameterization $\alpha : \mathcal{T} \to [1, \infty]$ such that $\alpha(\tau_0) = 1$. Then for any complex Borel measure, the function $g_\rho : \mathcal{T}^o \to \mathbf{C}$ defined by*

$$g_\rho(\tau) = \int_{\mathcal{T}} \alpha(\sigma \wedge \tau)\, d\rho(\sigma) \tag{7.19}$$

is a complex tree potential, and

$$(\mathcal{M}, \|\cdot\|) \ni \rho \mapsto g_\rho \in (\mathcal{P}, \|\cdot\|)$$

is an isometry. When $g \in \mathcal{P}$, we shall denote by Δg the unique complex Borel measure such that $g_{\Delta g} = g$.

Definition 7.51. *We call Δ the* Laplace operator*; if g is a complex tree potential, then Δg is the* Laplacian *of g. Given a complex Borel measure $\rho \in \mathcal{M}$, the function g_ρ in (7.19) is called* the potential of ρ .

The following formulas relate a complex tree potential and its Laplacian.

Proposition 7.52. *Let $g \in \mathcal{P}$, and $\rho \in \mathcal{M}$ such that $\rho = \Delta g$. Then*

(i) $\rho\{\sigma \geq \tau\} = \frac{dg}{d\alpha}(\tau)$ for every $\tau \in \mathcal{T}^o$;
(ii) $\rho(U(\vec{v})) = D_{\vec{v}}g$ for every tangent vector $\vec{v}$ not represented by τ_0;
(iii) $\rho(U(\vec{v})) = D_{\vec{v}}g + g(\tau_0)$ for every tangent vector $\vec{v}$ represented by τ_0;
(iv) $\rho\{\tau\} = \Delta(\tau; g)$ for every $\tau \in \mathcal{T}^o$.

Proof (Theorem 7.50). For any complex measure ρ it is easy to check from both equations (7.14) and (7.19) that $g_\rho = If_\rho$ and $f_\rho = \delta g_\rho$. As $\delta : (\mathcal{P}, \|\cdot\|) \to (\mathcal{N}, TV)$ is an isometry, Theorem 7.50 follows from Theorem 7.38. □

Remark 7.53. In the absence of branching, Δ reduces to the usual Laplacian on the real line: if $\mathcal{T} = [1, \infty]$ is parameterized by $\alpha(x) = x$, then $\Delta = -\frac{d^2}{dx^2}$.

Remark 7.54. Our definition of Δ also generalizes the Laplace operator on (rooted) simplicial trees as presented in e.g. [Car].

To see this, we make use of the discussion in Section 3.1.7. We hence view a rooted simplicial tree as a rooted (nonmetric) **N**-tree $\mathcal{T}_{\mathbf{N}}$ and equip it with its canonical parameterization $\alpha : \mathcal{T}_{\mathbf{N}} \to \mathbf{N}$ with $\alpha(\tau_0) = 1$. Let $\mathcal{T}_{\mathbf{R}}$ be the associated rooted nonmetric **R**-tree obtained by "adding edges". Finally let $\mathcal{T}$ be the completion of $\mathcal{T}_{\mathbf{R}}$ (if $\mathcal{T}_{\mathbf{N}}$ is finite, then $\mathcal{T} = \mathcal{T}_{\mathbf{R}}$). We view $\mathcal{T}_{\mathbf{N}}$ as a subset of $\mathcal{T}$ and equip $\mathcal{T}$ with a parameterization $\alpha : \mathcal{T} \to [1, \infty]$ extending the one on $\mathcal{T}_{\mathbf{N}}$.

Any complex valued function $g : \mathcal{T}_{\mathbf{N}} \to \mathbf{C}$ extends uniquely as a function $g : \mathcal{T}^o \to \mathbf{C}$ which is affine (in the parameterization α) on the edges of $\mathcal{T}^o$.

If $\mathcal{T}_{\mathbf{N}}$ is finite, then g is always a complex tree potential whose Laplacian Δg is atomic, supported on $\mathcal{T}_{\mathbf{N}}$, and

$$\Delta g\{\tau\} = -\sum_{\sigma \sim \tau} (g(\sigma) - g(\tau)) \text{ if } \tau \neq \tau_0 \tag{7.20}$$

$$\Delta g\{\tau_0\} = g(\tau_0) - \sum_{\sigma \sim \tau_0} (g(\sigma) - g(\tau_0)) \tag{7.21}$$

Here the sums are over all $\sigma \in \mathcal{T}_{\mathbf{N}}$ adjacent to τ and τ_0, respectively.

If $\mathcal{T}_{\mathbf{N}}$ is infinite, the situation is more complicated. Suffice it to say that g is a complex tree potential on $\mathcal{T}$ if the quantities in the right hand sides of (7.20) and (7.21) are all nonnegative. In that case the Laplacian Δg is a positive measure on $\mathcal{T}$ whose restriction to $\mathcal{T}^o$ is a sum of (at most) countably many atoms at points in $\mathcal{T}_{\mathbf{N}}$. The masses at these points are still given by (7.20)-(7.21). It is quite possible, however, for Δg to be nonzero and even nonatomic on $\mathcal{T} \setminus \mathcal{T}^o$.

7.8 Atomic Measures

A measure $\rho \in \mathcal{M}$ is *atomic* if $\rho = \sum_{j=1}^{n} c_j \tau_j$, where $n < \infty$, $c_j \in \mathbf{C}$ and $\tau_j \in \mathcal{T}$ (recall that we identify a point in $\mathcal{T}$ with the corresponding point mass in $\mathcal{M}^+$).

The representation Theorems 7.38 and 7.50 immediately yield

Proposition 7.55. *Pick $f \in \mathcal{N}$. Then $df \in \mathcal{M}$ is atomic iff there exists a finite subset $F \subset \mathcal{T}$ containing the root τ_0, such that:*

(i) f is zero outside the finite tree $\bigcup_{\tau \in F} [\tau_0, \tau]$;

(ii) f is constant on any segment in $\mathcal{T}$ not containing any point in F.

Proposition 7.56. *Pick $g \in \mathcal{P}$. Then $\Delta g \in \mathcal{M}$ is atomic iff there exists a finite subset $F \subset \mathcal{T}$ containing the root τ_0, such that:*

(i) g is constant on any segment that intersects the finite subtree $\bigcup_{\tau \in F}[\tau_0, \tau]$ in at most one point;
(ii) g is an affine function of the parameterization α on any segment in $\mathcal{T}$ not containing any point in F in its interior.

7.9 Positive Tree Potentials

Our next goal is to describe the image of the set of positive measures respectively in the set of functions of bounded variation, and in the set of complex tree potentials (Theorem 7.61 below).

In fact, we have already identified functions $f \in \mathcal{N}$ giving rise to positive measures $df \in \mathcal{M}^+$ as belonging to the cone $\mathcal{N}^+$ of nonnegative, left continuous, strongly decreasing functions on $\mathcal{T}^o$ (Theorem 7.38).

7.9.1 Definition

We now wish to describe the preimage of $\mathcal{M}^+$ in $\mathcal{P}$ under the Laplace operator.

Definition 7.57. *A function $g : \mathcal{T}^o \to \mathbf{R}$ is called a* positive tree potential, *or simply* tree potential *(on $\mathcal{T}$), if the following conditions are satisfied:*

(P1) g is nonnegative, increasing, and concave along totally ordered segments;
(P2) if $\tau \neq \tau_0$, then $\sum_{\vec{v} \in T\tau} D_{\vec{v}} g \leq 0$;
(P3) $\sum_{\vec{v} \in T\tau_0} D_{\vec{v}} g \leq g(\tau_0)$.

We denote by $\mathcal{P}^+$ the set of all positive tree potentials on $\mathcal{T}$.

Note that (P1) implies that the directional derivative $D_{\vec{v}} g$ is well defined for every tangent vector $\vec{v}$ at any point $\tau \in \mathcal{T}^o$. Moreover, $D_{\vec{v}} g \geq 0$ except if $\vec{v}$ is represented by τ_0, in which case $D_{\vec{v}} = -\frac{dg}{d\alpha} \leq 0$. Hence the series in (P2) and (P3) are well-defined.

If $\tau \in \mathcal{T}^o$, then (P2) and (P3) imply that $D_{\vec{v}} g = 0$ for all but countably many tangent vectors $\vec{v} \in T\tau$. Moreover, if $\vec{v}$ is not represented by τ_0 and $D_{\vec{v}} g = 0$, then (P1) implies that g is constant in the open set $U(\vec{v})$. Hence conditions (P1)-(P3) are quite strong.

Proposition 7.58. *Every positive tree potential $g \in \mathcal{P}^+$ can be written $g = If$ for some $f \in \mathcal{N}^+$ (see (7.16)). As a consequence, every positive tree potential is a complex tree potential, i.e. $\mathcal{P}^+ \subset \mathcal{P}$.*

Proof. Consider a positive tree potential $g \in \mathcal{P}^+$. It follows from (P1) that the left derivative of g is defined at any point. Thus we may define $f : \mathcal{T}^o \to \mathbf{R}$ by (7.10), i.e. $f(\tau_0) = g(\tau_0)$ and $f(\tau) = \frac{dg}{d\alpha}(\tau)$ for $\tau \neq \tau_0$. It then follows from (P1) that f is nonnegative and left continuous. We will show that f is also strongly decreasing so that $f \in \mathcal{N}^+ \subset \mathcal{N}$. By definition of $\mathcal{P}$, this will show that g is a complex tree potential.

Hence consider a finite, nonempty set $E \subset \mathcal{T}^o$ without relations and $\tau \in \mathcal{T}^o$ such that $\tau < \sigma$ for all $\sigma \in E$. We have to show that $f(\tau) \geq \sum_E f$.

First assume that $\sigma_1 \wedge \sigma_2 = \tau$ for any two distinct elements $\sigma_1, \sigma_2 \in E$. (This is true if E has only one element!) Let V be the set of tangent vectors at τ represented by the elements of E. Then (P1) implies that $\sum_E f \leq \sum_V D_{\vec{v}} g$. If $\tau \neq \tau_0$, then we conclude from (P2) that

$$\sum_E f - f(\tau) \leq \sum_V D_{\vec{v}} g - \frac{dg}{d\alpha}(\tau) \leq \sum_{T\tau} D_{\vec{v}} g \leq 0.$$

If instead $\tau = \tau_0$, then (P3) gives

$$\sum_E f - f(\tau) \leq \sum_V D_{\vec{v}} g - g(\tau_0) \leq \sum_{T\tau} D_{\vec{v}} g - g(\tau_0) \leq 0.$$

In the general case we proceed by induction on the number of elements in E. By the previous step we may assume that there exists $\tau' > \tau$ and a decomposition $E = E' \cup E''$, where E' and E'' are disjoint, E' has at least two elements (E'' could be empty), $\sigma_1' \wedge \sigma_2' = \tau'$ for all distinct elements $\sigma_1', \sigma_2' \in E'$, and $\sigma' \wedge \sigma'' < \tau'$ whenever $\sigma' \in E'$ and $\sigma'' \in E''$. Then $E'' \cup \{\tau'\}$ has no relations and by the inductive hypothesis and the first step we obtain

$$f(\tau) \geq f(\tau') + \sum_{E''} f \geq \sum_{E'} f + \sum_{E''} f = \sum_E f.$$

This concluses the proof. □

Let us mention here some continuity properties of positive tree potentials with respect to the weak topology.

Lemma 7.59. *Any positive tree potential on $\mathcal{T}$ is (weakly) lower semicontinuous and restricts to a continuous tree potential on any finite subtree.*

Proof. Let $f := \delta g \in \mathcal{N}^+$. For each $n \in \mathbf{N}^*$, define $\mathcal{S}_n := \{f > 1/n\}$. This is a finite tree. Set $f_n := \mathbf{1}_{\mathcal{S}_n} f$, and $g_n := I f_n$. Then (P1) implies that the potential g_n is continuous on the finite tree $\mathcal{S}_n$. As it is locally constant outside $\mathcal{S}_n$, it is weakly continuous on $\mathcal{T}$. The sequence f_n increases pointwise towards f. By integration g_n increases pointwise to g as $n \to \infty$. Thus g is lower semicontinuous. □

However, positive tree potentials are not necessarily continuous as the following example on the valuative tree shows.

Example 7.60. We work in the valuative tree $\mathcal{V}$. Fix local coordinates (x, y), set $\rho = \sum_{n\geq 1} n^{-2}\nu_{y+nx}$ and let $g = g_\rho$ be the associated positive tree potential given by (7.16). If $\nu_n = \nu_{y+nx,n^3}$, then $\nu_n \to \nu_{\mathfrak{m}}$ weakly but $g(\nu_n) > n \to \infty > g(\nu_{\mathfrak{m}})$.

7.9.2 Jordan Decompositions

We now show that the three positive cones $\mathcal{M}^+$, $\mathcal{N}^+$ and $\mathcal{P}^+$ are isomorphic and deduce the existence of Jordan decompositions of real-valued elements in $\mathcal{M}$, $\mathcal{N}$ and $\mathcal{P}$.

Theorem 7.61. *The isomorphisms*

$$d : \mathcal{N} \to \mathcal{M} \quad \text{and} \quad \Delta : \mathcal{P} \to \mathcal{M}$$

restrict to bijections

$$d : \mathcal{N}^+ \to \mathcal{M}^+ \quad \text{and} \quad \Delta : \mathcal{P}^+ \to \mathcal{M}^+.$$

Proof. The fact that $d : \mathcal{N}^+ \to \mathcal{M}^+$ is a bijection follows from Theorem 7.38. To prove that $\Delta : \mathcal{P}^+ \to \mathcal{M}^+$ is also a bijection, we prove that $g \in \mathcal{P}^+$ iff f belongs to $\mathcal{N}^+$ where g and f are related by (7.16). That f belongs to $\mathcal{N}^+$ if $g \in \mathcal{P}^+$ follows from Proposition 7.58. Conversely, pick $f \in \mathcal{N}^+$, and define $g(\tau) := f(\tau_0) + \int_{\tau_0}^{\tau} f(\sigma)d\alpha(\sigma)$. As f is non-decreasing and non-negative it is clear that g satisfies (P1). By Proposition 7.40 (iv), we have $d(\tau; f) = df\{\tau\}$ which is non-negative as df is a positive measure. Recall from (7.12) that $d(\tau; f) = -\sum_{\vec{v}\in T\tau} f(\vec{v})$ when $\tau \neq \tau_0$, and $= f(\tau_0) - \sum_{\vec{v}\in T\tau} f(\vec{v})$ when $\tau = \tau_0$, and that $D_{\vec{v}}g = f(\vec{v})$. Thus (P2) and (P3) are satisfied, so $g \in \mathcal{P}^+$, completing the proof. □

Using the Jordan decomposition of a real measure into positive measures and the isomorphisms in Theorems 7.38, 7.50, 7.61, we infer from the result above:

Proposition-Definition 7.62.

- *Any real Borel measure ρ is the difference of two positive measures $\rho = \rho_+ - \rho_-$. These measures are uniquely determined by the condition $\|\rho\| = \|\rho_+\| + \|\rho_-\|$.*
- *Any real-valued function $f \in \mathcal{N}$ is the difference of two functions $f = f_+ - f_-$, where $f_\pm \in \mathcal{N}^+$. These functions are uniquely determined by the condition $TV(f) = TV(f_+) + TV(f_-)$.*
- *Any real-valued complex tree potential $g \in \mathcal{P}$ is the difference of two positive tree potentials $g = g_+ - g_-$, with $g_\pm \in \mathcal{P}^+$. This decomposition is uniquely determined by the condition $\|g\| = \|g_+\| + \|g_-\|$.*

Any of these decompositions above is called the Jordan decomposition *of ρ, f or g respectively.*

Recall that in the case of measures, ρ_+ and ρ_- are also characterized by the fact that their support are disjoint in the sense there are Borel sets $E_\pm$ such that $\rho_+(E_+) = \rho_-(E_-) = 1$, and $\rho_+(E_-) = \rho_-(E_+) = 0$.

For further reference we note two inequalities satisfied by positive tree potentials.

Proposition 7.63. *For any positive tree potential g on $\mathcal{T}$ and any $\tau \in \mathcal{T}^o$ we have*

$$(\Delta g)\{\sigma \geq \tau\} \leq \frac{g(\tau)}{\alpha(\tau)} \leq \|\Delta g\| = g(\tau_0),$$

with equality (in either inequality) iff Δg is supported on $\{\sigma \geq \tau\}$.

Proof. This follows easily from $g(\tau) = g(\tau_0) + \int_{\tau_0}^{\tau} \Delta g\{\sigma' \geq \sigma\}\, d\alpha(\sigma)$ and $\|\Delta g\| = g(\tau_0)$. □

7.10 Weak Topologies and Compactness

We have so far considered $\mathcal{M}$, $\mathcal{N}$ and $\mathcal{P}$ with topologies induced by natural norms on these three spaces. It is important in applications to consider weaker topologies, in which these spaces, or at least subspaces of them, are compact. Here we shall show how to accomplish this in the cones $\mathcal{M}^+$, $\mathcal{N}^+$ and $\mathcal{P}^+$.

Recall that the weak topology on $\mathcal{M}^+$ is defined in terms of convergence: $\rho_k \to \rho$ iff $\int \varphi\, d\rho_k \to \int \varphi\, d\rho$ for any (weakly) continuous φ on $\mathcal{T}$.

We define the weak topology on $\mathcal{N}^+$ in terms of convergent sequences as follows: $f_k \to f$ iff $f_k(\tau_0) \to f(\tau_0)$, and $f_k \to f$ pointwise on $\mathcal{T} \setminus \{\tau_0\}$ except at (at most) countably many points. Note that this is a well-defined Hausdorff topology as $f_1 = f_2$ outside countably many points implies $f_1 = f_2$ on $\mathcal{T} \setminus \{\tau_0\}$. A weak limit in $\mathcal{N}^+$ is hence uniquely determined.

Finally the weak topology on $\mathcal{P}^+$ is defined in terms of pointwise convergence: $g_k \to g$ iff $g_k(\tau) \to g(\tau)$ for any $\tau \in \mathcal{T}^o$.

Theorem 7.64. *The maps*

$$\mathcal{N}^+ \xrightarrow{d} \mathcal{M}^+ \quad \textit{and} \quad \mathcal{P}^+ \xrightarrow{\Delta} \mathcal{M}^+$$

are homeomorphisms in the weak topology.

As the set of positive measures of mass 1 is compact, so are its images in $\mathcal{N}^+$ and $\mathcal{P}^+$. This remark has ramifications for the structure of the cones $\mathcal{N}^+$ and $\mathcal{P}^+$. Here we only mention an application to positive tree potentials.

Corollary 7.65. *The space of positive tree potentials normalized by $g(\tau_0) = 1$ is compact in the topology of pointwise convergence. Moreover:*

(i) from any sequence $(g_n)_1^\infty$ of positive tree potentials such that $\sup g_n(\tau_0) < +\infty$, one can extract a subsequence g_{n_k} converging pointwise;

(ii) if $(g_i)_{i \in I}$ *is any family of positive tree potentials, then* $g = \inf_i g_i$ *is also a positive tree potential;*

(iii) if $(g_n)_1^\infty$ *is an increasing sequence of positive tree potentials such that* $\sup g_n(\tau_0) < +\infty$*, then* $g = \sup_n g_n$ *is also a positive tree potential.*

Remark 7.66. As the properties above indicate, positive tree potentials play a role similar to that of concave functions on the real line, or superharmonic functions on the unit disk in the complex plane.

Proof. The first assertion is a consequence of the compactness of the set of positive measures with bounded mass. The second statement is proved using (P1)-(P3) in the same way as the fact that the family of nonnegative concave functions on $[0, \infty[$ are closed under infima. The same is true of (iii). Notice that $\sup g_n(\tau_0) < \infty$ implies $\sup_n g_n(\tau) \leq \alpha(\tau) \sup_n g(\tau_0) < \infty$ for all $\tau \in \mathcal{T}^o$: see Proposition 7.63. The details are left to the reader. □

Proof (Theorem 7.64). Thanks to Theorems 7.38, 7.50 and 7.61, both maps d, Δ are bijective. To complete the proof, we need to show that these maps and their inverses are weakly continuous. For sake of simplicity we shall only prove that they are sequentially continuous. We leave to the reader to check that they are indeed continuous, using the language of nets, or of filters.

Let us first prove that d is a homeomorphism. By Theorem 7.38, d is an isometry. It is hence sufficient to prove that its restriction to $\mathcal{M}^+(1)$ the set of positive measures of mass 1, induces a homeomorphism onto its image $\mathcal{N}^+(1)$, which consists of functions in $\mathcal{N}^+$ with $f(\tau_0) = 1$. The set $\mathcal{M}^+(1)$ is weakly compact, and $\mathcal{N}^+(1)$ is Hausdorff (see above). We thus only need to prove that $I = d^{-1}$ is weakly continuous.

So consider a sequence of positive measures ρ_n of mass 1 converging weakly to ρ. Write $f_n = I\rho_n$, $f = I\rho$. It is clear that $f_n(\tau_0) = \rho_n\{\mathcal{T}\} = 1 \to f(\tau_0)$. The following result generalizes Proposition 7.19 in [Fo] and exemplifies the idea that the quantity $d(\tau; f)$ measures the discontinuity of f at τ:

Lemma 7.67. *We have* $f_n(\tau) \to f(\tau)$ *for all* $\tau \in \mathcal{T}^o$ *with* $d(\tau; f) = 0$.

In view of Lemma 7.30, this result implies that $f_n \to f$ on $\mathcal{T}^o$ except on a countable subset. Thus $f_n \to f$ weakly, which completes the proof that d is a homeomorphism.

To prove that Δ is a homeomorphism, we proceed in the same way. Note that $\Delta^{-1}\rho = g_\rho = \int \alpha(\sigma \wedge \cdot)\, d\rho(\sigma)$ for any measure ρ. It is sufficient to prove that $g_{\rho_n} \to g_\rho$ pointwise when $\rho_n \to \rho$ weakly. But for any $\tau \in \mathcal{T}^o$, then

$$g_{\rho_n}(\tau) = \int \alpha(\sigma \wedge \tau)\, d\rho_n(\sigma) \to \int \alpha(\sigma \wedge \tau)\, d\rho(\sigma) = g_\rho(\tau)$$

since the function $\sigma \mapsto \alpha(\sigma \wedge \tau)$ on $\mathcal{T}^o$ is weakly continuous (Corollary 7.9). This completes the proof. □

Proof (Lemma 7.67). Fix $\tau \in \mathcal{T}^0$, $\tau \neq \tau_0$. First pick an increasing sequence $\tau_k < \tau$ converging towards τ. Let φ_k be a continuous increasing function, with values in $[0,1]$, 1 on $\{\sigma \geq \tau\}$, and 0 on $\{\sigma \not\geq \tau_k\}$. Then

$$\limsup_n f_n(\tau) \leq \limsup_n \int \varphi_k \, d\rho_n = \int \varphi_k \, d\rho \leq f(\tau_k).$$

As f is left continuous, we infer $\limsup f_n(\tau) \leq f(\tau)$.

Now assume $d(\tau; f) = 0$. Fix $\varepsilon > 0$. Pick finitely many tangent vectors $\{\vec{v}_j\}$ at τ, not representing τ_0, such that $\sum f(\vec{v}_j) \geq f(\tau) - \varepsilon$. Pick sequences $(\tau_{jk})_{k=1}^{\infty}$ decreasing to τ such that τ_{jk} represents $\vec{v}_j$ for all j, k. Also pick continuous increasing functions φ_{jk} with values in $[0,1]$, 1 on $\{\sigma \geq \tau_{jk}\}$ and vanishing outside $U(\vec{v}_j)$. Set $\varphi_k = \sum_j \varphi_{jk}$. Then

$$\liminf_n f_n(\tau) \geq \liminf_n \int \varphi_k \, d\rho_n = \int \varphi_k \, d\rho \geq \sum_j f(\tau_{jk}).$$

Letting $k \to \infty$ yields $\liminf_n f_n(\tau) \geq \sum_j f(\vec{v}_j) \geq f(\tau) - \varepsilon$, so as $\varepsilon \to 0$ we obtain $\liminf_n f_n(\tau) \geq f(\tau)$. Hence $\lim f_n(\tau) = f(\tau)$ and we are done. □

7.11 Restrictions to Subtrees

It is often important in applications to consider the restriction of functions and measures to subtrees, as well as extensions from a subtree to the larger tree.

Let $\mathcal{S}$ be a complete subtree of $\mathcal{T}$ (as $\mathcal{T}$ is a rooted tree, we assume that the root τ_0 is contained in $\mathcal{S}$). Denote by $p = p_{\mathcal{S}} : \mathcal{T} \to \mathcal{S}$ the retraction map defined by $p_{\mathcal{S}}(\tau) := \max[\tau_0, \tau] \cap \mathcal{S}$ and by $\imath = \imath_{\mathcal{S}} : \mathcal{S} \to \mathcal{T}$ the inclusion map. By Lemma 7.6 and Lemma 7.8, these are both continuous.

When $f : \mathcal{T}^o \to [0, \infty)$ belongs to $\mathcal{N}^+$, its restriction to $\mathcal{S}^o$ is an element of $\mathcal{N}^+(\mathcal{S})$. (Here it is important to regard the elements of $\mathcal{N}^+(\mathcal{S})$ as functions on $\mathcal{S}^o$ and extend them by zero on $\mathcal{S} \setminus \mathcal{S}^o$.) Conversely, if $f \in \mathcal{N}^+(\mathcal{S})$, we may extend it to a function in $\mathcal{N}^+$ by declaring it to be zero outside $\mathcal{S}^o$. Clearly the composition $\mathcal{N}^+(\mathcal{S}) \to \mathcal{N}^+(\mathcal{T}) \to \mathcal{N}^+(\mathcal{S})$ is the identity.

In the case of tree potentials the situation is only slightly more complicated. Let $g : \mathcal{T}^o \to [0, \infty)$ be a positive tree potential. It is straightforward to verify from (P1)-(P3) that the restriction $\imath^* g$ of g to $\mathcal{S}^o$ is a positive tree potential on $\mathcal{S}$. Conversely, if g is a positive tree potential on $\mathcal{S}$, then the *minimal extension* of g to $\mathcal{T}^o$ given by $p^* g = g \circ p$ is a positive tree potential on $\mathcal{T}$. Again the composition $\mathcal{P}^+(\mathcal{S}) \to \mathcal{P}^+(\mathcal{T}) \to \mathcal{P}^+(\mathcal{S})$ equals the identity.

Finally we consider measures. If ρ is a positive measure on $\mathcal{S}$, then its pushforward $\imath_* \rho$, i.e. the extension of ρ by zero, is a positive measure on $\mathcal{T}$. Conversely, if ρ is a positive measure on $\mathcal{T}$, then the pushforward $p_* \rho$ is a positive measure on $\mathcal{S}$. This time, too, the composition $\mathcal{M}^+(\mathcal{S}) \to \mathcal{M}^+(\mathcal{T}) \to \mathcal{M}^+(\mathcal{S})$ is the identity.

Proposition 7.68. *All the mappings above are continuous and respect the homeomorphisms between $\mathcal{N}^+$, $\mathcal{P}^+$ and $\mathcal{M}^+$ given by Theorem 7.64.*

In particular, $\mathcal{N}^+(\mathcal{S})$, $\mathcal{P}^+(\mathcal{S})$ and $\mathcal{M}^+(\mathcal{S})$) are retracts of $\mathcal{N}^+(\mathcal{T})$, $\mathcal{P}^+(\mathcal{T})$ and $\mathcal{M}^+(\mathcal{T})$), respectively.

Proof. That the mappings respect the isomorphisms is straightforward to verify and is left to the reader. It is immediate from the definition of the weak topology on $\mathcal{N}^+$ that the two mappings between $\mathcal{N}^+(\mathcal{T})$ and $\mathcal{N}^+(\mathcal{S})$ are continuous. Hence continuity holds also in the cones $\mathcal{P}^+$ and $\mathcal{M}^+$. □

Remark 7.69. The image measure $\rho_\mathcal{S} \in \mathcal{M}^+(\mathcal{S})$ of a measure $\rho \in \mathcal{M}^+(\mathcal{T})$ can be written as

$$\rho_\mathcal{S} = \rho|_\mathcal{S} + \sum_{\tau \in \mathcal{S}} \left(\sum_{\vec{v} \in T\tau \setminus T_\mathcal{S}\tau} \rho(U(\vec{v})) \right) \tau,$$

where $T\tau$ and $T_\mathcal{S}\tau$ denote the tangent spaces of τ in $\mathcal{T}$ and $\mathcal{S}$, respectively, and where $U(\vec{v})$ as usual denotes the open subset of $\mathcal{T}$ consisting of points in $\mathcal{T} \setminus \{\tau\}$ represented by $\vec{v}$.

7.12 Inner Products

We have constructed three isomorphic Banach spaces $\mathcal{M}$, $\mathcal{N}$ and $\mathcal{P}$ associated to a given complete, parameterized tree. In this section we wish to equip these spaces with inner products. As we show in Chapter 8, there are several interesting interpretations of these inner products when working on the valuative tree $\mathcal{V}$. First, general complex atomic measures on $\mathcal{V}$ can be viewed as cohomology classes on the voûte étoilée, and the intersection product on $\mathcal{M}$ agrees with the cup product on cohomology. Second, positive atomic measures ρ on $\mathcal{V}$ with integer coefficients correspond to integrally closed ideals ρ_I in the ring R, and the intersection product $\rho_I \cdot \rho_J$ gives the mixed multiplicity $e(I, J)$. An analytic version of the latter result is studied in [FJ1], where the intersection product is instead interpreted as a mixed Monge-Ampère mass at the origin.

We start by showing how to define the inner products on the positive cones $\mathcal{M}^+$, $\mathcal{N}^+$ and $\mathcal{P}^+$. The definition relies on a parameterization of the tree even in the case of measures and functions of bounded variation. In order to extend this inner product in the complex case, we need to impose some integrability conditions. The upshot is that the inner products are well defined on subspaces $\mathcal{M}_0$, $\mathcal{N}_0$ and $\mathcal{P}_0$, and turn these into isometric pre-Hilbert spaces. Finally, we compare the topologies induced by the inner products with the strong and weak topologies and show that the pre-Hilbert spaces are not complete.

7.12.1 Hausdorff Measure

The inner products on (subspaces) of $\mathcal{N}$ and $\mathcal{P}$ are defined by suitable integrals over $\mathcal{T}^o$ with respect to (one-dimensional) *Hausdorff measure*, whose definition we now recall.

We fix here a complete tree $\mathcal{T}$ with an increasing parameterization $\alpha : \mathcal{T} \to [1, \infty]$ with $\alpha(\tau_0) = 1$. The latter restricts to a parameterization $\alpha : \mathcal{T}^o \to [1, \infty[$ of the subtree $\mathcal{T}^o$. Consider the metric on $\mathcal{T}^o$ associated to the parameterization as in Section 3.1, i.e. $d(\tau, \tau') = \alpha(\tau) + \alpha(\tau') - 2\alpha(\tau \wedge \tau')$, and let λ be the corresponding one-dimensional Hausdorff measure:

$$\lambda\{A\} := \lim_{\delta \to 0} \inf \left\{ \sum_1^\infty \operatorname{diam}(E_i) \; ; \; A \subset \bigcup_1^\infty E_i, \; \operatorname{diam}(E_i) \leq \delta \right\} \tag{7.22}$$

for any subset $A \subset \mathcal{T}^o$. Here we use the convention that $\inf \emptyset = \infty$ so that $\lambda(A) = \infty$ if A cannot be covered by countably many sets of diameter $\leq \delta$ for any $\delta > 0$. From the remarks following Proposition 10.20 in [Fo] it follows that λ restricts to a measure on the Borel σ-algebra generated by the open sets in the topology associated to the metric d. As the latter topology is at least as strong as the weak topology (see Proposition 5.5) we conclude that λ restricts to a (positive, weak) Borel measure on $\mathcal{T}^o$. We can therefore integrate Borel measurable functions against λ. In particular we can integrate the functions $\mathcal{N}^+$ and $\mathcal{P}^+$.

Notice that, in general, the mass of λ is infinite. That $\lambda\{\mathcal{T}\} < +\infty$ implies in particular that $\mathcal{T}$ has at most countably many ends and is bounded for the metric d introduced above. Note, moreover, that if $\mathcal{T}$ has no branch points, then λ is isomorphic to Lebesgue measure on the interval $\alpha(\mathcal{T}) \subset \mathbf{R}$.

7.12.2 The Positive Case

The inner product is defined first on the set of Dirac masses, and then extended to $\mathcal{M}^+$ by bilinearity.

Definition 7.70. *If $\tau, \tau' \in \mathcal{T}$, then we define $\tau \cdot \tau' := \alpha(\tau \wedge \tau') \in [1, \infty]$.*

Note that this definition depends on the choice of the parameterization.

Remark 7.71. This definition has a natural interpretation in the case when $\mathcal{T}$ is the valuative tree $\mathcal{V}$: then $\nu_C \cdot \nu_{C'} = \frac{C \cdot C'}{m(C)m(C')}$ for any two irreducible curves C, C'. See Section 3.8.

Definition 7.72. *If $\rho, \rho' \in \mathcal{M}^+$, then we define $\rho \cdot \rho' \in [1, \infty]$ by*

$$\rho \cdot \rho' := \iint_{\mathcal{T} \times \mathcal{T}} \tau \cdot \tau' \, d\rho(\tau) d\rho'(\tau'). \tag{7.23}$$

Notice that by Fubini's theorem and by (7.19) we have

$$\rho \cdot \rho' = \int_{\mathcal{T}} g \, d\rho' = \int_{\mathcal{T}} g' \, d\rho, \tag{7.24}$$

where $g, g' \in \mathcal{P}^+$ are the positive tree potentials associated to ρ and ρ', respectively. Here it is important to consider the natural extension of these potentials from $\mathcal{T}^o$ to $\mathcal{T}$ for the formula to work also when ρ or ρ' charges an end.

The inner product is defined on $\mathcal{N}^+$, the cone of nonnegative, left continuous, strongly decreasing functions on $\mathcal{T}^o$, by

$$\langle f, f' \rangle := f(\tau_0) f'(\tau_0) + \int_{\mathcal{T}^o} f f' \, d\lambda, \tag{7.25}$$

where λ is 1-dimensional Hausdorff measure as above. Finally we define an inner product on $\mathcal{P}^+$, the cone of positive tree potentials, by passing to $\mathcal{N}^+$ (see (7.16) and (7.17)):

$$\langle g, g' \rangle := g(\tau_0) g'(\tau_0) + \int_{\mathcal{T}^o} \frac{dg}{d\alpha} \frac{dg'}{d\alpha} \, d\lambda. \tag{7.26}$$

Theorem 7.73. *The isomorphisms*

$$d : \mathcal{N}^+ \to \mathcal{M}^+ \quad \text{and} \quad \Delta : \mathcal{P}^+ \to \mathcal{M}^+$$

given in Theorem 7.61 preserve the inner products defined above.

The proof relies on the following lemma of independent interest.

Lemma 7.74. *Let $\rho \in \mathcal{M}^+$ be any positive Borel measure, and let $f = f_\rho$ be its associated functions in $\mathcal{N}^+$. Then one can find a sequence of positive atomic measures ρ_n tending weakly to ρ, such that $f_n := f_{\rho_n}$* increases *pointwise (and in fact uniformly) to f.*

Proof (Theorem 7.73). Pick positive measures $\rho, \rho' \in \mathcal{M}^+$ and let $f, f' \in \mathcal{N}^+$ and $g, g' \in \mathcal{P}^+$ be their preimages under d and Δ, respectively. It is clear from the definition that $\langle g, g' \rangle = \langle f, f' \rangle$. Hence it suffices to show that $\rho \cdot \rho' = \langle f, f' \rangle$.

First suppose ρ, ρ' are Dirac masses at τ and τ' respectively. Then $\rho \cdot \rho' = \tau \cdot \tau' = \alpha(\tau \wedge \tau')$. Moreover, f and f' are the characteristic functions of the segments $[\tau_0, \tau] \cap \mathcal{T}^o$ and $[\tau_0, \tau'] \cap \mathcal{T}^o$, respectively, so

$$\langle f, f' \rangle = f(\tau_0) f'(\tau_0) + \int_{\mathcal{T}^o} f f' \, d\lambda = 1 + \int_{\tau_0}^{\tau \wedge \tau'} d\alpha = \alpha(\tau \wedge \tau').$$

By bilinearity we conclude that $\rho \cdot \rho' = \langle f, f' \rangle$ holds when ρ and ρ' are positive atomic measures.

In the general case, we rely on Lemma 7.74, and find positive atomic measures ρ_n, ρ'_m whose corresponding functions $f_n, f'_m \in \mathcal{N}^+$ increase to f and f' respectively. Note that by integration it follows that g_n, g'_m increase to g and g' respectively. By monotone convergence, it is clear that

$$\lim_{m,n \to \infty} \langle f_n, f'_m \rangle = \lim f_n(\tau_0) f'_m(\tau_0) + \lim \int f_n f'_m \, d\lambda = \langle f, f' \rangle.$$

On the other hand, monotone convergence applied to g_n and g'_m gives

$$\rho_n \cdot \rho'_m = \int g_n \, d\rho'_m \overset{n\to\infty}{\longrightarrow} \int g \, d\rho'_m = \int g'_m \, d\rho \overset{m\to\infty}{\longrightarrow} \int g' \, d\rho = \rho \cdot \rho'.$$

By what precedes, $\rho_n \cdot \rho'_m = \langle f_n, f'_m \rangle$ for all n, m. Hence $\rho \cdot \rho' = \langle f, f' \rangle$, which completes the proof. □

Proof (Lemma 7.74). Suppose first that the support of f is included in a finite tree $\mathcal{S}$. For each n, pick a finite subset $B_n \subset \mathcal{S}$ containing the root, all branch points, and all ends of $\mathcal{S}$, and such that $\rho(I) \leq 1/n$ for any connected component I of $\mathcal{S} \setminus B_n$. For $\tau \in B_n$ set $m(\tau) = \rho\{\tau\} + \sum_I \rho(I)$, where the sum is over all $I \in \mathcal{I}$ having τ as a *left* endpoint. Now define the positive atomic measure ρ_n by $\rho_n = \sum_{\tau \in B} m(\tau)\tau$. In other words, we slide the mass on each segment I to its left endpoint. Set $f_n = I\rho_n$. Clearly ρ_n has the same mass as ρ, hence $f_n(\tau_0) = f(\tau_0)$. From the construction we have, for $\tau \in \mathcal{S}$:

$$f(\tau) - f_n(\tau) = \rho\{\sigma \geq \tau\} - \rho_n\{\sigma \geq \tau\} \in [0, 1/n].$$

By choosing $B_n \subset B_{n+1}$ for all n, we get that f_n increases uniformly to f.

In the general case, the set $\mathcal{S}_n := \{f > 1/n\}$ is a finite tree. Define $f_n := f \times \mathbf{1}_{\mathcal{S}_n}$. This is a sequence of functions in $\mathcal{N}^+$ increasing uniformly to f, and supported on a finite tree. We conclude by a diagonal argument. □

7.12.3 Properties

It is clear that the inner product on $\mathcal{N}^+$ (and on $\mathcal{P}^+$) defined by (7.25) satisfies the Cauchy-Schwartz inequality. An immediate consequence of Theorem 7.73 above is

Corollary 7.75. *The inner product defined on $\mathcal{M}^+$ satisfies the Cauchy-Schwartz inequality:*

$$(\rho \cdot \rho')^2 \leq (\rho \cdot \rho)(\rho' \cdot \rho').$$

For further reference, we note that in general the intersection product is not (weakly) continuous on $\mathcal{M}^+ \times \mathcal{M}^+$, and not even on the set of Dirac masses $\mathcal{T} \times \mathcal{T}$. An example on the valuative tree $\mathcal{T} = \mathcal{V}$ is given by $\nu_n = \nu_{y-nx,2}$ in local coordinates (x, y). Here $\nu_n \cdot \nu_n = 2$ but $\nu_n \to \nu_{\mathfrak{m}}$ and $\nu_{\mathfrak{m}} \cdot \nu_{\mathfrak{m}} = 1$.

It is also clear that the inner product is not continuous for the strong topology induced by the mass norm on $\mathcal{M}^+$. The self-intersection of a Dirac mass at an end with infinite parameter is infinite; see also Example 7.84. Note however that $\rho \cdot \rho' \geq \operatorname{mass}\rho \cdot \operatorname{mass}\rho'$.

Proposition 7.76. *The intersection product on $\mathcal{M}^+ \times \mathcal{M}^+$ is (weakly) lower semicontinuous: if $\rho_n \to \rho$, $\rho'_n \to \rho'$ weakly, then $\liminf \rho_n \cdot \rho'_n \geq \rho \cdot \rho'$.*

The same results are also true in $\mathcal{P}^+$ and $\mathcal{N}^+$

Proof. We may suppose the mass of all measures $\rho, \rho', \rho_n, \rho'_n$ is equal to 1. Write $f = I\rho$, $f' = I\rho'$. The functions $f, f' \in \mathcal{N}^+$ are uniformly bounded from above by 1. Note that $\rho \cdot \rho' = 1 + \int f f' \, d\lambda = 1 + \int_0^1 \lambda\{f f' > t\} \, dt$ by Theorem 7.73. Fix $\varepsilon > 0$ arbitrarily small, and k sufficiently large such that $1 + \int_{1/k}^1 \lambda\{f f' > t\} \, dt \geq \rho \cdot \rho' - \varepsilon$ if $\rho \cdot \rho' < +\infty$, or $\geq \varepsilon^{-1}$ if $\rho \cdot \rho' = +\infty$. As $\{f f' > 1/k\} \subset \{f > 1/k\} \cup \{f' > 1/k\}$, we may find a finite tree $\mathcal{S}_k$ containing $\{f f' > 1/k\}$ (see Corollary 7.26).

Define $f_n := I\rho_n$, $f'_n := I\rho'_n$. As ρ_n tends weakly to ρ, f_n tends to f λ-almost everywhere. These functions are bounded uniformly from above by $1 = \sup \|\rho_n\|$, hence $f_n \to f$ in $L^2(\mathcal{S}_k)$ also. The same is true for f'_n. We hence have

$$\rho_n \cdot \rho'_n \geq 1 + \int_{\mathcal{S}_k} f_n f'_n \, d\lambda \overset{n\to\infty}{\longrightarrow} 1 + \int_{\mathcal{S}_k} f f' \, d\lambda.$$

When $\rho \cdot \rho'$ is finite, the right term is greater than $\rho \cdot \rho' - \varepsilon$. When $\rho \cdot \rho' = \infty$, it is greater than ε^{-1}. In any case, we conclude that $\liminf \rho_n \cdot \rho'_n \geq \rho \cdot \rho'$ by letting $\varepsilon \to 0$. □

7.12.4 The Complex Case

We now wish to extend the previous definitions and results to the complex case. In order to do this, we have to impose suitable integrability restrictions.

Definition 7.77.

- *A complex measure ρ belongs to $\mathcal{M}_0$ iff its total variation measure $|\rho|$ satisfies $|\rho| \cdot |\rho| < \infty$.*
- *A function $f \in \mathcal{N}$ belongs to $\mathcal{N}_0$ iff the function T_f, defined in (7.11), satisfies $\langle T_f, T_f \rangle < +\infty$.*
- *A function $g \in \mathcal{P}$ belongs to $\mathcal{P}_0$ iff δg belongs to $\mathcal{N}_0$.*

Remark 7.78. Note that any atomic measure supported on the set $\{\alpha < \infty\}$ lies in $\mathcal{M}_0$. In particular, in the valuative tree $\mathcal{V}$ (parameterized by skewness), any atomic measure supported on quasimonomial valuations lies in $\mathcal{M}_0$.

Let us now describe how to define inner products on the sets $\mathcal{M}_0, \mathcal{N}_0, \mathcal{P}_0$.

Proposition 7.79. *For any measures $\rho, \rho' \in \mathcal{M}_0$ the function $(\tau, \tau') \mapsto \tau \cdot \tau'$ lies in $L^1(\rho \otimes \overline{\rho'})$. One can thus define*

$$\rho \cdot \rho' := \iint_{\mathcal{T} \times \mathcal{T}} \tau \cdot \tau' \, d(\rho \otimes \overline{\rho'})(\tau, \tau'). \tag{7.27}$$

Proof. We have $\tau \cdot \tau' \in L^1(\rho \otimes \overline{\rho'})$ iff $\tau \cdot \tau' \in L^1(|\rho| \otimes |\rho'|)$. By the Cauchy-Schwartz inequality (Corollary 7.75), the latter condition is satisfied when $\tau \cdot \tau' \in L^1(|\rho| \otimes |\rho|) \cap L^1(|\rho'| \otimes |\rho'|)$, i.e. when $\rho, \rho' \in \mathcal{M}_0$. □

Proposition 7.80. *For any $f, f' \in \mathcal{N}_0$, one has $f\overline{f}' \in L^1(\lambda)$. One can thus define*

$$\langle f, f' \rangle := f(\tau_0)\,\overline{f'(\tau_0)} + \int_{\mathcal{T}^o} f\,\overline{f'}\,d\lambda. \tag{7.28}$$

As a consequence, when $g, g' \in \mathcal{P}_0$, we can set

$$\langle g, g' \rangle := g(\tau_0)\,\overline{g'(\tau_0)} + \int_{\mathcal{T}^o} \frac{dg}{d\alpha}\,\overline{\frac{dg'}{d\alpha}}\,d\lambda. \tag{7.29}$$

Proof. Pick $f, f' \in \mathcal{N}_0$. Then $\langle T_f, T_f \rangle < \infty$, $\langle T_{f'}, T_{f'} \rangle < \infty$, so that $\langle T_f, T_{f'} \rangle < \infty$ by Corollary 7.75. But $|f\overline{f}'| \le T_f T_{f'}$, implying $f\overline{f}' \in L^1(\lambda)$. □

Remark 7.81. The space $\mathcal{M}_0$ can be characterized in a slightly different way. Namely, one can show that $\rho \in \mathcal{M}_0$ iff $\tau \cdot \tau' \in L^1(\rho \otimes \overline{\rho})$. In a similar way, a real-valued function $f \in \mathcal{N}_0$ iff $\langle f_1, f_1 \rangle < \infty$, $\langle f_2, f_2 \rangle < \infty$, where $f = f_1 - f_2$ is the Jordan decomposition of f (see Proposition 7.62).

On the other hand, a function $f \in \mathcal{N}$ may be in $L^2(\lambda)$ but not in $\mathcal{N}_0$, as the following example shows.

Example 7.82. Pick an end $\tau_\infty \in \mathcal{T}$ with $\alpha(\tau_\infty) = \infty$ (assuming such an end exists) and define increasing sequences $(\tau_k)_1^\infty$ and $(\tau'_n)_1^\infty$ by $\tau_k, \tau'_n < \tau_\infty$, $\alpha(\tau'_n) = 2^n$ and $\alpha(\tau_k) = 2^n + 2^{-n}$. Set $\rho_1 = \sum_1^\infty 2^{-n/2}\tau_k$ and $\rho_2 = \sum_1^\infty 2^{-n/2}\tau'_n$. Finally let $f_1 = d^{-1}(\rho_1)$, $f_2 = d^{-1}(\rho_2)$ and $f = f_1 - f_2$. Then it is straightforward to see that $f \in L^2(\lambda)$ but $f_1, f_2 \notin L^2(\lambda)$, so that $f \notin \mathcal{N}_0$.

Theorem 7.83. *The three spaces $\mathcal{M}_0$, $\mathcal{N}_0$ and $\mathcal{P}_0$ are vector spaces, and the natural inner product defined in (7.27), (7.28), (7.29) endow them with a pre-Hilbert space structure. Further, the maps $d : \mathcal{N} \to \mathcal{M}$, $\Delta : \mathcal{P} \to \mathcal{M}$ restrict to bijections*

$$d : \mathcal{N}_0 \to \mathcal{M}_0 \quad \textit{and} \quad \Delta : \mathcal{P}_0 \to \mathcal{M}_0$$

which preserve the inner products.

Proof. It is clear by definition that δ maps $\mathcal{P}_0$ bijectively onto $\mathcal{N}_0$ and preserves the inner product. Hence we need only consider $\mathcal{N}_0$ and $\mathcal{M}_0$.

First note that d maps $\mathcal{N}_0$ onto $\mathcal{M}_0$. Indeed:

$$f \in \mathcal{N}_0 \Leftrightarrow \langle T_f, T_f \rangle < \infty \overset{\text{thm 7.73}}{\Longleftrightarrow} d(T_f) \cdot d(T_f) < \infty$$
$$\overset{\text{prop 7.42}}{\Leftrightarrow} |df| \cdot |df| < \infty \Leftrightarrow df \in \mathcal{M}_0.$$

It is also clear that $\mathcal{N}_0$ is a vector space because $T_{f+f'} \le T_f + T_{f'}$ for any functions $f, f' \in \mathcal{N}$. Hence $\mathcal{M}_0$ is also a vector space.

We are thus reduced to proving that $\rho \cdot \rho' = \langle I\rho, I\rho' \rangle$ for any $\rho, \rho' \in \mathcal{M}_0$. Note that by linearity, we need only check that $\rho \cdot \rho = \langle I\rho, I\rho \rangle$ for all $\rho \in \mathcal{M}_0$.

First assume that ρ is a real measure. Let $\rho = \rho_+ - \rho_-$ be the Jordan decomposition of ρ. Thus $\rho_\pm$ are positive measures and $|\rho| = \rho_+ + \rho_-$. It is clear that $\rho \in \mathcal{M}_0$ implies $\rho_\pm \cdot \rho_\pm < \infty$. By Theorem 7.73, we infer that

$$\rho \cdot \rho = \rho_+ \cdot \rho_+ + \rho_- \cdot \rho_- - 2\rho_+ \cdot \rho_- = \langle f_+, f_+ \rangle + \langle f_-, f_- \rangle - 2\langle f_+, f_- \rangle = \langle f, f \rangle.$$

If $\rho = \rho_1 + i\rho_2$ is a complex Borel measure in $\mathcal{M}_0$, with ρ_i real Borel measures, then $|\rho_j| \le |\rho|$, implying $\rho_j \in \mathcal{M}_0$ for $j = 1, 2$. By what precedes, $\rho_j \cdot \rho_j = \langle f_j, f_j \rangle$, for $j = 1, 2$, where $f_j = f_{\rho_j}$. This gives

$$\rho \cdot \rho = \rho_1 \cdot \rho_1 + \rho_2 \cdot \rho_2 = \langle f_1, f_1 \rangle + \langle f_2, f_2 \rangle = \langle f, f \rangle,$$

and completes the proof. □

7.12.5 Topologies and Completeness

We end this section by briefly discussing the topology on the three pre-Hilbert spaces $\mathcal{M}_0$, $\mathcal{N}_0$ and $\mathcal{P}_0$ induced by the associated norms. Namely, we give examples showing that the spaces are not complete with respect to these norms, and that the topologies are not comparable to the strong topologies in general.

Example 7.84. Assume that α is unbounded on $\mathcal{T}$ and pick a sequence $(\tau_n)_1^\infty$ in $\mathcal{T}^o$ with $\alpha(\tau_n) = n^2$. Then $\rho_n := n^{-1}\tau_n$ tends to zero strongly but $\rho_n \cdot \rho_n = n$. On the other hand, if α is bounded on $\mathcal{T}$, say $\alpha \le C$, then it is clear that $\rho \cdot \rho \le C\|\rho\|^2$.

Example 7.85. Pick any strictly increasing sequence $(\tau_n)_1^\infty$ such that $\alpha(\tau_n)$ is bounded and set $\rho_n = \sum_{j=1}^{2n}(-1)^j \tau_{2n+j}$. Then clearly $\rho_n \in \mathcal{M}_0$ and

$$\rho_n \cdot \rho_n = \sum_{j=1}^{n} (\alpha(\tau_{2n+2j}) - \alpha(\tau_{2n+2j-1})) < \alpha(\tau_{4n}) - \alpha(\tau_{2n+1}) \to 0$$

as $n \to \infty$. On the other hand, ρ_n has total variation $2n$ so $\rho_n \not\to 0$ strongly.

Notice that the fact that $\|\rho_n\|$ is unbounded implies that ρ_n does not tend to zero weakly either.

A small modification of the same example shows that $\mathcal{M}_0$ (and hence $\mathcal{N}_0$ and $\mathcal{P}_0$) is not complete.

Example 7.86. Let $(\tau_n)_1^\infty$ be as in Example 7.85 and set $\rho_n = \sum_1^{2n}(-1)^{j-1}\tau_j$. Then $\rho_n \in \mathcal{M}_0$ and

$$(\rho_{n+m} - \rho_n) \cdot (\rho_{n+m} - \rho_n) = \sum_{j=1}^{2m} (-1)^j \alpha(\tau_{2n+j}) \le \alpha(\tau_{2n+2m}) - \alpha(\tau_{2n+1}) \to 0$$

as $n, m \to \infty$. Hence $(\rho_n)_1^\infty$ is a Cauchy sequence, but there is no $\rho \in \mathcal{M}$ such that $(\rho_n - \rho) \cdot (\rho_n - \rho) \to 0$. (Notice that ρ_n has total variation $2n$.)

Remark 7.87. In fact, the completion of $\mathcal{N}_0$ is naturally isomorphic to $\mathbf{C} \oplus L^2(\mathcal{T}^o)$.

8

Applications of the Tree Analysis

This chapter is devoted to applications of the tree analysis developed in the previous chapter. We shall use measures on the valuative tree $\mathcal{V}$ to describe singularities of ideals in Section 8.1, and cohomology classes of the voûte étoilée in Section 8.2. Further applications, to singularities of plurisubharmonic functions and to the dynamics of fixed point germs $f : (\mathbf{C}^2, 0) \circlearrowleft$, will be explored in forthcoming papers: see [FJ1], [FJ2], [FJ3].

Let us describe in more detail the content of this chapter.

We first attach to any ideal $I \subset R$ a *tree transform* $g_I : \mathcal{V}_{\mathrm{qm}} \to \mathbf{R}_+$, by setting $g_I(\nu) = \min\{\nu(\phi) \, ; \, \phi \in I\}$. This is a positive tree potential in the sense of Section 7.9, hence the Laplacian of g_I is a well-defined positive measure $\rho_I = \Delta g_I$, called the *tree measure* of I. We characterize the positive measures on $\mathcal{V}$ that are tree measures of ideals (Theorem 8.2). Any tree measure ρ_I for an ideal $I \subset R$ is atomic, and its support coincides with the set of Rees valuations of I when I is primary. This gives a tree-theoretic approach to the Rees valuations of a primary ideal and to Zariski's factorization of integrally closed ideals (complete ideals in Zariski's terminology).

We next introduce the *voûte étoilée* $\mathfrak{X}$. This space was first defined by Hironaka [Hi] in a quite general context. In our setting, $\mathfrak{X}$ has a simpler description than in the general case, and can be viewed as the total space of the set of all blowups above the origin. Our aim is to describe the cohomology of this space, that is, the sheaf cohomology $H^2(\mathfrak{X}, \mathbf{C})$. In doing so, we were much inspired by the monograph of Hubbard-Papadopol [HP], where the toric case was described in detail.

Let us summarize our approach. The cohomology $H^2(\mathfrak{X}, \mathbf{C})$ is a natural complex vector space, endowed with an intersection form coming from the cup product. We describe in Section 8.2.4 a natural map sending a cohomology class $\omega \in H^2(\mathfrak{X}, \mathbf{C})$ to a function g_ω defined on $\mathcal{V}_{\mathrm{qm}}$, the set of quasimonomial valuations. As we show, g_ω is always a complex tree potential in the sense of Section 7.6, and its Laplacian $\rho_\omega = \Delta g_\omega$ is a complex atomic measure supported $\mathcal{V}_{\mathrm{div}}$, the set of divisorial valuations. Thus ρ_ω belongs to the subspace $\mathcal{M}_0$ of complex measures on which we defined a inner product in

Section 7.12. We show that the mapping $\omega \mapsto \rho_\omega$ gives an isometric embedding of $H^2(\mathfrak{X}, \mathbf{C})$ into $\mathcal{M}_0$ (Theorem 8.30). Finally we identify the images inside $\mathcal{M}_0$ of the subspaces $H^2(\mathfrak{X}, \mathbf{C})$, $H^2(\mathfrak{X}, \mathbf{Z})$ and $H^2(\mathfrak{X}, \mathbf{R})$, as well as two natural positive cones in $H^2(\mathfrak{X}, \mathbf{R})$ (Theorem 8.33).

8.1 Zariski's Theory of Complete Ideals

8.1.1 Basic Properties

We define the *tree transform* of an ideal $I \subset R$ as the function $g_I : \mathcal{V}_{\text{qm}} \to \mathbf{R}$ given by

$$g_I(\nu) = \nu(I) = \min\{\nu(\phi) \; ; \; \phi \in I\}. \tag{8.1}$$

We will show that g_I is a positive tree potential in the sense of Definition 7.57, where the valuative tree $\mathcal{V}$ is parameterized by *skewness*. Moreover, we will characterize all positive tree potentials that are of this form.

Let us start with the case of a principal ideal. For $\phi \in \mathfrak{m}$, set $g_\phi(\nu) = \nu(\phi)$.

Lemma 8.1. *The function $g_\phi : \mathcal{V}_{\text{qm}} \to [1, \infty)$ is a positive tree potential with the property that if $\vec{v}$ is any tangent vector in $\mathcal{V}$, then $D_{\vec{v}} g_\phi$ is an integer divisible by $m(\vec{v})$. Further, if ϕ is irreducible, then $\Delta g_\phi = m(\phi)\nu_\phi$.*

Proof. By additivity in $\mathcal{P}^+$ and unique factorization in R we may assume that ϕ is irreducible. Then Proposition 3.25 and Definition 7.70 imply that

$$g_\phi(\nu) = m(\phi)\, \alpha(\nu \wedge \nu_\phi) = m(\phi)\, \nu \cdot \nu_\phi,$$

hence (7.19) shows that $g_\phi = g_\rho$, where $\rho = m(\phi)\nu_\phi$. If $\vec{v} \in T\nu$ is a tangent vector, then $D_{\vec{v}} g_\phi = 0$ unless $\nu \in [\nu_\mathfrak{m}, \nu_\phi]$ and $\vec{v}$ is represented by either $\nu_\mathfrak{m}$; or ν_ϕ, in which case $D_{\vec{v}} g_\phi = \pm m(\phi)$ and $m(\vec{v})$ is a factor of $m(\phi)$. □

We let $\mathcal{M}^+_\mathcal{I}$ be the set of positive measures $\rho \in \mathcal{M}^+$ of the form

$$\rho = \sum_{i=1}^{s} n_i b_i \nu_i. \tag{8.2}$$

Here $1 \le s < \infty$, n_i are positive integers, ν_i is a divisorial or curve valuation, and $b_i = b(\nu_i)$ is the generic multiplicity of ν_i if ν_i is divisorial and $b_i = m(\nu_i)$ is the multiplicity of ν_i if ν_i is a curve valuation.

To any positive measure $\rho \in \mathcal{M}^+$ we associate an ideal $I_\rho \subset R$ by

$$I_\rho = \{\phi \in R \; ; \; g_\phi \ge g_\rho\}. \tag{8.3}$$

If $\rho = b\nu$, i.e. $i = 1$ and $n_1 = 1$ in (8.2), then we write $I_\nu = I_\rho$.

Theorem 8.2. *The tree transform g_I of any ideal $I \subset R$ is a positive tree potential whose Laplacian $\rho_I = \Delta g_I$ is a positive measure in $\mathcal{M}^+_\mathcal{I}$ of mass $m(I)$. Conversely, if $\rho \in \mathcal{M}^+_\mathcal{I}$, then the ideal $I = I_\rho$ has associated measure $\rho_I = \rho$.*

Definition 8.3. *We call the positive measure ρ_I the* tree measure *of I.*

Remark 8.4. If R' is any subring of $R = \mathbf{C}[[x, y]]$ whose completion equals R, and I is any ideal in R', the curve valuations in (8.2) (if any) are associated to elements in R'. In particular, when R' is the ring of convergent power series, any such curve valuation is analytic.

Remark 8.5. If I, J are ideals, then $g_{IJ} = g_I + g_J$, and $g_{I+J} = \min\{g_I, g_J\}$. Moreover $g_{I \cap J}$ is the smallest positive tree potential dominating $\max\{g_I, g_J\}$.

One can rephrase the first two of these properties nicely in the terminology of semi-rings. The set of ideals in R defines a semi-ring, with addition $I+J$, and multiplication $I \cdot J$. We may endow $\mathbf{R}_+$ with its tropical semi-ring structure: addition is given by $\min\{a, b\}$, and multiplication by $a + b$. This induces a semi-ring structure on the set of functions $\mathcal{V}_{\mathrm{qm}} \to R_+$, and in particular to the set $\mathcal{P}^+$ of positive tree potentials. The properties above assert that the map $I \mapsto g_I$ is a semi-ring homomorphism.

Remark 8.6. Fix a composition of blowups $\pi \in \mathfrak{B}$, and let E_i be its exceptional components. It is a classical problem to characterize the collections of integers r_i which appear as multiplicities of some ideal I i.e. such that $r_i = \mathrm{div}_{E_i}(\pi^* I)$ for all i. A necessary and sufficient condition is that the r_i's satisfy certain *proximity inequalities* (or relations). For principal ideals this goes back to Enriques [En] (see [Cas] for a more recent presentation). The case of general ideals was treated by Lejeune-Jalabert [Le] and Lipman in [Li]. Theorem 8.2 can be viewed as a translation of these proximity relations into the tree language.

Proof (Theorem 8.2). That $g_I = \inf_{\phi \in I} g_\phi$ is a positive tree potential is a consequence of Lemma 8.1 and the fact that $\mathcal{P}^+$ is closed under infima (Corollary 7.65). Clearly ρ_I has mass $g_I(\nu_{\mathfrak{m}}) = m(I)$.

Let $S \subset I$ be a finite set of generators for I. Then $g_I = \min_{\phi \in S} g_\phi$. This implies that g_I is supported on the smallest subtree of $\mathcal{V}$ containing $\nu_{\mathfrak{m}}$ and any ν_ψ, where ψ ranges over the irreducible factors of the elements of S. This is a finite subtree $\mathcal{S}$. Moreover, it follows from Lemma 8.1 that on any segment in $\mathcal{S}$ parameterized by skewness, g_I is a piecewise affine function with integer slopes. Thus ρ_I is an atomic measure supported on valuations that are either ends or branch points in $\mathcal{S}$, or regular points in $\mathcal{S}$ where g_I fails to be locally affine: see Proposition 7.56. From the integer slope property we conclude that $\rho_I = \sum_{i=1}^r \tilde{n}_i \nu_i$, where ν_i are divisorial (i.e. have rational skewness) or curve valuations and $\tilde{n}_i$ are positive integers.

We have to show that b_i divides $\tilde{n}_i$. For this, it suffices to show that if ν is a curve or divisorial valuation, then $\rho_I\{\nu\}$, which we now know is an integer, is divisible by $b(\nu)$ in the case of a divisorial valuation, and by $m(\nu)$ in the case of a curve valuation.

If ν is a curve valuation, then on $[\nu_{\mathfrak{m}}, \nu[$ $\mu \mapsto \mu(\phi)$ is an affine function of skewness with slope in $m(\nu)\mathbf{N}$ as $\mu \to \nu$. Thus so is g_I, which implies that $\rho_I\{\nu\} \in m(\nu)\mathbf{N}$.

If ν is divisorial we have to work a bit harder. We may assume that $b(\nu) > 1$, so that in particular $\nu \neq \nu_{\mathfrak{m}}$. The proof relies on the following lemma.

Lemma 8.7. *Let $\nu \neq \nu_{\mathfrak{m}}$ be a divisorial valuation with approximating sequence $\nu_{\mathfrak{m}} = \nu_0 < \nu_1 < \cdots < \nu_g < \nu_{g+1} = \nu$ as in Proposition 3.44 and let $\phi \in \mathcal{C}$. Assume that ν_ϕ and $\nu_{\mathfrak{m}}$ represent the same tangent vector at ν. Then $\nu(\phi) \in \sum_{i=1}^g \mathbf{N} m_i \alpha_i$.*

We continue the proof of the theorem. Recall that $\rho_I\{\nu\} = -\sum_{\vec{v}\in T\nu} D_{\vec{v}} g_I$. Lemma 8.1 implies that $D_{\vec{v}} g_I \in m(\vec{v})\mathbf{N}$ for every $\vec{v} \in T\nu$. If $b(\nu) = m(\nu)$ then $m(\vec{v}) = b(\nu)$ for every $\vec{v}$ and we are done, so suppose that $b(\nu) > m(\nu)$.

Let $\vec{v}_- \in T\nu$ be the tangent vector represented by $\nu_{\mathfrak{m}}$. There is then a unique tangent vector $\vec{v}_+ \in T\nu$, $\vec{v} \neq \vec{v}_-$ such that $m(\vec{v}) = m(\nu)$; for all other $\vec{v}$ we have $m(\vec{v}) = b(\nu)$. It hence suffices to show that $D_{\vec{v}_+} g_I + D_{\vec{v}_-} g_I \in b(\nu)\mathbf{N}$. Moreover, we may find $\psi_\pm \in S$ such that $g_I(\mu) = \mu(\psi_\pm)$ as $\mu \to \nu$, $\mu \in U(\vec{v}_\pm)$. Then $\nu(\psi_+) = \nu(\psi_-)$ and we have to show that $D_{\vec{v}_+} g_{\psi_+} + D_{\vec{v}_-} g_{\psi_-} \in b(\nu)\mathbf{N}$.

Write $\psi_\pm = \psi'_\pm \psi''_\pm \psi'''_\pm$, where $\psi'_\pm$ ($\psi''_\pm$) is the product of all irreducible factors representing $\vec{v}_-$ ($\vec{v}_+$). Then

$$0 = \nu(\psi_+) - \nu(\psi_-) = \nu(\psi'_+) - \nu(\psi'_-) + \alpha(\nu)(m(\psi''_+\psi'''_+) - m(\psi''_-\psi'''_-)).$$

By Lemma 8.7 we get that $\alpha(\nu)(m(\psi''_+\psi'''_+) - m(\psi''_-\psi'''_-)) \in \sum_{i=1}^g \mathbf{N} m_i \alpha_i$. This implies that $m(\psi''_+\psi'''_+) - m(\psi''_-\psi'''_-) \in b(\nu)\mathbf{N}$ by Proposition 3.53. But we always have $m(\psi'''_\pm) \in b(\nu)\mathbf{N}$ so we conclude that $m(\psi''_+) - m(\psi''_-) \in b(\nu)\mathbf{N}$. Finally this gives

$$D_{\vec{v}_+} g_{\psi_+} + D_{\vec{v}_-} g_{\psi_-} = m(\psi''_+) - m(\psi''_-) - m(\psi'''_-) \in b(\nu)\mathbf{N},$$

which completes the proof that $\rho_I \in \mathcal{M}^+_{\mathcal{I}}$.

Conversely, if $\rho \in \mathcal{M}^+_{\mathcal{I}}$, define $I = I_\rho$ by (8.3) and let $\mathcal{S}$ be the support of g_ρ (thus $\mathcal{S}$ is a finite tree: see Section 7.6). Clearly $g_I = \inf_{\phi\in I} g_\phi \geq g_\rho$. For the reverse inequality we pick irreducible elements $\psi_{ij} \in \mathfrak{m}$, $1 \leq i \leq s$, $1 \leq j \leq n_i$ as follows. Write $\mu_{ij} = \nu_{\psi_{ij}}$. If ν_i is a curve valuation, then $\mu_{ij} = \nu_i$ for all j. If ν_i is divisorial, then $\mu_{ij} > \nu_i$, $m(\mu_{ij}) = b_i$ and μ_{ij} represent distinct tangent vectors at ν_i, none of which is in $T_{\mathcal{S}}\nu_i$.

Write $\psi = \prod_{i,j} \psi_{ij}$. It is then straightforward to verify that $g_\psi = g_\rho$ on $\mathcal{S}$. Hence $\psi \in I$. Now consider $\nu \notin \mathcal{S}$ and let $\nu_0 = \max \mathcal{S} \cap [\nu_{\mathfrak{m}}, \nu]$. If $\nu_0 \neq \nu_i$ for all i, then $g_\psi(\nu) = g_\psi(\nu_0) = g_\rho(\nu_0) = g_\rho(\nu)$ for any choice of ψ. If $\nu_0 = \nu_i$, then we pick ψ_{ij} such that no μ_{ij} represent the same tangent vector as ν_0 at ν_i. Again $g_\psi(\nu) = g_\rho(\nu)$. Hence $g_I = g_\rho$, which implies $\rho_I = \rho$. The proof is complete. □

Proof (Lemma 8.7). Set $\mu = \nu_\phi \wedge \nu$ and write $\mu \in [\nu_j, \nu_{j+1}[$ for some $0 \leq j \leq g$. Then $b(\mu)$ divides $m(\phi)$ and $b(\mu)\alpha(\mu) \in \sum_{i\leq j} \mathbf{N} m_i \alpha_i \subset \sum_{i=1}^g \mathbf{N} m_i \alpha_i$. This proves the lemma as $\nu(\phi) = m(\phi)\alpha(\mu)$. □

8.1.2 Normalized Blowup

The proof of Theorem 8.2 was based on tree arguments. We now follow [Te2], and use the classical normalized blowup of an ideal to describe the geometric structure of an ideal of the form I_ρ.

Recall that an ideal I is *primary* iff $I \supset \mathfrak{m}^n$ for some $n \geq 1$. When I is primary, denote by $\pi : X \to (\mathbf{C}^2, 0)$ the normalization of the blowup along I. The exceptional components E_i of π are associated to divisorial valuations called the *Rees valuations* of I.

Proposition 8.8. *Let $\rho = \sum n_i b_i \nu_i \in \mathcal{M}_\mathcal{I}$. Then $I_\rho = \prod I_{\nu_i}^{n_i}$. Moreover, I_ρ is primary iff all ν_i are divisorial, in which case the latter coincide with the Rees valuations of I.*

Proof. Pick $\rho \in \mathcal{M}_\mathcal{I}^+$, and write $I = I_\rho$. Assume first that I is primary. Then $I \supset \mathfrak{m}^n$ for some n so the tree transform of I is bounded by n. Hence all the valuations ν_i are divisorial. Let us show that they coincide with the Rees valuations of I, and that $I = \prod I_{\nu_i}^{n_i}$.

Let $\pi : X \to (\mathbf{C}^2, 0)$ be the normalization of the blowup along I, and E_i be the set of exceptional components of π. Let $(\psi_0, \cdots, \psi_N)$ be a finite set of generators for I. For $a \in \mathbf{C}^{N+1}$ write $\psi_a = \sum a_i \psi_i$, and let V_a be the strict transform of $\psi_a^{-1}(0)$ by π.

The structure of V_a is described in [Te2, p. 332]. We may find integers $n'_i \geq 1$ such that outside a proper closed Zariski subset $Z \subset \mathbf{P}^N$, V_a is a union of smooth curves V_a^{ij}, $1 \leq j \leq n'_i$. Moreover each V_a^{ij} intersects E_i transversely at a smooth point and $V_a \cap V_b = \emptyset$ for $a \neq b \in Z$.

Let μ_i be the divisorial valuation associated to E_i, ν_{ij}^a the valuation associated to the irreducible curve $\pi(V_a^{ij})$, and $\psi_{ij} \in \mathfrak{m}$ the irreducible element attached to $\pi(V_a^{ij})$. Pick $\phi \in \mathfrak{m}$ irreducible. Then $g_{\mu_i}(\phi)$ is equal to $b(\mu_i)$ times the order of vanishing of $\pi^*\phi$ along E_i. On the other hand, $\nu_{ij}^a(\phi)$ is equal to $m(V_a^{ij})^{-1}$ times the order of vanishing of $\phi \circ h(t)$ where $t \to h(t) \in \mathbf{C}^2$ is a parameterization of $\pi(V_a^{ij})$. As V_a^{ij} is smooth and transverse to E_i, $\pi(V_a^{ij})$ is a curvette for ν_{ij}^a in the sense of Section 6.6.1. By Proposition 6.31, its multiplicity equals $b(\mu_i)$. The parameterization h can be obtained by composing a parameterization of V_a^{ij} with π, hence $\nu_{ij}^a(\phi)$ is equal $b(\mu_i)$ times the intersection product of $\pi^*\phi^{-1}(0)$ with V_a^{ij}. Whence $\nu_{ij}^a(\phi) \geq g_{\mu_i}(\phi)$, with equality when the strict transform of $\phi^{-1}(0)$ does not intersect V_a^{ij}. In particular, $\psi_a^{ij} \in I_{\mu_i}$ for all j and all $a \notin Z$. When $a \neq b$, $V_a^{ij} \cap V_b^{ij} = \emptyset$ hence $g_{\mu_i}(\phi) = \min\{\nu_{ij}^a(\phi), \nu_{ij}^b(\phi)\}$, so that $g_{\mu_i} = \min\{\nu_{ij}^a \; ; \; a \neq Z, \, 1 \leq j \leq n'_i\}$. We infer that the tree transform $\min\{g_{\psi_a} \; ; \; a \notin Z\}$ coincides with the positive tree potential associated to the measure $\sum n'_i b(\mu_i)\mu_i$.

But I is generated by a finite number of sufficiently generic elements ψ_{a_i}, $a_i \notin Z$ so we conclude that $\sum n_i b_i \nu_i = \sum n'_i b(\mu_i)\mu_i$. In particular the ν_i's are divisorial and are exactly the Rees valuation of I.

We choose $a_i \neq Z$ such that ψ_{a_i} belongs to $\prod I_{\mu_i}^{n'_i} = \prod I_{\nu_i}^{n_i}$, hence $I \subset \prod I_{\nu_i}^{n_i}$. Conversely $\psi \in \prod I_{\nu_i}^{n_i}$ implies $\nu(\psi) \geq \sum n_i g_{I_{\nu_i}}(\nu) = g_\rho(\nu)$, hence $I = \prod I_{\nu_i}^{n_i}$.

Suppose now I is not primary. We will show that $I = \prod I_{\nu_i}^{n_i}$, and that some ν_i are curve valuations. Write $\rho_I = \sum n_i b_i \nu_i \in \mathcal{M}_{\mathcal{I}}^+$. Suppose ν_i is divisorial for $i \leq r$ and a curve valuation, $\nu_i = \nu_{\phi_i}$ for $i > r$. Define $\rho' = \sum_{i \leq r} n_i b_i \nu_i$, and $I' = I_{\rho'}$. We have proved that I' is a primary ideal equal to $\prod_{i \leq r} I_{\nu_i}^{n_i}$.

Pick $\psi \in I' \cdot \prod_{i>r} (\phi_i)^{n_i}$. The tree transform of ν_ψ clearly dominates g_I, hence $\psi \in I$. Conversely suppose $g_\psi(\nu) \geq g_I(\nu)$ for all ν. Then letting $\nu \to \nu_{\phi_i}$ we infer that the mass of Δg_ψ at ν_{ϕ_i} is greater than n_i hence $\phi_i^{n_i}$ divides ψ. We may then write $\psi = \psi' \prod_{i>r} \phi_i^{n_i}$, and clearly $g_{\psi'} \geq g_{I'}$, whence $\psi' \in I'$. We have proved $I = I' \prod_{i>r} (\phi_i)^{n_i} = \prod_i I_{\nu_i}^{n_i}$.

This concludes the proof. □

8.1.3 Integral Closures

The mapping $I \mapsto \rho_I$ is not injective in general. For instance, the ideals $\langle x^2, y^2 \rangle$ and $\langle x^2, xy, y^2 \rangle$ both have tree measure $2\nu_{\mathfrak{m}}$. However, the lack of injectivity can be well understood. Recall that the *integral closure* $\overline{I}$ of I is the set of $\phi \in R$ such that $\phi^n + a_1 \phi^{n-1} + \cdots + a_n = 0$ for some $n \geq 1$ and $a_i \in I^i$. Then $\overline{\overline{I}} = \overline{I}$ and I is *integrally closed* if $\overline{I} = I$. We have the following classical result (see [ZS2, p. 350]), rephrased in our language.

Proposition 8.9. *For any ideal $I \subset R$ we have $\overline{I} = \{\phi \in R \; ; \; g_\phi \geq g_I\}$.*

Remark 8.10. Fix an ideal I and $\phi \in \mathfrak{m}$. Suppose we want to show that $\phi \in \overline{I}$. By Proposition 8.9 we must show that $g_\phi \geq g_I$. Since g_I is locally constant outside $\mathcal{T}_I = \operatorname{supp} g_I$ it suffices to show that $g_\phi \geq g_I$ on $\mathcal{T}_I$. Write $\rho_I = \sum n_i b_i \nu_i$ as before. Assume that ν_i is divisorial for $i \leq r$ and a curve valuation for $r < i \leq s$. Then g_I is affine and g_ϕ concave on any segment in $\mathcal{T}_I \setminus \operatorname{supp} \rho_I$. Hence it suffices to check that $\nu_i(\phi) \geq \nu_i(I)$ for $i \leq r$ and $\lim \alpha(\nu)^{-1}(\nu(\phi) - \nu(I)) \geq 0$ as $\nu \to \nu_i$ for $i > r$.

In particular, if I is primary, then it suffices to check that $g_\phi \geq g_I$ at the Rees valuations of I (see e.g. [Te2, p.333]).

Corollary 8.11. *If $\rho \in \mathcal{M}_{\mathcal{I}}^+$, then I_ρ is integrally closed.*

Let $\mathcal{I}$ be the set of integrally closed ideals in R. Theorem 8.2 and Corollary 8.11 imply that the mapping $\rho \mapsto I_\rho$ gives a bijection between $\mathcal{M}_{\mathcal{I}}^+$ and $\mathcal{I}$, with inverse is given by $I \mapsto \rho_I$. Now $g_{IJ} = g_I + g_J$ for any ideals I, J, hence $\rho_{IJ} = \rho_I + \rho_J$. We get from this that if I is integrally closed and $\rho = \rho_I = \sum n_i b_i \nu_i$, then $I = I_\rho = I_{\nu_1}^{n_1} \cdots I_{\nu_s}^{n_s}$ by Proposition 8.8. This leads to the following result that incorporates Zariski's celebrated decomposition of complete (=integrally closed) ideals (see [ZS2]).

Theorem 8.12. *The set $\mathcal{I}$ of integrally closed ideals in R is a semigroup under multiplication and the mappings*

$$\mathcal{I} \ni I \mapsto \rho_I \in \mathcal{M}^+_{\mathcal{I}} \qquad \text{and} \qquad \mathcal{M}^+_{\mathcal{I}} \ni \rho \mapsto I_\rho \in \mathcal{I}$$

define inverse semigroup isomorphisms.

In particular, every $I \in \mathcal{I}$ has a unique factorization $I = I^{n_1}_{\nu_1} \cdots I^{n_s}_{\nu_s}$, where n_i are positive integers and ν_i are divisorial or curve valuations.

8.1.4 Multiplicities

Suppose I and J are primary ideals in R. We define the *mixed multiplicity* $e(I, J) \in \mathbf{N}^*$ following [Te1, Chapter I, §2] (when $I = J$ this gives the *multiplicity* $e(I)$ of I in the classical sense). For all $n, m \geq 0$, the vector space $R/(I^n J^m)$ is finite dimensional over $\mathbf{C}$, and we write $\dim_k(R/(I^n J^m)) = \frac{1}{2}n^2 e(I) + nme(I, J) + \frac{1}{2}m^2 e(J) + O(n + m)$. The multiplicity $e(I, J)$ coincides with the multiplicity of their respective integral closures (e.g. [Te1, Chapter 0, §0.6]).

Here we show how to compute mixed multiplicities in terms of the associated tree measures. Recall from Section 7.12 the definition of the inner product on the cone $\mathcal{M}^+$ of positive Borel measures on $\mathcal{V}$. We then have

Theorem 8.13. *For any primary ideals I, J we have*

$$e(I, J) = \rho_I \cdot \rho_J = \iint_{\mathcal{V} \times \mathcal{V}} \mu \cdot \nu \, d\rho_I(\mu) d\rho_J(\nu) = \sum_{i,j} n_i b_i m_j c_j \, \mu_i \cdot \nu_j, \qquad (8.4)$$

where $\rho_I = \sum_1^p n_i b_i \mu_i$ and $\rho_J = \sum_1^q m_j c_j \nu_j$ are the measures associated to I and J, respectively (see Proposition 8.8), and b_i, c_j are the generic multiplicities of ν_i and μ_j respectively.

Remark 8.14. In [FJ1] we shall prove an analytic version of Theorem 8.13 where I and J are replaced by plurisubharmonic functions u and v, and where $e(I, J)$ is replaced by the mass of the mixed Monge-Ampère measure $dd^c u \wedge dd^c v$ at the origin.

Proof. We may assume that I and J are integrally closed. Fix a finite set of generators for I and J, say $I = \langle \phi_i \rangle$, $J = \langle \psi_j \rangle$. By [Te1, Chapter I, §2] $e(I, J)$ is equal to the intersection multiplicity of $\{\phi_\alpha = \sum \alpha_i \phi_i = 0\}$ with $\{\psi_\beta = \sum \beta_j \psi_j = 0\}$ if the coefficients (α_i), (β_j) are sufficiently generic.

Introduce a composition of blowups π dominating the blowups of both ideals I and J, i.e. such that all Rees valuations of I and J are determined by some exceptional divisor of π. Decompose the strict transform of $\{\phi_\alpha = 0\}$ into irreducible components V^{is}_α, $1 \leq i \leq p$, $1 \leq s \leq n_i$ as in the proof of Proposition 8.8. Each V^{is}_α corresponds to a curve valuation μ^{is}_α dominating μ_i, of multiplicity $b_i = b(\mu_i)$, and the tangent vectors defined by all μ^{is}_α at μ_i are distinct. We do the same decomposition of the strict transform of $\{\psi_\beta = 0\}$ as a union of irreducible germs W_β^{jt}, with associated valuations ν_β^{jt}. We note that if $\mu_i = \nu_j$ for some i, j, then for sufficiently generic α, β, the tangent vectors of μ_β^{jt} and ν_α^{is} at $\mu_i = \nu_j$ are all distinct. We infer $V_\alpha^{is} \cdot W_\beta^{jt} = b_i c_j \, \mu_i \cdot \nu_j$. Formula (8.4) now follows immediately by bilinearity since

$$\{\phi_\alpha = 0\} \cdot \{\psi_\beta = 0\} = \sum_{i,j,s,t} V_\alpha^{is} \cdot W_\beta^{jt} = \sum_{i,j} b_i c_j n_i m_j \, \mu_i \cdot \nu_j.$$

The proof is complete. □

8.2 The Voûte Étoilée

We now turn to the voûte étoilée $\mathfrak{X}$. Our objective is to analyze its cohomology $H^2(\mathfrak{X}, \mathbf{C})$ in terms of complex Borel measures on the valuative tree. We start by giving the definition and basic properties of $\mathfrak{X}$, of its cohomology $H^2(\mathfrak{X}, \mathbf{C})$, and of the inner product on cohomology induced from the cup product. In Section 8.2.4 we then show how each cohomology class ω defines a function $g_\omega : \mathcal{V}_{\mathrm{qm}} \to \mathbf{C}$. The function g_ω turns out to be a complex tree potential in the sense of Chapter 7, so its Laplacian $\rho_\omega = \Delta g_\omega$ is a well-defined complex Borel measure on $\mathcal{V}$. In fact, the map $\omega \to \rho_\omega$ gives an isomorphism between $H^2(\mathfrak{X}, \mathbf{C})$ and the set of complex atomic measure supported on divisorial valuations. Moreover, as we prove in Section 8.2.5, this isomorphism preserves the inner product in the sense that $-\omega \cdot \omega' = \rho_\omega \cdot \rho_\omega$. We end the chapter by describing the images of various subsets of $H^2(\mathfrak{X}, \mathbf{C})$ under the isomorphism: see Theorem 8.33.

Throughout the section, we shall make essential use of the fundamental isomorphism from Chapter 6 between the valuative tree and the universal dual graph.

8.2.1 Definition

We shall denote by $\mathfrak{B}$ the set of all blowups above the origin in $\mathbf{C}^2$. We let X_π be the total space of a fixed element $\pi \in \mathfrak{B}$ so that $\pi : X_\pi \to (\mathbf{C}^2, 0)$. We saw in Chapter 6 that $\mathfrak{B}$ forms an inverse system.

Definition 8.15. *The voûte étoilée is the projective limit*

$$\mathfrak{X} := \operatorname{proj\,lim}_{\pi \in \mathfrak{B}} X_\pi .$$

Each X_π is an algebraic variety, hence $\mathfrak{X}$ is naturally a pro-algebraic variety. We endow it with the topology induced by the product topology from the natural embedding of $\mathfrak{X}$ into the product $\prod X_\pi$. There is a natural proper projection map $\mathfrak{X} \to (\mathbf{C}^2, 0)$. The space $\mathfrak{X}$ is not algebraic (nor even a topological manifold): we shall see that its second cohomology group is infinite dimensional.

Let us quickly indicate why our definition is equivalent to the usual one, given for instance in [Hi]. Any "étoile"[1] in the sense of Hironaka has a well-defined "center" in X_π for any $\pi \in \mathfrak{B}$: take the intersection of all images $\varpi(U)$ over all (ϖ, U) in the étoile. As we are in dimension two, this center is always a (closed) point. This gives a natural map from the set of "étoiles" to $\mathfrak{X}$, which is easily seen to be bijective, and also bicontinuous.

8.2.2 Cohomology

Consider $\pi, \pi' \in \mathfrak{B}$ such that $\pi' = \pi \circ \varpi$ for some modification ϖ. The map ϖ induces a map between cohomology groups $\varpi^* : H^2(X_\pi, \mathbf{C}) \to H^2(X_{\pi'}, \mathbf{C})$.[2] By [Br, Corollary 14.6] (see also [Ki]), the sheaf cohomology of the constant sheaf $\mathbf{C}$ on the projective limit $\mathfrak{X}$ is equal to

$$H^2(\mathfrak{X}, \mathbf{C}) = \operatorname{inj}\lim_{\pi \in \mathfrak{B}} H^2(X_\pi, \mathbf{C}). \tag{8.5}$$

The choice of the base field is essentially irrelevant in the sequel. One may replace $\mathbf{C}$ by $\mathbf{R}$ or $\mathbf{Q}$ or even $\mathbf{Z}$. We restrict our attention to H^2 as all the other cohomology groups are easy to compute.

In order to proceed further, we need to describe more precisely the cohomology $H^2(X_\pi, \mathbf{C})$ and the pullback ϖ^* discussed above. We shall use some of the notation from Chapter 6. Specifically, we let Γ^*_π be the (partially ordered) set consisting of irreducible components of the exceptional divisor $\pi^{-1}(0)$.

The following result is classical; we refer to [GH, p.473-474] for a proof.

Proposition 8.16.

- *Each* $E \in \Gamma^*_\pi$ *induces a natural class* $[E] \in H^2(X_\pi, \mathbf{C})$. *The cohomology group* $H^2(X_\pi, \mathbf{C})$ *is isomorphic to the direct sum of* $\mathbf{C}[E]$ *over* $E \in \Gamma^*_\pi$.
- *Suppose* $\varpi : X_{\pi'} \to X_\pi$ *is the blowup at* one point $p \in \pi^{-1}(0)$, *and let* $E := \varpi^{-1}\{p\}$. *Then* $\varpi^* : H^2(X_\pi, \mathbf{C}) \to H^2(X_{\pi'}, \mathbf{C})$ *is an injective map and we have*
$$H^2(X_{\pi'}, \mathbf{C}) = \varpi^* H^2(X_\pi, \mathbf{C}) \oplus \mathbf{C}[E].$$

Remark 8.17. If $\pi \in \mathfrak{B}$ and $E \in \Gamma^*_\pi$ (i.e. E is an irreducible component of $\pi^{-1}(0)$) then there are three natural objects associated to E: a cohomology class $[E] \in H^2(X_\pi, \mathbf{C})$; an element E in the universal dual graph Γ^*; and a divisorial valuation $\nu_E \in \mathcal{V}_{\mathrm{div}}$. The last two of these are independent of π (as long as $E \in \Gamma^*_\pi$) but the cohomology class *does* depend on π. Indeed, if

[1] An étoile is a collection of finite compositions of local blowups, an element is hence a map $\varpi : U \to (\mathbf{C}^2, 0)$, where U is some analytic space.

[2] When a class ω is represented by a smooth form α, $\varpi^*\omega$ is the class of $\varpi^*\alpha$.

$\pi' = \pi \circ \varpi$ for some modification ϖ, then the image of $[E]$ in $H^2(X_{\pi'}, \mathbf{C})$ corresponds to the *total* transform of E whereas the image of E in $\Gamma^*_{\pi'}$ is the *strict* transform.

From Proposition 8.16 and (8.5) we infer:

- The natural map $\imath_\pi : H^2(X_\pi, \mathbf{C}) \to H^2(\mathfrak{X}, \mathbf{C})$ is injective for any $\pi \in \mathfrak{B}$.
- For each element $\omega \in H^2(\mathfrak{X}, \mathbf{C})$, there exists $\pi \in \mathfrak{B}$, and $\omega_\pi \in H^2(X_\pi, \mathbf{C})$ such that $\imath_\pi \omega_\pi = \omega$. Moreover, $\omega_\pi = \sum_{E \in \Gamma^*_\pi} a(E)[E]$, where $a(E) \in \mathbf{C}$.
- Two elements $\omega_1 \in H^2(X_{\pi_1}, \mathbf{C})$ and $\omega_2 \in H^2(X_{\pi_2}, \mathbf{C})$ determine the same element $\omega \in H^2(\mathfrak{X}, \mathbf{C})$ iff there exists $\pi \in \mathfrak{B}$ and modifications $\varpi_1 : X_\pi \to X_{\pi_1}$, $\varpi_2 : X_\pi \to X_{\pi_2}$ such that $\varpi_1^* \omega_1 = \varpi_2^* \omega_2$.

Corollary 8.18. *The set $H^2(\mathfrak{X}, \mathbf{C})$ is an infinite dimensional vector space.*

8.2.3 Intersection Product

Each complex vector space $H^2(X_\pi, \mathbf{C})$ is endowed with a natural hermitian form, the cup product. If $E, E' \in \Gamma^*_\pi$, then $[E] \cdot [E']$ is by definition the intersection product of the curves E and E'. For two arbitrary elements in $H^2(X_\pi, \mathbf{C})$, we have $(\sum a_i[E_i]) \cdot (\sum b_j[E_j]) := \sum a_i \bar{b}_j \, E_i \cdot E_j$.

The map ϖ also induces a push-forward map $\varpi_* : H^2(X_{\pi'}, \mathbf{C}) \to H^2(X_\pi, \mathbf{C})$.[3] It is a basic fact that $\varpi^* \omega \cdot \omega' = \omega \cdot \varpi_* \omega'$ for $\omega \in H^2(X_\pi, \mathbf{C})$, $\omega' \in H^2(X_{\pi'}, \mathbf{C})$. Moreover $\varpi_* \varpi^* = \mathrm{id}$ as ϖ is birational. From these two facts one immediately deduces

Proposition 8.19. *Suppose $\varpi : X_{\pi'} \to X_\pi$ is the blowup at* one point $p \in \pi^{-1}(0)$, *and let $E := \varpi^{-1}(p)$. Then $\varpi^* : H^2(X_\pi, \mathbf{C}) \to H^2(X_{\pi'}, \mathbf{C})$ is an isometric embedding and we have the orthogonal direct sum decomposition*

$$H^2(X_{\pi'}, \mathbf{C}) = \varpi^* H^2(X_\pi, \mathbf{C}) \perp \mathbf{C}[E].$$

Moreover, $[E] \cdot [E] = -1$.

Corollary 8.20.

- *The cup product is a negative definite hermitian form on $H^2(X_\pi, \mathbf{C})$.*
- *For any modification $\varpi : X_{\pi'} \to X_\pi$, the map $\varpi^* : H^2(X_\pi, \mathbf{C}) \to H^2(X_{\pi'}, \mathbf{C})$ is an isometric embedding.*

Remark 8.21. It follows from Proposition 8.16, and the definition of $H^2(\mathfrak{X}, \mathbf{C})$ as an injective limit, that $H^2(\mathfrak{X}, \mathbf{C})$ is in fact generated by classes $[E]$ of exceptional divisors with self-intersection -1. More precisely, it is generated by classes ω of the following form: $\omega = \imath_\pi[E]$, where $E \in \Gamma^*$ and $\pi \in \mathfrak{B}$ is minimal such that $E \in \Gamma^*_\pi$.

[3] As ϖ is proper, $\varpi_* T$ is even defined for any current T on $X_{\pi'}$.

Remark 8.22. Proposition 8.19 is also true when replacing $\mathbf{C}$ by the ring of integers $\mathbf{Z}$. By decomposing $\pi \in \mathfrak{B}$ into a sequence of point blowups, we infer the existence of a basis $F_1, \ldots, F_n$ of $H^2(X_\pi, \mathbf{Z})$ (as a $\mathbf{Z}$-module) such that $F_i \cdot F_j = -\delta_{ij}$.

Now we can define the intersection product on $H^2(\mathfrak{X}, \mathbf{C})$. Pick two elements $\omega, \omega' \in H^2(\mathfrak{X}, \mathbf{C})$. By Proposition 8.16, we can find $\pi \in \mathfrak{B}$ and $\omega_\pi, \omega'_\pi \in H^2(X_\pi, \mathbf{C})$ such that $\imath_\pi \omega_\pi = \omega$ and $\imath_\pi \omega'_\pi = \omega'$. We set

$$\omega \cdot \omega' := \omega_\pi \cdot \omega'_\pi.$$

By Corollary 8.20, this number does not depend on the choice of $\pi \in \mathfrak{B}$.

Corollary 8.23. *The intersection product is a negative definite hermitian form on $H^2(\mathfrak{X}, \mathbf{C})$.*

8.2.4 Associated Complex Tree Potentials

Let us associate a function $g_\omega : \mathcal{V}_{\mathrm{qm}} \to \mathbf{C}$ to each cohomology class $\omega \in H^2(\mathfrak{X}, \mathbf{C})$.

Fix $\pi \in \mathfrak{B}$, and irreducible components $E, F \subset \pi^{-1}\{0\}$ (i.e. $E, F \in \Gamma^*_\pi$). We define a function $g_{[E]} : \Phi(\Gamma^*_\pi) \to \mathbf{Q}$ by

$$g_{[E]}(\nu_F) = \begin{cases} b(\nu_E)^{-1} & \text{if } F = E \\ 0 & \text{otherwise.} \end{cases}$$

Here $b(\nu_E)$ is the generic multiplicity of the divisorial valuation ν_E and $\Phi : \Gamma^* \to \mathcal{V}_{\mathrm{div}}$ denotes the isomorphism between the universal dual graph and the valuative tree as in Theorem 6.22.

We then define $g_{\omega_\pi} : \Phi(\Gamma^*_\pi) \to \mathbf{C}$ for any $\omega_\pi \in H^2(X_\pi, \mathbf{C})$ by linearity. If $\pi' \in \mathfrak{B}$ and $\pi' = \pi \circ \varpi$ for some modification ϖ, then the function $g_{\varpi^*\omega_\pi}$ is defined on $\Phi(\Gamma^*_{\pi'}) \supset \Phi(\Gamma^*_\pi)$ and restricts to g_{ω_π} on $\Phi(\Gamma^*_\pi)$.

Now fix a cohomology class $\omega \in H^2(\mathfrak{X}, \mathbf{C})$, and a divisorial valuation ν. Pick $\pi \in \mathfrak{B}$ such that $\imath_\pi \omega_\pi = \omega$ with $\omega_\pi \in H^2(X_\pi, \mathbf{C})$ and $\nu = \nu_E$ for some $E \in \Gamma^*_\pi$. We set

$$g_\omega(\nu_E) := g_{\omega_\pi}(\nu_E).$$

By what precedes, this does not depend on the choice of π.

Theorem 8.24. *The function $g_\omega : \mathcal{V}_{\mathrm{div}} \to \mathbf{C}$ extends (uniquely) to a complex tree potential $g_\omega : \mathcal{V}_{\mathrm{qm}} \to \mathbf{C}$ whose associated Laplacian $\rho_\omega = \Delta g_\omega$ is a complex atomic measure on $\mathcal{V}$ supported on divisorial valuations.*

Remark 8.25. As before, we are using the parameterization of $\mathcal{V}$ by skewness when talking about complex tree potentials.

Remark 8.26. We shall later show that any complex atomic measure ρ supported on divisorial valuations is of the form $\rho = \rho_\omega$ for a unique $\omega \in H^2(\mathfrak{X}, \mathbf{C})$.

In view of Remark 8.21, Theorem 8.24 follows by linearity from the following more precise result:

Proposition 8.27. *Consider a cohomology class $\omega \in H^2(\mathfrak{X}, \mathbf{C})$ of the form $\omega = \imath_\pi[E]$, where $\pi \in \mathfrak{B}$, $E \in \Gamma_\pi^*$. Assume that E has self-intersection -1. Then $g_\omega : \mathcal{V}_{\mathrm{div}} \to \mathbf{C}$ extends to a complex tree potential whose Laplacian $\rho = \Delta g_\omega$ is an atomic measure given as follows:*

(i) if $E = E_0$ is the exceptional divisor obtained by blowing up the origin once, then $b(E) = 1$ and

$$\rho = \nu_{\mathfrak{m}};$$

(ii) if E intersects a unique $E' \in \Gamma_\pi^*$, i.e. if E is obtained by blowing up a free point on E', then $b(E) = b(E')$ and

$$\rho = b(\nu_E)\nu_E - b(\nu_{E'})\nu_{E'};$$

iii) if E intersects $E', E'' \in \Gamma_\pi^*$, i.e. if E is obtained by blowing up the intersection point $E' \cap E''$, then $b(E) = b(E') + b(E'')$ and

$$\rho = b(\nu_E)\nu_E - b(\nu_{E'})\nu_{E'} - b(\nu_{E''})\nu_{E''}.$$

In order to prove this proposition, we shall use the following two results, whose proofs are postponed until the end of the section.

Lemma 8.28. *Fix $\pi \in \mathfrak{B}$ and a point $p \in \pi^{-1}(0)$. Let ϖ be the blowup of p with exceptional divisor $F = \varpi^{-1}(p)$. Consider $\omega \in H^2(\mathfrak{X}, \mathbf{C})$ and assume that $\omega = \imath_\pi\omega_\pi$ for some $\omega_\pi \in H^2(X_\pi, \mathbf{C})$. Then the following hold:*

(a) if p is a free point, i.e. p belongs to a unique $F' \in \Gamma_\pi^*$, then

$$g_\omega(\nu_F) = g_\omega(\nu_{F'}); \tag{8.6}$$

(b) if p is a satellite point, i.e. $p = F' \cap F''$ for $F', F'' \in \Gamma_\pi^*$, then

$$g_\omega(\nu_F) = \frac{b_{F'}}{b_{F'} + b_{F''}} g_\omega(\nu_{F'}) + \frac{b_{F''}}{b_{F'} + b_{F''}} g_\omega(\nu_{F''}). \tag{8.7}$$

Lemma 8.29. *Consider $\omega \in H^2(\mathfrak{X}, \mathbf{C})$ and pick $\pi \in \mathfrak{B}$ such that $\omega = \imath_\pi\omega_\pi$ for some $\omega_\pi \in H^2(X_\pi, \mathbf{C})$. Assume that Γ_π^* has more than one element and pick two adjacent elements $F', F'' \in \Gamma_\pi^*$ (i.e. F' intersects F''). Then g_ω is an affine function of skewness on the segment $[\nu_{F'}, \nu_{F''}]$ in $\mathcal{V}_{\mathrm{qm}}$.*

Proof (Proposition 8.27). We only need to prove the formulas for ρ as the expressions for the generic multiplicities $b(\nu_E)$ are known from Chapter 6.

In (i) we need to show that $g_\omega(\nu) = 1$ for every divisorial valuation $\nu = \nu_F$. This is clear for $F = E_0$. We now proceed by induction on the "length" of ν, i.e. the number of elements in Γ_π^*, where $\pi \in \mathfrak{B}$ is minimal such that $F \in \Gamma_\pi^*$. If this length is one, then $F = E_0$ and we are done. Otherwise we may apply Lemma 8.28. The inductive assumption gives $g_\omega(\nu_{F'}) = 1$ in case (a) and $g_\omega(\nu_{F'}) = g_\omega(\nu_{F''}) = 1$ in case (b). In both cases we get $g_\omega(\nu_F) = 1$, completing the proof of (i).

The proof in cases (ii) and (iii) is similar to that in case (i). By Lemma 8.29, g_ω is an affine function of skewness on all segments $[\nu_{F'}, \nu_{F''}]$, where F' and F'' are adjacent vertices in Γ_π^*. Moreover, $g_\omega(\nu_E) = 1$ and $g_\omega(\nu_F) = 0$ for every $F \in \Gamma_\pi^*$, $F \neq E$. This determines the value of g_ω on the smallest subtree $\mathcal{S} \subset \mathcal{V}_{\mathrm{qm}}$ containing all valuations ν_F, $F \in \Gamma_\pi^*$. Proving the formulas for ρ in (ii) or (iii) then amounts to showing that g_ω is locally constant outside $\mathcal{S}$.

Pick a divisorial valuation $\nu \notin \mathcal{S}$ and define $\nu_0 = \max\{\mu \in \mathcal{S} \; ; \; \mu \leq \nu\}$. We need to prove that $g_\omega(\nu) = g_\omega(\nu_0)$. The valuations ν_0 and ν are both divisorial, say $\nu_0 = \nu_{F_0}$ and $\nu = \nu_F$ for some $F_0, F \in \Gamma^*$. Moreover, F is the last exceptional divisor obtained by blowing up a sequence of infinitely near points $p_1, \dots, p_n$, starting with a point $p_1 \in F_0$. Let F_i be the exceptional divisor obtained by blowing up p_i, and write $\nu_i = \nu_{F_i}$. The key point is now that p_1 is a *free* point on F_0. Thus $g_\omega(\nu_1) = g_\omega(\nu_0)$ by Lemma 8.28. Inductively, we obtain from (a) or (b) in the same lemma that $g_\omega(\nu_i) = g_\omega(\nu_0)$ for $1 \leq i \leq n$. But $\nu_n = \nu$ so we are done. □

Proof (Lemma 8.28). Both proofs are analogous; we only treat (b). In the basis of $H^2(X_\pi, \mathbf{C})$ consisting of classes of irreducible components of $\pi^{-1}(0)$, write $\omega_\pi = c'[F'] + c''[F''] + \dots$ with $c', c'' \in \mathbf{C}$. By definition, $g_\omega(\nu_{F'}) = c'/b_{F'}$ and $g_\omega(\nu_{F''}) = c''/b_{F''}$. In $H^2(X_{\varpi\circ\pi}, \mathbf{C})$, we have $\varpi^*\omega_\pi = (c' + c'')[F] + \dots$. Together with $b_F = b_{F'} + b_{F''}$, this gives (8.7). □

Proof (Lemma 8.29). Assume that $\nu_{F'} < \nu_{F''}$. The divisorial valuations in the segment $]\nu_{F'}, \nu_{F''}[$ are of the form ν_F, where $F \in \Gamma^*$ is obtained by a finite sequence of blowups at satellite points, starting with the point $F' \cap F''$. By induction it therefore suffices to show that g_ω is an affine function on the totally ordered set $\{\nu_{F'}, \nu_F, \nu_{F''}\}$, where F is the exceptional divisor obtained by blowing up the point $F' \cap F''$ once.

If (a', b') and (a'', b'') are the Farey weights of F' and F'', respectively, then the Farey weight of F is $(a' + a'', b' + b'')$. We thus obtain from Lemma 8.28 and Corollary 6.39:

$$\frac{g_\omega(\nu_F) - g_\omega(\nu_{F'})}{\alpha(\nu_F) - \alpha(\nu_{F'})} = \frac{g_\omega(\nu_F) - g_\omega(\nu_{F'})}{\frac{1}{bb'}} = \frac{\frac{b' g_\omega(\nu_{F'}) + b'' g_\omega(\nu_{F''})}{b'+b''} - g_\omega(\nu_{F'})}{\frac{1}{b'(b'+b'')}}$$
$$= \frac{g_\omega(\nu_{F''}) - g_\omega(\nu_{F'})}{\frac{1}{b'b''}} = \frac{g_\omega(\nu_{F''}) - g_\omega(\nu_{F'})}{\alpha(\nu_{F''}) - \alpha(\nu_{F'})}.$$

This completes the proof. □

8.2.5 Isometric Embedding

As we saw above, we have a natural intersection product on the cohomology $H^2(\mathfrak{X}, \mathbf{C})$. In Section 7.12 we showed that there is a subspace $\mathcal{M}_0$ of the spaces of complex Borel measures on which we have a well-defined inner product.

If $\omega \in H^2(\mathfrak{X}, \mathbf{C})$, then the measure ρ_ω is atomic and supported on $\mathcal{V}_{\mathrm{div}}$. This implies that $\rho_\omega \in \mathcal{M}_0$: see Remark 7.78.

Theorem 8.30. *The map $\omega \mapsto \rho_\omega$ gives an isometric embedding of $H^2(\mathfrak{X}, \mathbf{C})$ into $\mathcal{M}_0$ in the following sense: for any classes $\omega, \omega' \in H^2(\mathfrak{X}, \mathbf{C})$, we have*

$$-\omega \cdot \omega' = \rho_\omega \cdot \rho_{\omega'}. \tag{8.8}$$

Here $\omega \cdot \omega'$ denotes the intersection product on $H^2(\mathfrak{X}, \mathbf{C})$ as in Section 8.2.3 and $\rho_\omega \cdot \rho_{\omega'}$ the inner product on $\mathcal{M}_0$ as in Section 7.12.

Proof. Instead of proving (8.8) we shall prove the equivalent formula

$$-\omega \cdot \omega' = g_\omega(\nu_{\mathrm{m}})\,\overline{g_{\omega'}(\nu_{\mathrm{m}})} + \int_{\mathcal{V}_{\mathrm{qm}}} \frac{dg_\omega}{d\alpha}\,\overline{\frac{dg_{\omega'}}{d\alpha}}\,d\lambda; \tag{8.9}$$

see Theorem 7.83. Fix $\pi \in \mathfrak{B}$, and $E, E' \in \Gamma_\pi^*$. Let us first prove (8.9) for $\omega = \imath_\pi[E]$ and $\omega' = \imath_\pi[E']$ under suitable assumptions.

(1) First suppose E and E' do not intersect, i.e. E and E' are not adjacent elements in Γ_π^*. By Proposition 8.27 above, the function g_ω is supported on the union of segments $[\nu_E, \nu_F[$ for all $F \in \Gamma_\pi^*$ with $E \cap F \neq \emptyset$. The analogous assertion holds for $g_{\omega'}$. Hence the product $g_\omega \overline{g_{\omega'}}$ is identically zero everywhere on $\mathcal{V}_{\mathrm{qm}}$. This proves (8.9) in this case.

(2) Now suppose $E \cap E' \neq \emptyset$ but $E \neq E'$, i.e. E and E' are adjacent elements in Γ_π^*. Then $-\omega \cdot \omega' = -[E] \cdot [E'] = -1$. By the same argument as before, the product $g_\omega \overline{g_{\omega'}}$ is identically zero except on the segment $I =]\nu_E, \nu_{E'}[$. Assume $\nu_E > \nu_{E'}$ and let (a_E, b_E) and $(a_{E'}, b_{E'})$ be the Farey weights of ν_E and $\nu_{E'}$, respectively.

The function g_ω is affine on I taking the value b_E^{-1} at ν_E and 0 at $\nu_{E'}$. Using Corollary 6.39, the (left) derivative with respect to skewness of g_ω on I is equal to

$$\frac{dg_\omega}{d\alpha} = \frac{g_\omega(\nu_E) - g_\omega(\nu_{E'})}{\alpha_E - \alpha_{E'}} = \frac{b_E^{-1} - 0}{b_E^{-1} b_{E'}^{-1}} = b_{E'}.$$

A similar computation shows that $\frac{dg_{\omega'}}{d\alpha} = -b_E$. The λ-length of I is given by $1/(b_E b_{E'})$ so again (8.9) holds.

(3) If π is the blowup of the origin and $E = E' = E_0$ is the exceptional divisor of π, then $-\omega \cdot \omega' = +1$, $g_\omega = g_{\omega'}$ is constant equal to 1 on $\mathcal{V}_{\mathrm{qm}}$, and (8.9) is immediate.

(4) Suppose π contains more than one blowup, that $E = E' \in \Gamma_\pi^*$ has self-intersection -1, and that E is obtained by blowing up a free point on a (unique) exceptional component F with Farey weight (a, b). Then $\nu_E > \nu_F$, the Farey weight of E is $(a+1, b)$, the multiplicity on $I =]\nu_F, \nu_E[$ is constant equal to b, and $\lambda(I) = b^{-2}$. The function $g_\omega = g_{\omega'}$ is affine on I, sends ν_E to b^{-1} and ν_F to 0, is locally constant outside I, and vanishes at ν_{m}. Whence

$$g_\omega(\nu_{\mathrm{m}})\,\overline{g_\omega(\nu_{\mathrm{m}})} + \int_{\mathcal{V}_{\mathrm{qm}}} \frac{dg_\omega}{d\alpha}\,\overline{\frac{dg_\omega}{d\alpha}}\,d\lambda = 0 + \left(\frac{1/b}{1/b^2}\right)^2 \frac{1}{b^2} = +1.$$

This proves (8.9) in this case.

(5) Now suppose $E = E'$ has self-intersection -1 and is obtained by blowing up a satellite point lying at the intersection of two divisors $F, F' \in \Gamma_\pi^*$ with Farey weights (a, b) and (a', b'), respectively. We may suppose $\nu_F > \nu_{F'}$. The valuation ν_E has Farey weight $(a+a', b+b')$ by definition. Corollary 6.39 implies that the segment $I =]\nu_F, \nu_E[$ has λ-length $\alpha(\nu_F) - \alpha(\nu_E) = 1/b(b+b')$ whereas the segment $I' =]\nu_E, \nu_{F'}[$ has length $1/b'(b+b')$.

The function $g_\omega = g_{\omega'}$ is affine on I and I', locally constant outside these two segments, and $g_\omega(\nu_{\mathrm{m}}) = g_\omega(\nu_F) = g_\omega(\nu_{F'}) = 0$, $g_\omega(\nu_E) = 1/(b+b')$. A direct computation shows

$$g_\omega(\nu_{\mathrm{m}})\,\overline{g_\omega(\nu_{\mathrm{m}})} + \int_{\mathcal{V}_{\mathrm{qm}}} \frac{dg_\omega}{d\alpha}\,\overline{\frac{dg_{\omega'}}{d\alpha}}\,d\lambda = \frac{b^2}{b(b+b')} + \frac{b'^2}{b'(b+b')} = +1.$$

This gives (8.9).

We can now prove Theorem 8.30 in full generality. Pick $\pi \in \mathfrak{B}$ minimal such that $\omega = \imath_\pi(\omega_\pi)$ and $\omega' = \imath_\pi(\omega'_\pi)$ for classes $\omega_\pi, \omega'_\pi \in H^2(X_\pi, \mathbf{C})$. We proceed by induction on the number $N(\pi)$ of elements in Γ_π^*. When $N(\pi) = 1$, (8.9) follows from (3).

For the inductive step, consider $\pi \in \mathfrak{B}$ with $N(\pi) > 1$. Then we may write $\pi = \pi' \circ \varpi$, where $N(\pi') = N(\pi) - 1$ and ϖ is the blowup at a point on $(\pi')^{-1}(0)$. Let E denote the exceptional divisor of ϖ. By linearity, and by Corollary 8.20, it suffices to check (8.9) for $\omega_\pi = [E]$ and $\omega'_\pi \in \varpi^* H^2(X_{\pi'}, \mathbf{C})$; and for $\omega_\pi = \omega'_\pi = [E]$. The last case is taken care of by (4) and (5) and the first one reduces to (1) or (2), again using linearity of both sides of (8.9). □

8.2.6 Cohomology Groups

We have shown that the assignment $\omega \mapsto \rho_\omega$ gives an isometric embedding of $H^2(\mathfrak{X}, \mathbf{C})$ into $\mathcal{M}_0$. We now wish to describe the image of $H^2(\mathfrak{X}, \mathbf{C})$ as well as of some of its subsets.

Definition 8.31. *We let $H^2_+(\mathfrak{X}) \subset H^2(\mathfrak{X}, \mathbf{R})$ be the set of cohomology classes ω which can be written $\omega = \imath_\pi \omega_\pi$, with $\omega_\pi = \sum_{E \in \Gamma_\pi^*} a(E)[E]$ where $a(E) \geq 0$. Such classes are called* pseudo-effective.

Definition 8.32. *We let $H^2_{\mathrm{an}}(\mathfrak{X}) \subset H^2(\mathfrak{X}, \mathbf{R})$ be the dual cone of $H^2_+(\mathfrak{X})$, that is, the set of classes $\omega \in H^2(\mathfrak{X}, \mathbf{R})$ for which $\omega \cdot \omega' \leq 0$ for all $\omega' \in H^2_+(\mathfrak{X})$. Such classes are called* anti-numerically effective *(or antinef for short).*

Theorem 8.33. *The map $\omega \to \rho_\omega$ induces an isometry between*

(i) $H^2(\mathfrak{X}, \mathbf{C})$ *and the set of complex atomic measures supported on divisorial valuations;*

(ii) $H^2(\mathfrak{X}, \mathbf{R})$ *and the set of real atomic measures supported on divisorial valuations;*

(iii) $H^2(\mathfrak{X}, \mathbf{Z})$ *and the set of real atomic measures* ρ *supported on divisorial valuations such that* $\rho\{\nu\} \in b(\nu)\mathbf{Z}$ *for every* $\nu \in \mathcal{V}_{\mathrm{div}}$*;*

(iv) $H^2_+(\mathfrak{X})$ *and the set of real atomic measures* ρ *supported on divisorial valuations such that* $\rho \cdot \rho' \geq 0$ *for all positive measures* ρ'*;*

(v) $H^2_{\mathrm{an}}(\mathfrak{X})$ *and the set of positive atomic measures supported on divisorial valuations.*

Remark 8.34. Thus the image of $H^2(\mathfrak{X}; \mathbf{Z}) \cap H^2_{\mathrm{an}}(\mathfrak{X})$ is exactly the set of tree measures ρ_I for primary ideals $I \subset R$. See Theorem 8.2 and Proposition 8.8. This fact is reminiscent of Lefschetz' theorem of realization of cohomology classes $H^2(X, \mathbf{Z}) \cap H^{1,1}(X)$ by divisors on a projective variety. On the voûte étoilée analytic ideals play the role of (effective) divisors.

Proof (Theorem 8.33). We have already seen in Theorem 8.30 that $\omega \mapsto \rho_\omega$ is an isometry and in particular injective. Let us first prove (iii). Then (i) and (ii) follow by linearity, whereas (iv)-(v) will be proved below.

Denote by $\mathcal{M}_{\mathbf{Z}}$ the set of real atomic measures supported on divisorial valuations whose mass on a divisorial valuation ν is always an integer multiple of the generic multiplicity of ν. Then $\mathcal{M}_{\mathbf{Z}}$ is an abelian subgroup of $\mathcal{M}_0$. We have to show that $\mathcal{M}_{\mathbf{Z}}$ is the image of $H^2(\mathfrak{X}; \mathbf{Z})$.

First pick $\omega \in H^2(\mathfrak{X}, \mathbf{Z})$. Let us show that $\rho_\omega \in \mathcal{M}_{\mathbf{Z}}$. By linearity it suffices to do this in the case $\omega = \imath_\pi[E]$, where $E \in \Gamma^*_\pi$ and $\pi \in \mathfrak{B}$ is minimal such that $E \in \Gamma^*_\pi$: see Remark 8.21. But then the conclusion is immediate from Proposition 8.27.

Conversely, let us show that any measure $\rho \in \mathcal{M}_{\mathbf{Z}}$ can be obtained as ρ_ω for some $\omega \in H^2(\mathfrak{X}, \mathbf{Z})$. Again by linearity, it is sufficient to prove this for $\rho = b(\nu)\nu$, where $\nu = \nu_E$ is divisorial. Pick $\pi \in \mathfrak{B}$ minimal such that $E \in \Gamma^*_\pi$. We proceed by induction on the number of elements in Γ^*_π, i.e. the minimal number of blowups necessary to create the exceptional divisor E.

If $E = E_0$ is the exceptional divisor obtained by blowing up the origin once, then $\nu_E = \nu_{\mathfrak{m}}$, $b(\nu_E) = 1$ and we conclude by Proposition 8.27 (i).

Now assume Γ^*_π has more than one element. Then either E intersects a unique $E' \in \Gamma^*_\pi$ or two distinct $E', E'' \in \Gamma^*_\pi$. We will consider only the second of these cases, the first one being almost identical. We may apply the inductive hypothesis to $\nu_{E'}$ and $\nu_{E''}$ and find $\omega', \omega'' \in H^2(\mathfrak{X}, \mathbf{Z})$ such that $\rho_{\omega'} = b(\nu_{E'})\nu_{E'}$ and $\rho_{\omega''} = b(\nu_{E''})\nu_{E''}$. Set $\omega = \omega' + \omega'' + b(E)\imath_\pi[E]$. We get from Proposition 8.27 (iii) that

$$\rho_\omega = \rho_{\omega'} + \rho_{\omega''} + b(\nu_E)\nu_E - b(\nu_{E'})\nu_{E'} - b(\nu_{E''})\nu_{E''} = b(\nu_E)\nu_E.$$

This completes the induction step and hence the proof of (iii).

As noted above (i) and (ii) follow by linearity. In order to prove (iv) it suffices to show that $\omega \in H^2_+(\mathfrak{X})$ iff $g_\omega \geq 0$ on $\mathcal{V}_{\mathrm{div}}$. Indeed, (7.24) shows that $\rho_\omega \cdot \rho' = \int_{\mathcal{V}} g_\omega \, d\rho'$ for any positive measure ρ'.

That $g_\omega \geq 0$ whenever $\omega \in H^2_+(\mathfrak{X})$ is a direct consequence of the definition of g_ω. For the converse, consider $\omega \notin H^2_+(\mathfrak{X})$. Pick $\pi \in \mathfrak{B}$ and $\omega_\pi \in H^2(X_\pi, \mathbf{R})$ such that $\omega = \imath_\pi \omega_\pi$. Write $\omega_\pi = \sum a_i[E_i]$ with $E_i \in \Gamma^*_\pi$ and $a_i \in \mathbf{R}$. As $\omega \notin H^2_+(\mathfrak{X})$, one of these real numbers is negative, say $a_1 < 0$. Then $g_\omega(\nu_{E_1}) = a_1/b(\nu_1) < 0$.

This completes the proof of (iv). Finally, (v) follows by duality from (iv) and from Theorem 8.30. □

Appendix

We end this monograph with an appendix containing complements to results already proved in the main body of the text.

Appendix A is devoted to infinitely singular valuations. Specifically it contains a list of properties that each characterizes a valuation as being infinitely singular. We also give a few constructions of infinitely singular valuations.

In Appendix B we give different characterizations of the tree tangent space at a divisorial valuation.

It is a fascinating fact that there are many paths to the valuative tree. We summarize in Appendix C the classification of Krull valuations on R from the different points of view emphasized in the monograph.

In order to help the reader familiar with invariants and terminology used to describe plane curve singularities (as in Zariski [Za3]), we give a short dictionary between this terminology and ours. This is done in Appendix D, where we also explain how the Eggers tree of a reduced curve singularity can be naturally embedded in the valuative tree.

We conclude in Appendix E by discussing the importance of the various assumptions we made on the ring R. Our standing assumption that R be the ring of formal power series in two complex variables is clearly unnecessarily restrictive. As the discussion shows, our method applies, for instance, to the ring of holomorphic germs at the origin in $\mathbf{C}^2$ and to the local ring at a smooth point on a surface over an algebraically closed field.

A Infinitely Singular Valuations

The infinitely singular valuations in $\mathcal{V}$ are numerous, but also the most complicated to describe. For the convenience of the reader we gather in one place all the characterizations of infinitely singular valuations that we have seen so far. Most of what we present here will also figure in the classification tables in Appendix C but we feel it is useful to have the information spelled out in more detail. In addition, we present a couple of explicit constructions of infinitely singular valuations.

As before we work with the ring R of formal power series in two complex variables, although as explained in Appendix E this is in fact unnecessarily restrictive.

A.1 Characterizations

The following result characterizes infinitely singular valuations from many different points of view.

Proposition A.1. *For a valuation $\nu \in \mathcal{V}$ the following properties are equivalent and characterize ν as being infinitely singular:*

(i) *the multiplicity $m(\nu)$ is infinite;*
(ii) *the approximating sequence $(\nu_i)_0^g$ of ν is infinite, i.e. $g = \infty$;*
(iii) *the value semigroup $\nu(R^*)$ is not finitely generated;*
(iv) *the numerical invariants of ν are given as follows:* $\operatorname{rat.rk} \nu = \operatorname{rk} \nu = 1$ *and* $\operatorname{tr.deg} \nu = 0$;
(v) *for some (or, equivalently, any) choice of local coordinates (x, y), the SKP $[(U_j); (\tilde{\beta}_j)]$ defined by ν has infinite length and $n_j \geq 2$ infinitely often (or, equivalently, $m(U_j) \to \infty$);*
(vi) *for any choice of local coordinates (x, y), some (or, equivalently, any) choice of extension $\hat{\nu}$ to a valuation on the ring $\bar{k}[[y]]$, is of the form $\hat{\nu} = \operatorname{val}[\hat{\phi}; \hat{\beta}]$, with $\hat{\phi} = \sum_1^\infty a_j x^{\hat{\beta}_j}$, $\hat{\beta} = \lim \hat{\beta}_j$, and the $\hat{\beta}_j$'s have unbounded denominators, i.e. $\hat{\phi} \notin \hat{k}$;*
(vii) *the sequence of infinitely near points associated to ν is of Type 1 in Definition 6.7, i.e. it contains both infinitely many free and infinitely many satellite points.*

Proof. As already noted, all the above characterizations can be read off from results already proved in the monograph. Specifically, our definition of an infinitely singularly valuation was exactly (v), for a fixed choice of coordinates. Theorem 2.28 asserted that this definition is equivalent to (iv) and hence independent on the choice of local coordinates. The equivalences of (i), (ii) follow from Proposition 3.37 and (iii) from Proposition 3.54. Further (vi) is a consequence of Theorem 4.17 and (vii) of Corollary 6.19 and Theorem 6.22. □

Remark A.2. In addition we have seen that the infinitely singular valuations are ends in the valuative tree, but this is not a characterizing property as the curve valuations are also ends.

A.2 Constructions

Next we outline a couple of constructions of infinitely singular valuations, starting with the construction of an infinitely singular valuation with prescribed skewness or thinness. See also Lemma 5.16.

Proposition A.3. *Pick a quasimonomial valuation ν, and any real number $\alpha > \alpha(\nu)$ (resp. $A > A(\nu)$). Then there exists an infinitely singular valuation $\mu > \nu$ with $\alpha(\mu) = \alpha$ (resp. with $A(\mu) = A$).*

Proof. We may assume that ν is divisorial. Indeed, otherwise replace ν by a divisorial valuation $\nu' > \nu$ such that $\alpha(\nu') < \alpha$ (resp. $A(\nu') < A$).

Let us first construct an infinitely singular valuation with prescribed skewness. Our strategy is to extend the approximating sequence $(\nu_i)_0^g$ of ν to an infinite approximating sequence $(\nu_i)_0^\infty$. By convention, $\nu_{g+1} = \nu$. Pick a curve valuation $\mu_{g+2} > \nu$ with $m(\mu_{g+2}) = b(\nu)$, and consider a divisorial valuation ν_{g+2} in the segment $]\nu, \mu_{g+2}[$. It follows from (3.13) that the set of divisorial valuations in $]\nu, \mu_{g+2}[$ with generic multiplicity equal to $b(\nu)$ is a discrete set. We may therefore pick ν_{g+2} in this segment such that $b(\nu_{g+2}) > m(\nu_{g+2}) = b(\nu)$ and $\alpha - 1/2 > \alpha(\nu_{g+2}) > \alpha(\nu)$.

Inductively, given $(\nu_i)_1^{g+k}$ we construct ν_{g+k+1} divisorial with $\nu_{g+k+1} > \nu_{g+k}$, $b(\nu_{g+k+1}) > m(\nu_{g+k+1}) = b(\nu_{g+k})$ and $\alpha - 1/2^k > \alpha(\nu_{g+k+1}) > \alpha(\nu_{g+k})$. Then $(\nu_i)_0^\infty$ defines an approximating sequence for an infinitely singular valuation μ, satisfying $\mu > \nu$ and $\alpha(\mu) = \alpha$.

For finding an infinitely singular valuation with prescribed thinness we may follow the argument in the proof of Lemma 5.16. However, let us instead show how to use Puiseux series, exploiting the analysis and notation of Chapter 4. Fix local coordinates (x, y) such that $\nu(y) \geq \nu(x) = 1$ and pick a valuation $\hat{\nu} \in \widehat{\mathcal{V}}_x$ whose image in $\mathcal{V}$ under the restriction map equals ν. Then $\hat{\nu} = \mathrm{val}[\hat{\phi}; \hat{\beta}]$ for a Puiseux series $\hat{\phi} = \sum_1^q a_j x^{\hat{\beta}_j}$ with $\hat{\beta}_j < \hat{\beta}_{j+1} < \hat{\beta}$, and all coefficients $a_j \neq 0$. By Theorem 4.17, $\hat{\beta} = A(\nu) - 1 < A - 1$. Since ν is divisorial, $\hat{\beta}$ is rational. Set $\hat{\beta}_{q+1} = \hat{\beta}$ and pick an arbitrary sequence of strictly increasing rational numbers $(\hat{\beta}_j)_{q+2}^\infty$ such that $\hat{\beta}_{q+2} > \hat{\beta}$ and $\hat{\beta}_j \to A - 1$. Define $\hat{\psi} = \sum_1^\infty a_j x^{\hat{\beta}_j}$, where, say, $a_j = 1$ for $j > q$. Set $\hat{\mu} = \mathrm{val}[\hat{\psi}; A - 1]$. Then $\hat{\mu}$ is of special type and has Puiseux parameter $A - 1$. Moreover, $\hat{\mu} > \hat{\nu}$. Let $\mu \in \mathcal{V}$ be the image of $\hat{\mu}$ under the restriction map. Then Theorem 4.17 implies that μ is infinitely singular, $\mu > \nu$ and $A(\mu) = A$. ☐

Remark A.4. By slightly modifying the first part of the proof above we can construct an infinitely singular valuation with prescribed skewness $t \in (1, \infty)$ and infinite thinness. The reason is that we may choose the ν_i's inductively to have very high multiplicity. Thus m_i grows very fast and this allows for skewness α_i to increase to $t \in (1, \infty)$ whereas thinness A_i tends to infinity. The details are left to the reader.

Remark A.5. Similarly one can construct a sequence $(\nu_n)_0^\infty$ of infinitely singular valuations with $\alpha(\nu_n) \to 1$ and $A(\nu_n) \geq 3$. Here is an outline of the construction. Fix a large integer n. Pick a monomial divisorial valuation μ' with $\alpha(\mu') = 1 + 1/n$. Then $b(\mu') = n$. Then pick μ divisorial with $\mu > \mu'$, μ representing a generic tangent vector at ν' and $\alpha(\mu) - \alpha(\mu') = 1/n$. Then $\alpha(\mu) = 1 + 2/n$ but $A(\mu) = 3 + 1/n > 3$. Now use Lemma 5.16 to replace μ by an infinitely singular valuation ν.

B The Tangent Space at a Divisorial Valuation

Recall from Section 3.1 that in a general nonmetric tree $\mathcal{T}$, the (tree) tangent space $T\tau$ at a point $\tau \in \mathcal{T}$ is the set of equivalence classes $\mathcal{T} \setminus \{\tau\}/\sim$, where $\sigma \sim \sigma'$ iff the two segments $]\tau, \sigma]$ and $]\tau, \sigma']$ intersect.

We saw in Section 3.2 that branch points of the valuative tree correspond to divisorial valuations. Our goal in this appendix is to describe the tangent space at a divisorial valuation from several points of view.

Theorem B.1. *Let $\nu = \nu_E \in \mathcal{V}$ be a divisorial valuation, associated to some exceptional component $E \subset \pi^{-1}(0)$, where π is a composition of point blowups. Then there exist natural bijections between the following four sets:*

- *$\mathcal{T}_1$: tree tangent vectors at ν (see Section 3.1.2);*
- *$\mathcal{T}_2$: Krull valuations μ satisfying $R_\mu \subsetneq R_\nu$ (see Sections 1.3 and 1.6);*
- *$\mathcal{T}_3$: sequences of infinitely near points $(p_j)_0^\infty$ of Type 3 (see Section 6.2), such that the divisorial valuation ν_n associated to the truncated sequence $(p_j)_0^n$ converges (weakly) to ν as $n \to \infty$;*
- *$\mathcal{T}_4$: points on the exceptional component E.*

The Krull valuations in $\mathcal{T}_2$ are exactly the exceptional curve valuations of the form $\nu_{E,p}$: see Lemma 1.5.

The proof of this result occupies the rest of Appendix B. We shall construct injective maps $\Psi_1 : \mathcal{T}_1 \to \mathcal{T}_2$, $\Psi_2 : \mathcal{T}_2 \to \mathcal{T}_3$, $\Psi_3 : \mathcal{T}_3 \to \mathcal{T}_4$ and $\Psi_4 : \mathcal{T}_4 \to \mathcal{T}_1$. and prove that the composition $\Psi_3 \circ \Psi_2 \circ \Psi_1 \circ \Psi_4$ is the identity.

Fix a divisorial valuation $\nu \in \mathcal{V}$ for the rest of the proof. Also fix a composition $\pi \in \mathfrak{B}$ of blowups such that $\nu = \nu_E$ for some exceptional component $E \subset \pi^{-1}(0)$.

The construction of $\Psi_1 : \mathcal{T}_1 \to \mathcal{T}_2$ is interesting as it allows us to interpret tangent vectors at ν in terms of directional derivatives. Let us first prove a preliminary result.

Lemma B.2. *Let $\nu \in \mathcal{V}$ be divisorial, $\varepsilon > 0$, and $[0, \varepsilon] \ni t \mapsto \nu_t \in \mathcal{V}$ a (weakly) continuous map with $\nu_0 = \nu$ such that $\frac{d}{dt}\alpha(\nu_t)$ is a nonzero constant on $]0, \varepsilon[$. Consider $\mathbf{R} \times \mathbf{R}$ with the lexicographic order. Then the function*

$$\mu(\psi) = \left(\nu(\psi), \left. \frac{d}{dt} \right|_{t=0} \nu_t(\psi) \right)$$

defines a centered Krull valuation on R, with valuation ring satisfying $R_\mu \subsetneq R_\nu$.

Proof. Clearly $\mu(\phi\psi) = \mu(\phi) + \mu(\psi)$ for $\phi, \psi \in R$. Further, $\nu_t(\phi + \psi) \geq \min\{\nu_t(\phi), \nu_t(\psi)\}$ for any t. If strict inequality holds at $t = 0$, then $\mu(\phi+\psi) \geq \min\{\mu(\phi), \mu(\psi)\}$ by the lexicographic ordering. Otherwise $t \mapsto \nu_t(\phi + \psi)$, and $t \mapsto \min\{\nu_t(\phi), \nu_t(\psi)\}$ are affine functions near $t = 0$, coinciding at $t = 0$. Then the slope of the latter cannot exceed the slope of the former, and this implies that $\mu(\phi + \psi) \geq \min\{\mu(\phi), \mu(\psi)\}$. This shows that μ is a Krull valuation. It is immediate that $R_\mu \subsetneq R_\nu$. □

Consider a tree tangent vector $\vec{v} \in \mathcal{T}_1$ at $\nu \in \mathcal{V}$. This is represented by a segment that can be parameterized by skewness. Lemma B.2 gives us a Krull valuation $\mu := \Psi_1(\vec{v})$ which satisfies $R_\mu \subsetneq R_\nu$, i.e. $\mu \in \mathcal{T}_2$. It does not depend on the choice of the segment, as two segments defining the same tree tangent vector intersect in a one-sided neighborhood of ν, hence define the same Krull valuation.

To show that Ψ_1 is injective, take two different tangent vectors $\vec{v}_1 \neq \vec{v}_2$, and choose $\phi \in \mathfrak{m}$ irreducible such that ϕ represents $\vec{v}_1$ but not $\vec{v}_2$. Then it is straightforward to verify that $\Psi_1(\vec{v}_1)(\phi) \neq \Psi_1(\vec{v}_2)(\phi)$.

The construction of $\Psi_2 : \mathcal{T}_2 \to \mathcal{T}_3$ is done as follows. For any Krull valuation μ, we let $\Psi_2(\mu)$ be the sequence of infinitely near points $(p_j)_0^\infty$ constructed in Section 1.7. Let us recall the construction. The point p_0 is the origin in $\mathbf{C}^2$. We let $\tilde{\pi}_0 : X_0 \to (\mathbf{C}^2, 0)$ be the blowup at p_0 with exceptional divisor E_0. Then $p_1 \in E_0$ is defined to be the center of the valuation μ in X_0. Inductively we construct a sequence of points $p_{j+1} \in X_j$, with $\tilde{\pi}_j : X_j \to X_{j-1}$ being the blowup at p_j, and we write E_j for the exceptional divisor of $\tilde{\pi}_j$. The point p_{j+1} is chosen to be the center of μ inside X_j. In particular, $p_{j+1} \in E_j$. Write $\pi_j = \tilde{\pi}_0 \circ \cdots \circ \tilde{\pi}_j$.

As $R_\mu \subsetneq R_\nu$, the center of μ is always included in the center of ν. As ν is divisorial, there exists $j_0 \geq 1$ such that the center of ν in X_j is a point for $j = 0, \ldots, j_0 - 1$, and an irreducible component of $\pi_j^{-1}(0)$ for $i \geq j_0$. In X_{j_0}, it is given by E_{j_0}. In X_j for $j > j_0$, it is the strict transform of E_{j_0} by $\tilde{\pi}_j \circ \cdots \circ \tilde{\pi}_{j_0+1}$. We shall write E_{j_0} for this strict transform, too. This shows that for all $j > j_0$, p_{j+1} has to be the intersection point of E_{j_0} and E_j. The sequence $(p_j)_0^\infty$ is thus of Type 3. By Theorem 6.11, the divisorial valuation ν_n associated with the truncated sequence $(p_j)_0^n$, converges to $\nu_{j_0} = \nu$ as $n \to \infty$. This shows that $\Psi_2(\mu) \in \mathcal{T}_3$.

It is clear that Ψ_2 is injective by Theorem 1.10.

Now pick a sequence of infinitely near points $(p_j)_0^\infty \in \mathcal{T}_3$ of Type 3. As before denote by $\tilde{\pi}_j : X_j \to X_{j-1}$ the blowup at p_j and E_j the exceptional divisor of $\tilde{\pi}_j$. The identity map id : $(\mathbf{C}^2, 0) \to (\mathbf{C}^2, 0)$ lifts to a rational map $\mathrm{id}_j : X_j \to X$ for all $j \geq 0$, where $\pi : X \to (\mathbf{C}^2, 0)$ was fixed at the beginning of the proof. As $(p_j)_0^\infty$ is of Type 3, there exists an index j_0 such that p_{j+1} is the intersection point of E_j with the strict transform of E_{j_0} (which we shall also denote by E_{j_0}) by $\tilde{\pi}_j \circ \cdots \circ \pi_{j_0+1}$ for all $j > j_0$. The divisorial valuation associated to E_j converges to ν_{j_0} (Theorem 6.11). As $\nu_{j_0} = \nu$, id_j sends E_{j_0} bijectively onto E for $j \geq j_0$. On the other hand, for j large enough, id_j is regular at p_{j+1}. The point $q := \mathrm{id}_j(p_{j+1})$ equals $\mathrm{id}_j(E_{j_0} \cap E_{j+1})$ hence lies in $\mathrm{id}_j(E_{j_0}) = E$. It is also clear that $\mathrm{id}_{j+1}(p_{j+2}) = \mathrm{id}_{j+1}(\pi_{j+1}(p_{j+2})) = \mathrm{id}_j(p_{j+1})$, so that q is independent of j. By definition we set $\Psi_3((p_j)_0^\infty) := q$.

To see that Ψ_3 is injective, choose two different sequences of infinitely near points $(p_j)_0^\infty, (p'_j)_0^\infty \in \mathcal{T}_3$. As the divisorial valuation associated to both sequences is the same, and equal to ν, it follows that $p_j = p'_j$ until $j = j_0$, where E_{j_0} is the exceptional component attached to ν. Being both of Type 3, the sequences are determined uniquely by p_{j_0+1} and p'_{j_0+1}, respectively. By assumption these two points are distinct. Now pick $j > j_0$, and let $\varpi_j : Y_j \to (\mathbf{C}^2, 0)$ be the composition of blowups at the points $p_0, \ldots, p_{j_0}$ and then at both $p_{j_0+1}, \ldots, p_j$ and $p'_{j_0+1}, \ldots, p'_j$. As before, the identity map $(\mathbf{C}^2, 0) \to (\mathbf{C}^2, 0)$ lifts as a map $\mathrm{id}_j : Y_j \to X$. For j large enough, id_j is regular at p_{j+1} and p'_{j+1}. These two distinct points belong to the strict transform of E_{j_0}, which we again denote by E_{j_0}. As id_j is a bijection of E_{j_0} onto E, we conclude that $\Psi_3((p_j)_0^\infty) \neq \Psi_3((p'_j)_0^\infty)$, i.e. that Ψ_3 is injective.

Finally, for a point $p \in E$, we define $\Psi_4(p)$ to be the tree tangent vector at ν defined by the divisorial valuation obtained by blowing-up p. Lemma 6.3 and Theorem 6.22 imply that Ψ_4 is injective.

The proof of Theorem B.1 will be complete, if we now show that the composition $\Psi_3 \circ \Psi_2 \circ \Psi_1 \circ \Psi_4$ equals the identity. Pick an element in $\mathcal{T}_4$, that is, a point $p \in E$. By definition, $\Psi_4(p)$ is the tree tangent vector at ν represented by the divisorial valuation ν_p associated to the blowup at p. The valuation $\mu := \Psi_1 \circ \Psi_4(p)$ is then a Krull valuation whose valuation ring is included in R_ν, hence the center of μ in X belongs to E. Pick an irreducible curve $C = \{\phi = 0\}$, $\phi \in \mathfrak{m}$, whose strict transform C' is smooth and intersects E transversely at p. Then ν_C and ν_p represent the same tree tangent vector at ν. The segment $[\nu, \nu_C] \cap [\nu, \nu_p]$ is thus nontrivial. We parameterize it by skewness: $t \mapsto \nu_{C, t+\alpha(\nu)}$. By definition, we have $\mu = (\nu, \frac{d}{dt}|_{t=0} \nu_{C, t+\alpha(\nu)})$. In particular, we infer that $\mu(\phi) = (\nu(\phi), c)$ with c *positive.*

Let $\tilde{\mu} = (\mathrm{div}_E, \frac{d}{d\tau}|_{\tau=0} \nu_{C', \tau})$. It defines an exceptional curve valuation $\tilde{\mu}$ which is centered at p. Its image by π_* is an exceptional curve valuation whose first component equals $\pi_* \mathrm{div}_E$ which is equivalent to ν. Thus $R_{\pi_* \tilde{\mu}} \subsetneq R_\nu$. It is also clear by definition that $(\pi_* \tilde{\mu})(\phi) = (\mathrm{div}_E(\pi^* \phi), c')$ with $c' > 0$.

Now pick a generic element $x \in \mathfrak{m}$ such that $\{x = 0\}$ is a smooth curve, $\nu_x \wedge \nu_C = \nu_{\mathfrak{m}}$, and the strict transform of $\{x = 0\}$ by π does not contain p. Then $\mu(x) = (\nu(x), 0)$, and $(\pi_* \tilde{\mu})(x) = (\mathrm{div}_E(\pi^* x), 0)$. Define $\psi := \phi^k / x^l$ where $\nu(\phi) k = l \nu(x)$ with $k, l \in \mathbf{N}$. Then ψ is an element of $\mathfrak{m}_\mu \cap \mathfrak{m}_{\pi_* \tilde{\mu}}$ with $\nu(\psi) = 0$. Remark 1.8 then implies $R_\mu = R_{\pi_* \tilde{\mu}}$. In particular, we conclude that the center of μ in X is precisely the center of $\tilde{\mu}$ which is equal to p.

At this stage, we have shown that the Krull valuation $\Psi_1 \circ \Psi_4(p)$ is an exceptional curve valuation centered at p, whose associated divisorial valuation is equal to ν. Now let $(p_j)_0^\infty := \Psi_2 \circ \Psi_1 \circ \Psi_4(p)$ be the sequence of infinitely near points associated to μ. Consider the sequence of blowups $\tilde{\pi}_j : X_j \to X_{j-1}$ as above. By construction, $\tilde{\pi}_j$ is the blowup at p_j, which is the center of μ in X_{j-1}. For j large enough, the identity map $(\mathbf{C}^2, 0) \to (\mathbf{C}^2, 0)$ lifts to a rational map $\mathrm{id}_j : X_j \to X$ which is regular at p_{j+1}. The image of the center of μ in X_j by id_j is the center of μ in X. Thus $\mathrm{id}_j(p_{j+1}) = p$ for j large enough. This proves that $\Psi_3 \circ \Psi_2 \circ \Psi_1 \circ \Psi_4(p) = p$, and concludes the proof.

C Classification

As we have shown in this monograph, centered Krull valuations on R can be interpreted in many different ways, each of which leads to a (full or partial) classification. Let us review the four approaches that we have considered.

The first (and classical) way to classify Krull valuations is through their value groups $\nu(K^*)$, in particular through the numerical invariants rk, rat. rk and tr. deg. In the two-dimensional case that we are concerned with in this monograph, this is feasible since Abhyankar's inequalities give strong restrictions on the values of these invariants.

A second way, and the one emphasized in the monograph at hand, is to identify Krull valuations with points or tangent vectors in the valuative tree $\mathcal{V}$. The non-metric tree structure on $\mathcal{V}$ then leads to a classification of Krull valuations into ends, regular points, branch points and tangent vectors. The valuative tree also comes with two natural parameterizations: skewness and thinness. By checking whether these are rational, irrational or infinite we obtain a third classification. Finally, we can classify valuations according to whether the multiplicity is finite or infinite.

Thirdly, we can take advantage of the isomorphism of $\mathcal{V}$ with the universal dual graph, as worked out in Chapter 6. Specifically, we can classify Krull valuations according to their associated sequences of infinitely near points, following the terminology of Spivakovsky: see Definition 6.7.

Finally, as explained in Chapter 4, any valuation in $\mathcal{V}$ extends to an element of the tree $\widehat{\mathcal{V}}_x$, i.e. a valuation on the ring of formal power series in one variable with Puiseux series coefficients. Even though this extension is not unique, we can classify valuations in the valuative tree using the tree structure on $\widehat{\mathcal{V}}_x$. This classification can then be rephrased in the terminology of Berkovich.

We list all of these classifications in three tables. Table C.1 contains the classification in terms of the numerical invariants (here we use the completeness of R—see Appendix E) and the associated sequences of infinitely near points.

In Table C.2 we instead focus on the tree structure of the valuative tree, specifically the nonmetric tree structure, skewness, thinness and multiplicity. For convenience we define the skewness (thinness) of a tangent vector in $\mathcal{V}$ to be the skewness (thinness) of the associated divisorial valuation.

Finally, in Table C.3 we present the classification using the Puiseux series approach. Notice that this table is a reproduction of Table 4.1 in Chapter 4, and that we used the relative valuative tree $\mathcal{V}_x$ instead of $\mathcal{V}$. Recall that the terminology for valuations in $\mathcal{V}$ and in $\mathcal{V}_x$ coincides except for div_x which is quasimonomial in $\mathcal{V}_x$, and corresponds to the curve valuation ν_x in $\mathcal{V}$.

A fifth classification is in terms of SKP's. We do not present this classification here as it depends on the choice of local coordinates. Instead we refer to Definition 2.23.

Classification		rk	rat.rk	tr.deg	Inf near pts
Quasim.	Divisorial	1	1	1	Type 0
	Irrational		2	0	Type 2
Curve	Nonexcept.	2	2	0	Type 4
	Exceptional				Type 3
Infinitely singular		1	1	0	Type 1

Table C.1. The table shows the classification of centered Krull valuations on R into five groups, and lists the three numerical invariants as well as the type of the associated sequence of infinitely near points as in Definition 6.7.

Classification		skewness	thinness	mult	Tree terminology
Quasim.	Divisorial	rational	rational	$< \infty$	Branch point
	Irrational	irrational	irrational		Regular point
Curve	Nonexcept.	∞	∞	$< \infty$	End
	Exceptional	rational	rational		Tangent vector
Infinitely singular		$\leq \infty$	$\leq \infty$	∞	End

Table C.2. Elements of the valuative tree. Here we show the classification of centered Krull valuations on R in terms of the tree structure on $\mathcal{V}$.

D Combinatorics of Plane Curve Singularities

The study of plane curve singularities is a rich and well developed subject with its own notation and terminology. Here we show how to interpret some classical invariants in terms of the valuative tree, specifically in terms of skewness, thinness and multiplicity.

We divide the study into two parts. In the first, we consider an irreducible formal curve with its associated invariants introduced by Zariski. In the second

<table>
<tr><th colspan="3">Valuations in $\mathcal{V}_x$</th><th colspan="2">Valuations in $\widehat{\mathcal{V}}_x$</th><th>Berkovich</th></tr>
<tr><td rowspan="2">Quasim.</td><td colspan="2">Divisorial</td><td rowspan="2">Finite type</td><td>Rational</td><td>Type 2</td></tr>
<tr><td colspan="2">Irrational</td><td>Irrational</td><td>Type 3</td></tr>
<tr><td rowspan="3">Not quasi-monomial</td><td colspan="2">Curve</td><td rowspan="2">Point type</td><td>$m < \infty$</td><td rowspan="2">Type 1</td></tr>
<tr><td rowspan="2">Inf sing</td><td>$A = \infty$</td><td>$m = \infty$</td></tr>
<tr><td>$A < \infty$</td><td colspan="2">Special type</td><td>Type 4</td></tr>
</table>

Table C.3. Here the classification of centered valuations on R is in terms of their preimages in the tree $\widehat{\mathcal{V}}_x$ and in terms of Berkovich's terminology.

part we extend this study to a (possibly reducible but reduced) formal curve and its associated Eggers tree.

In both cases, we shall see that all the classical invariants can be understood in terms of the subtree of the valuative tree whose end points are the curve valuations associated to the irreducible components of the curve. This serves to illustrate the idea that the valuative tree provides an efficient way of encoding singularities (in this case of plane curves).[4]

D.1 Zariski's Terminology for Plane Curve Singularities

Let us recall some notation from [Za3]. As before, R denotes the ring of formal power series in two complex variables. Consider an irreducible formal curve C.[5]

Zariski writes n for the multiplicity $m(C)$ and gives two equivalent sets of invariants for C. The first set is defined through Puiseux expansions. Pick local coordinates (x, y) such that the curve C is not tangent to $\{x = 0\}$. In these coordinates, there is a Puiseux parameterization $t \to (t^n, \sum a_j t^j)$ of C. Here $a_j = 0$ for $0 \leq j < n$. Define

- $\beta_1 = \min\{j \ ; \ a_j \neq 0, \ j \not\equiv 0 \mod n\}$;
- $e_1 = \gcd\{n, \beta_1\}$;
- $\beta_2 = \min\{j \ ; \ a_j \neq 0, \ j \not\equiv 0 \mod e_1\}$ (if $e_1 > 1$);
- $e_2 = \gcd\{n, \beta_1, \beta_2\}$ etc…

This process produces two finite sets of integers $(\beta_1, \ldots, \beta_g)$ and $(e_1, \ldots, e_g)$, but notice that the e_i's can be recovered from the β_i's and n. Set $e_0 = \beta_0 = n$ and $n_i = e_{i-1}/e_i$ for $i = 1, \ldots, g$. The set $(\beta_1/n, \ldots, \beta_g/n)$ is usually called the set of *generic characteristic exponents* of C.

The second set of invariants of C is defined through its value semigroup. This is by definition the collection of all intersection products $C \cdot D$ when D ranges over all formal curves. We let $\overline{\beta}_0, \ldots, \overline{\beta}_g \in \mathbf{N}^*$ be a minimal set

[4]We do not claim, however, that the valuative tree leads to any results on plane curve singularities that could not be proved by other means.

[5]Zariski actually writes X instead of C.

of generators of the semi-group of C, that is, $\overline{\beta}_0 = \min_D\{C \cdot D\} = n$, and, inductively, $\overline{\beta}_{i+1} = \min\{a = C \cdot D\ ;\ a \notin \sum_0^i \mathbf{N}\overline{\beta}_l\}$ for $i \geq 0$.

The β_i's (with n) determine the $\overline{\beta}_i$'s uniquely (and vice-versa) as follows: $\overline{\beta}_0 = n$, $\overline{\beta}_1 = \beta_1$ and

$$\overline{\beta}_i = n_{i-1}\overline{\beta}_{i-1} + \beta_i - \beta_{i-1} \quad \text{for } i = 2, \dots, g. \tag{D.1}$$

It is known that $(n; \beta_1, \dots, \beta_g)$ determines the equisingularity class of C, i.e. the topological type of the embedding of C in $\mathbf{C}^2$.

We now translate these invariants into the tree language that we have developed, specifically using skewness, thinness and multiplicity on the valuative tree. For this we consider the approximating sequence of ν_C as defined in Section 3.5:

$$\nu_{\mathfrak{m}} = \nu_0 < \nu_1 < \dots < \nu_g < \nu_{g+1} = \nu_C. \tag{D.2}$$

Here ν_i, $1 \leq i \leq g$, are divisorial valuations with strictly increasing multiplicities $m_i = m(\nu_i)$, such that the multiplicity is constant, equal to m_im on each segment $]\nu_{i-1}, \nu_i]$. See Figure 3.6. Thus the generic multiplicity of ν_i is $b(\nu_i) = m_{i+1}$, where $m_{g+1} = m(C)$.

Write α_i and A_i for the skewness and thinness of ν_i, respectively. Then we claim that we have the identification given by Table D.1.

Classical invariants	Tree language
n	$m(C)$
g	g
β_i/n	$A_i - 1$
n/e_i	$m_{i+1} = b_i$
n_i	b_i/m_i
$\overline{\beta}_i/n$	$\alpha_i m_i$

Table D.1. Dictionary expressing classical invariants of a curve singularity in terms of the multiplicity, skewness and thinness of the elements of the approximating sequence of the associated curve valuation.

An irreducible formal curve naturally defines an end in the valuative tree. This provides an embedding of a fundamentally discrete object (the ultrametric space $\mathcal{C}$ of local irreducible curves) into a "continuous" object (the valuative tree $\mathcal{V}$). The dictionary in Table D.1 shows that the *discrete* invariants for curves can be viewed as special cases of *continuous* invariants for valuations.

Let us now prove the validity of the correspondence given in Table D.1. As the invariants β_i and e_i are defined in terms of Puiseux series, it is natural to use the analysis in Chapter 4 to verify the first four entries in the table. Here we will freely apply the results from that chapter.

Thus define a Puiseux series by $\hat{\phi} = \sum a_j t^{j/n} \in \hat{k}$ and let $\hat{\nu}_{\hat{\phi}} = \mathrm{val}[\hat{\phi}; \infty]$ be the associated valuation in $\widehat{\mathcal{V}}_x$ of point type. Let ϕ be the minimal polynomial of $\hat{\phi}$ over k. Then $C = \{\phi = 0\}$ and $\hat{\nu}_{\hat{\phi}}$ is sent to the curve valuation $\nu_C \in \mathcal{V}_x$[6] by the restriction mapping $\Phi : \widehat{\mathcal{V}}_x \to \mathcal{V}_x$ induced by the inclusion $\mathbf{C}[[x, y]] \subset \bar{k}[[y]]$.

For $0 \le i \le g$, let

$$\hat{\nu}_i = \mathrm{val}\left[\sum_0^{\beta_i} a_j t^{j/n}; \beta_i/n\right] = \mathrm{val}\left[\sum_0^{\beta_i - 1} a_j t^{j/n}; \beta_i/n\right]$$

This defines an increasing sequence of rational valuations in $\widehat{\mathcal{V}}_x$. The multiplicity of $\hat{\nu}_i$ is given by

$$m(\hat{\nu}_i) = \mathrm{lcm}\left\{\frac{j}{n} \;;\; 0 < j < \beta_i, a_j \neq 0\right\} = n/\gcd\{\beta_1, \dots, \beta_{i-1}\} = n/e_{i-1}. \tag{D.3}$$

and the multiplicity is constant in the segment $]\hat{\nu}_{i-1}, \hat{\nu}_i]$. By Theorem 4.17 this implies that if $\nu_i = \Phi(\hat{\nu}_i)$ is the restriction of $\hat{\nu}_i$ to R, then $(\nu_i)_1^g$ is the approximating sequence of ν_C. The first and the second entries in Table D.1 are thus clear. The fourth entry follows from (D.3) and implies the fifth. Further, as $\hat{\nu}_i$ has Puiseux parameter $\hat{\beta}_i/n$, ν_i must have thinness $\hat{\beta}_i/n - 1$, which gives the third entry in the table.

Finally let us prove that $\overline{\beta}_i/n = m_i \alpha_i$ for $1 \le i \le g$. This can be seen from (D.1) and the relation between thinness and skewness. Indeed, we first have $\overline{\beta}_1/n = \beta_1/n = A_1 - 1 = \alpha_1 m_1$ as $m_1 = 1$. Suppose we have proved that $\overline{\beta}_{i-1}/n = \alpha_{i-1} m_{i-1}$. Then

$$\begin{aligned}\overline{\beta}_i = n_{i-1}\overline{\beta}_{i-1} + \beta_i - \beta_{i-1} &= n n_{i-1}\alpha_{i-1}m_{i-1} + n(A_i - A_{i-1}) = \\ &= n n_{i-1} m_{i-1}\alpha_{i-1} + n m_i(\alpha_i - \alpha_{i-1}) = n\alpha_i m_i.\end{aligned}$$

By induction, this completes the verification of Table D.1.

An alternative way of showing that $\overline{\beta}_i/n = a_i m_i$ is to use the fact that the value group of ν_C is given by $\nu_C(R^*) = \sum_{i=0}^g m_i \alpha_i \mathbf{N}$ (see Proposition 3.52) and the formula $m(C)\nu_C(\psi) = C \cdot \{\psi = 0\}$ for any $\psi \in \mathfrak{m}$.

D.2 The Eggers Tree

Let C be a reduced formal curve and let $C_1, \dots, C_r$ be its branches (i.e. irreducible components). The embedding of C in $\mathbf{C}^2$ is determined up to

[6] By our choice of coordinates, ν_C is a valuation in both $\mathcal{V}$ and $\mathcal{V}_x$.

topological conjugacy[7] by the generic characteristic exponents of each branch, and by the order of contact between them.

This numerical information can be encoded geometrically in the *Eggers tree* $\mathcal{T}_C$, first introduced by Eggers [Eg], and subsequently used in studies of plane curve singularities [Ga, Po]. Here we show that $\mathcal{T}_C$ has a natural interpretation inside the valuative tree $\mathcal{V}$.

Let us recall the definition of $\mathcal{T}_C$, essentially following [Po]. First assume C is irreducible, i.e. $r = 1$, and let $\beta_1/n, \dots, \beta_g/n$ be its generic characteristic exponents (with the notation in Appendix D.1). Also define $\beta_0 := n$ and $\beta_{g+1} := +\infty$. Then $\mathcal{T}_C$ is a simplicial tree whose vertices are exactly of the form $\{\beta_i/n\}_{0\le i\le g+1}$, and with exactly $g+1$ edges linking β_i/n to β_{i+1}/n for $0 \le i \le g$. By sending each vertex to the corresponding element in $[1, +\infty]$, we may also view $\mathcal{T}_C$ as the interval $[1, +\infty]$ with $g+2$ marked points $\{\beta_i/n\}_{0\le i\le g+1}$. Alternatively speaking, $\mathcal{T}_C$ is a rooted, nonmetric tree with a parameterization onto $[1, +\infty]$, still with $g+2$ marked points.

When $C = \bigcup C_j$ is reducible, $\mathcal{T}_C$ is constructed by patching together the trees $\mathcal{T}_{C_j}$ as follows. First define the coincidence order $K_{jj'} = K(C_j, C_{j'})$ between two branches C_j and $C_{j'}$ as follows. Fix local coordinates (x, y) such that all branches are transverse to $\{x = 0\}$, and let $\hat\phi = \sum a_l x^{\hat\beta_l}$, $\hat\phi' = \sum a'_l x^{\hat\beta'_l}$ be Puiseux series associated to C_j and $C_{j'}$, respectively. Let $(\hat\phi_l)_1^{n_j}$, $(\hat\phi'_{l'})_1^{n_{j'}}$ denote their orbits under the action of the Galois group $\mathrm{Gal}(\hat k/k)$ where k and $\hat k$ are the fields of Laurent and Puiseux series in x, respectively (see Section 4.3). Then

$$K_{jj'} := K(C_j, C_{j'}) := \max_{l,l'} \nu_\star(\hat\phi_l - \hat\phi'_{l'}),$$

where $\nu_\star$ is the valuation on $\hat k$ given by $\nu_\star(\sum a_j x^{\hat\beta_j}) := \min \hat\beta_j$. Note that $K_{jj'} \ge 1$.

The trees $\mathcal{T}_{C_j} \simeq [1, +\infty]$ and $\mathcal{T}_{C_{j'}} \simeq [1, +\infty]$ are now patched together at the point corresponding to $K_{jj'}$. In other words, $\mathcal{T}_C$ is the set of pairs (C_j, t) where $1 \le j \le r$ and $t \in [1, +\infty]$ modulo the relation $(C_j, s) = (C_{j'}, t)$ iff $s = t \le K_{jj'}$. Notice the similarity to the construction in Section 3.1.6. Thus $\mathcal{T}_C$ is naturally a rooted, nonmetric tree with a parameterization onto $[1, +\infty]$. The root of $\mathcal{T}_C$ is $(C_j, 1)$ and the maximal points of $\mathcal{T}_C$ are exactly of the form $(C_j, +\infty)$.

To be precise, $\mathcal{T}_C$ also comes with a finite number of marked points. These are exactly the points coming from marked points on the trees $\mathcal{T}_{C_j}$ (i.e. the points of the form $(C_j, \beta_{ji}/n_j)$) together with the branch points of $\mathcal{T}_C$ (i.e. the points of the form $(C_j, K_{jj'}) = (C_{j'}, K_{jj'})$ for $j \ne j'$). This marking allows us to recover $\mathcal{T}_C$ as a simplicial tree: the vertices are the marked points and the edges are open segments in $\mathcal{T}_C$ containing no marked points.

[7] i.e. the equisingularity type of C

Proposition D.1. *Let C be a reduced formal curve with irreducible components $C_1, \dots C_r$. Let n_j be the multiplicity of C_j, and let $\{\beta_{ji}/n_j\}_{i=0}^{g_j+1}$ be the set of (generic) characteristic exponents of C_j (by convention 1 and $+\infty$ belong to this set). Denote by $\nu_j = \nu_{C_j}$ the curve valuation associated to C_j and by $(\nu_{ji})_{i=1}^{g_j}$ the approximating sequence of ν_j.*

Define a map $\Psi : \mathcal{T}_C \to \mathcal{V}$ by sending (C_j, t) to the unique valuation in the segment $[\nu_{\mathrm{m}}, \nu_j]$ with thinness equal to $t+1$. Then Ψ gives an isomorphism of parameterized trees from $\mathcal{T}_C$ onto the subtree $\mathcal{V}_C = \cup_{1\le j\le r}\{\nu \le \nu_j\}$ of the valuative tree parameterized by thinness (minus one).

Moreover $\Psi(C_j, \beta_{ji}/n_j) = \nu_{ji}$ and $\Psi(C_j, K_{jj'}) = \nu_j \wedge \nu_{j'}$. In other words, the marked points on $\mathcal{T}_C$ are sent to the points on $\mathcal{V}_C$ consisting of the root, the ends, the branch points, and all regular points where the multiplicity function is not locally constant.

Proof. For an irreducible curve C, we have seen in the previous section that the generic characteristic exponents corresponds exactly to the values $-1 + A(\nu_i)$ where $(\nu_i)_1^g$ is the approximating sequence of the curve valuation ν_C. When C_j, $C_{j'}$ are two irreducible curves, then by Theorem 4.17, we have $K_{jj'} = -1 + A(\nu_j \wedge \nu_{j'})$. These two facts immediately imply the proposition. □

In the original definition of the Eggers tree [Eg], two types of edges were distinguished: dashed or plain; as well as two types of vertices: white or black. Using the isomorphism above, the different types of vertices and edges can be seen inside the tree $\mathcal{V}_C \subset \mathcal{V}$ as follows: the white vertices are exactly the curve valuations, and an edge $]\nu, \nu'[$ is dashed iff the multiplicity on the edge is (constant) equal to that of its left end point ν.

It is also possible to interpret the Eggers tree $\mathcal{T}_C$ in terms of the minimal desingularization of C. This was done by Popescu-Pampu: see [Po, Théorème 4.4.1. p.152]. Let us outline how to understand this interpretation using the analysis in Section 6.7.

Let $\pi : X \to (\mathbf{C}^2, 0)$ be the minimal desingularization of C, i.e. $\pi \in \mathfrak{B}$ is minimal with the property that the total transform $\pi^{-1}(C)$ has normal crossings in X. Let Γ_C^* be the dual graph of π. This is by definition a finite subset of $\Gamma^* \subset \Gamma$ and in particular an $\mathbf{N}$-tree. Enlarge Γ_C^* by adding the elements $C_j \in \Gamma$ corresponding to the irreducible components of C. The resulting set is still an $\mathbf{N}$-tree. Now remove from this set all points except for the following: the points C_j; the branch points; the root E_0. We obtain a set X_C, which by construction is a subset of the $\mathbf{R}$-tree $\mathcal{S}_C := \bigcup_j [E_0, C_j] \subset \Gamma$. Let us equip $\mathcal{S}_C$ with the Farey parameterization, from which we subtract a constant 1, and the multiplicity function induced from Γ. Then X_C is exactly the subset consisting of the root, the ends, the branch points, and all regular points where the multiplicity function (induced from Γ) is not locally constant.

Proposition D.1 and Theorem 6.22 now imply that $\mathcal{S}_C$ is isomorphic to the Eggers tree $\mathcal{T}_C$. More precisely, the fundamental isomorphism $\Phi : \Gamma \to \mathcal{V}$

restricts to an isomorphism of parameterized trees $\Phi : \mathcal{S}_C \to \mathcal{V}_C$, where $\mathcal{V}_C$ is defined in Proposition D.1. Thus the composition $\Psi^{-1} \circ \Phi : \mathcal{S}_C \to \mathcal{T}_C$ is an isomorphism of parameterized trees and maps X_C onto the marked points on $\mathcal{T}_C$.

We may also recover the minimal desingularization π_C of C from the Eggers tree $\mathcal{T}_C$ using the algorithm in Section 6.7, as $\mathcal{T}_C$ gives precisely the equisingularity type of the curve.

E What are the Essential Assumptions on the Ring R?

Throughout the monograph we have assumed that R is the ring of formal power series in two complex variables. An equivalent way of describing R is as an *equicharacteristic, complete, two-dimensional, regular local ring with residue field* **C**. Here we address the question of which of these conditions are actually necessary for the analysis to go through.

First of all, nothing changes if we replace the residue field **C** by any algebraically closed field k of characteristic zero. On the other hand, some assumptions are still crucial to our analysis like locality, regularity, and dimension two. Without these assumptions $\mathcal{V}$ needs not be a tree (see Remark 3.16 for instance).

Let us briefly discuss the assumption of completeness, and the assumptions on the residue field k.

E.1 Completeness

In this text we have always assume R to be a complete ring. Let us explain why this assumption is not essential.

Recall that Cohen's Theorem [ZS2, p.304] asserts that any complete, equicharacteristic, regular local ring is isomorphic to the ring of formal power series with coefficients in the residue field of the ring.

Proposition E.1. *Suppose R is an equicharacteristic, two-dimensional, regular local ring with algebraically closed residue field, and let $\hat{R}$ be its completion. Denote by $\mathcal{V}(R)$ the set of normalized centered valuations on R with values in $\overline{\mathbf{R}}_+$. Then the natural restriction map $\mathcal{V}(\hat{R}) \to \mathcal{V}(R)$ is a bijection.*

Proof. If (x, y) is a regular system of parameters for R, and k is the residue field of R, then the assumptions imply that $k[x, y] \subset R \subset k[[x, y]] = \hat{R}$. The assertion then follows immediately from Proposition 2.10. □

There is still one point where the assumption of completeness is important. It concerns the numerical invariants of a curve valuation.

Proposition E.2. *Let R be a (not necessarily complete) equicharacteristic, two-dimensional, regular local ring with algebraically closed residue field. Suppose $\nu \in \mathcal{V}(R)$ is a curve valuation associated to the irreducible curve V. Then* $\operatorname{tr.deg} \nu = 0$, *and*

- $\mathrm{rat.rk}\,\nu = \mathrm{rk}\,\nu = 2$ *when* $V = \{\phi = 0\}$ *for some* $\phi \in R$;
- $\mathrm{rat.rk}\,\nu = \mathrm{rk}\,\nu = 1$ *otherwise.*

In particular when R is complete, we are always in the first case. When R is the ring of *convergent* power series, the first case appears exactly when V is a formal curve, and the second when V is an analytic curve.

Proof. Embed R in its completion $\hat{R}$. Pick $\phi \in \hat{R}$ such that the Krull valuation ν is given by $\nu(\psi) = (\nu_\phi(\tilde{\psi}), \mathrm{div}_\phi(\psi)) \in \mathbf{N} \times \mathbf{N}$ where $\psi = \phi^{\mathrm{div}_\phi(\tilde{\psi})}\hat{\psi}$ with ϕ not dividing $\tilde{\psi}$, and $\nu_\phi(\tilde{\psi})$ is equal to the intersection number between $\{\phi = 0\}$ and $\{\tilde{\psi} = 0\}$.

The set $I := \{\psi \in R \,;\, \mathrm{div}_\phi(\psi) > 0\}$ is a prime ideal in R. Its completion $I \cdot \hat{R}$ is included in the height one ideal $\{\psi \in \hat{R} \,;\, \mathrm{div}_\phi(\psi) > 0\}$, hence I has height zero or one. As R is a factorial domain [ZS2, p.312], we have either $I = (0)$, in which case $\mathrm{rk}\,\nu = \mathrm{rat.rk}\,\nu = 2$; or I is generated by an irreducible element $\phi' \in R$, in which case $\mathrm{rk}\,\nu = \mathrm{rat.rk}\,\nu = 1$. □

E.2 The Residue Field

We assumed the characteristic of the residue field of k to be zero for sake of convenience. But Weierstrass' preparation theorem holds on $k[[x, y]]$ without any assumption on the characteristic of k (see [ZS2, p.139]). Hence k may be taken to be an arbitrary algebraically closed field.

The fact that k is algebraically closed is used in an essential way in Corollary 2.25, hence in Theorem 2.29 which gives the description of valuations in terms of SKP's. Most results in the monograph rely on this description. It would be interesting to extend some of our results to the case when k is not algebraically closed (for instance when $k = \mathbf{R}$). We refer to [AGG] for some results in this direction.

Finally all methods presented here using SKP's or Puiseux series fail on a non-equicharacteristic ring. On the other hand, the geometric method in Chapter 6 based on the universal dual graph is likely to work. Again it would be interesting to extend our results to this more general setting.

References

[AA] Abhyankar, S.S., Assi, A.: Factoring the Jacobian. Singularities in algebraic and analytic geometry (San Antonio, TX, 1999), 1–10, Contemp. Math., 266, Amer. Math. Soc., Providence, RI (2000)

[AM] Abhyankar, S.S, Moh, T.T.: Newton-Puiseux expansion and generalized Tschirnhausen transformation. I, II. J. Reine Angew. Math. 260, 47–83 (1973)

[AGG] Aparicio, J.J., Granja, A., Sánchez-Giralda, T.: On proximity relations for valuations dominating a two-dimensional regular local ring. Rev. Mat. Iberoamericana 15, no. 3, 621–634 (1999)

[BR] Baker, M., Rumely, R.: Harmonic analysis on metrized graphs. Preprint.

[Be] Berkovich, V.G.: Spectral theory and analytic geometry over non-Archimedean fields. Mathematical Surveys and Monographs, 33. American Mathematical Society, Providence, RI (1990)

[Bo] Bourbaki, N.: Intégration. Hermann. Paris (1965)

[Br] Bredon, G.: Sheaf theory. Second edition. Graduate Texts in Mathematics, 170. Springer-Verlag, New York (1997)

[BT] Bruhat, F., Tits, J.: Groupes réductifs sur un corps local. Inst. Hautes Études Sci. Publ. Math. No. 41, 5–251 (1972)

[Cas] Casas-Alvero, E.: Infinitely near imposed singularities and singularities of polar curves. Math. Ann. 287, no. 3, 429–454 (1990)

[Car] Cartier, P.: Harmonic analysis on trees. Harmonic analysis on homogeneous spaces (Proc. Sympos. Pure Math., Vol. XXVI, Williams Coll., Williamstown, Mass., 1972), pp. 419–424. Amer. Math. Soc., Providence, RI (1973)

[De] Demailly, J.P.: Nombres de Lelong généralisés, théorèmes d'intégralité et d'analyticité. Acta Math. 159, no. 3-4, 153–169 (1987)

[Do] Doob, J.L.: Classical potential theory and its probabilistic counterpart. Grundlehren der Mathematischen Wissenschaften 262. Springer-Verlag (1984)

[Eg] Eggers, H.: Polarinvarianten und die Topologie von Kurvensingularitäten. Dissertation, Rheinische Friedrich-Wilhelms-Universität, Bonn. Bonner Mathematische Schriften, 147 (1982).

[En] Enriques, O.C.: Lezioni sulla teoria geometrica delle equazioni e delle funzioni algebriche (1915) (CM5 Zanichelli 1985), Libro IV.

[ELS] Ein, L., Lazarsfeld, R., Smith, K.E.: Uniform approximation of Abhyankar valuation ideals in smooth function fields. Amer. J. Math. 125, no. 2, 409–440 (2003)

[FJ1] Favre, C., Jonsson, M.: Valuative analysis of planar plurisubharmonic functions. Preprint.

[FJ2] Favre, C., Jonsson, M.: Valuations and multiplier ideals. Preprint.

[FJ3] Favre, C., Jonsson, M.: Eigenvaluations. Preprint.

[Fo] Folland, G.B.: Real Analysis. Wiley, New York (1984)

[Ga] Garcia Barroso, E.R.: Invariants des singularités de courbes planes et courbures des fibres de Milnor. Thèse, Univ. de la Laguna, Tenerife (Spain), LMENS-96-35, ENS (1996)

[GH] Griffiths, P., Harris, J.: Principles of algebraic geometry. Wiley Classics Library. John Wiley & Sons, Inc., New York (1994)

[Hi] Hironaka, H.: La voûte étoilée. Singularités à Cargèse (Rencontre Singularités en Géom. Anal., Inst. études Sci., Cargèse, 1972), pp. 415–440. Astérisque, Nos. 7 et 8, Soc. Math. France, Paris (1973)

[HP] Hubbard, J.H., Papadopol, P.: Newton's method applied to two quadratic equations in $\mathbf{C}^2$ viewed as a global dynamical system. Preprint. Available at http://www.math.sunysb.edu/cgi-bin/preprint.pl?ims00-01

[Ki] Kiyek, K.: Kohomologiegruppen und Konstantenreduktion in Funktionenkörpern. Invent. Math. 9, 99–120 (1969–70)

[La] Laufer, H.B.: Normal two dimensional singularities. Annals of mathematical studies. Princeton University Press (1971)

[Le] Lejeune-Jalabert, M.: Linear systems with infinitely near base conditions and complete ideals in dimension two. Singularity theory (Trieste, 1991), 345–369, World Sci. Publishing, River Edge, NJ (1995)

[Li] Lipman, J.:. Proximity inequalities for complete ideals in two-dimensional regular local rings. Commutative algebra: syzygies, multiplicities, and birational algebra (South Hadley, MA, 1992), 293–306. Contemp. Math. 159, Amer. Math. Soc., Providence, RI (1994)

[Ma] MacLane, S.: A construction for prime ideals as absolute values of an algebraic field. Duke M. J. 2, 363-395 (1936)

[MO] Mayer, J.C., Oversteegen, L.C.: A topological characterization of $\mathbf{R}$-trees. Trans. Amer. Math. Soc. 320:1, 395–415 (1990)

[Po] Popescu-Pampu, P.: Arbres de contacts des singularités quasi-ordinaires et graphe d'adjacences pour les 3-variétés réelles. Thèse, Université de Paris VII (2001)

[Ri] Rivera-Letelier, J.: Dynamique des fonctions rationnelles sur des corps locaux. Thèse de l'université d'Orsay (2000)

[Ru] Rudin, W.: Real and complex analysis. Third edition. McGraw-Hill Book Co., New York, (1987)

[Se] Serre, J.P.: Arbres, amalgames, SL_2. Astérisque, No. 46 (1977)

[Sp] Spivakovsky, M.: Valuations in function fields of surfaces. Amer. J. Math. 112, no. 1, 107–156 (1990)

[Te1] Teissier, B.: Cycles évanescents, sections planes et conditions de Whitney. Singularités à Cargèse, pp. 285–362. Astérisque, Nos. 7 et 8, Soc. Math. France, Paris (1973)

[Te2] Teissier, B.: Variétés polaires II. Multiplicités polaires, sections planes, et conditions de Whitney. Algebraic geometry (La Rábida, 1981), 314–491, Lecture Notes in Math., 961, Springer, Berlin (1982)

[Te3] Teissier, B.: Introduction to curve singularities. Singularity theory (Trieste, 1991), 866–893, World Sci. Publishing, River Edge, NJ (1995)

[Te4] Teissier, B.: Valuations, deformations, and toric geometry. Valuation theory and its applications, Vol. II (Saskatoon, SK, 1999), 361–459, Fields Inst. Commun., 33, Amer. Math. Soc., Providence, RI (2003)

[Va] Vaquié, M.: Valuations. In *Resolution of singularities*, Progress in Math., Vol. 181, Birkhäuser Verlag, 539–590 (2000)

[Za1] Zariski, O.: The reduction of the singularities of an algebraic surface. Ann. Math. 40, 639–689 (1939)

[Za2] Zariski, O.: The compactness of the Riemann manifold of an abstract field of algebraic functions. Bull. Amer. Math. Soc. 50, 683–691 (1944)

[Za3] Zariski, O.: Le problème des modules pour les branches planes. Course given at the Centre de Mathématiques de l'Ecole Polytechnique, Paris, October–November 1973. Hermann, Paris (1986)

[ZS1] Zariski, O., Samuel, P.: Commutative algebra. Vol. 1. Graduate Texts in Mathematics, No. 28. Springer-Verlag, New York-Heidelberg-Berlin, (1975)

[ZS2] Zariski, O., Samuel, P.: Commutative algebra. Vol. 2. Graduate Texts in Mathematics, No. 29. Springer-Verlag, New York-Heidelberg-Berlin (1975)

Index

Printing and Binding: Strauss GmbH, Mörlenbach

Lecture Notes in Mathematics

For information about Vols. 1–1679
please contact your bookseller or Springer

Vol. 1680: K.-Z. Li, F. Oort, Moduli of Supersingular Abelian Varieties. V, 116 pages. 1998.

Vol. 1681: G. J. Wirsching, The Dynamical System Generated by the 3n+1 Function. VII, 158 pages. 1998.

Vol. 1682: H.-D. Alber, Materials with Memory. X, 166 pages. 1998.

Vol. 1683: A. Pomp, The Boundary-Domain Integral Method for Elliptic Systems. XVI, 163 pages. 1998.

Vol. 1684: C. A. Berenstein, P. F. Ebenfelt, S. G. Gindikin, S. Helgason, A. E. Tumanov, Integral Geometry, Radon Transforms and Complex Analysis. Firenze, 1996. Editors: E. Casadio Tarabusi, M. A. Picardello, G. Zampieri. VII, 160 pages. 1998.

Vol. 1685: S. König, A. Zimmermann, Derived Equivalences for Group Rings. X, 146 pages. 1998.

Vol. 1686: J. Azéma, M. Émery, M. Ledoux, M. Yor (Eds.), Séminaire de Probabilités XXXII. VI, 440 pages. 1998.

Vol. 1687: F. Bornemann, Homogenization in Time of Singularly Perturbed Mechanical Systems. XII, 156 pages. 1998.

Vol. 1688: S. Assing, W. Schmidt, Continuous Strong Markov Processes in Dimension One. XII, 137 page. 1998.

Vol. 1689: W. Fulton, P. Pragacz, Schubert Varieties and Degeneracy Loci. XI, 148 pages. 1998.

Vol. 1690: M. T. Barlow, D. Nualart, Lectures on Probability Theory and Statistics. Editor: P. Bernard. VIII, 237 pages. 1998.

Vol. 1691: R. Bezrukavnikov, M. Finkelberg, V. Schechtman, Factorizable Sheaves and Quantum Groups. X, 282 pages. 1998.

Vol. 1692: T. M. W. Eyre, Quantum Stochastic Calculus and Representations of Lie Superalgebras. IX, 138 pages. 1998.

Vol. 1694: A. Braides, Approximation of Free-Discontinuity Problems. XI, 149 pages. 1998.

Vol. 1695: D. J. Hartfiel, Markov Set-Chains. VIII, 131 pages. 1998.

Vol. 1696: E. Bouscaren (Ed.): Model Theory and Algebraic Geometry. XV, 211 pages. 1998.

Vol. 1697: B. Cockburn, C. Johnson, C.-W. Shu, E. Tadmor, Advanced Numerical Approximation of Nonlinear Hyperbolic Equations. Cetraro, Italy, 1997. Editor: A. Quarteroni. VII, 390 pages. 1998.

Vol. 1698: M. Bhattacharjee, D. Macpherson, R. G. Möller, P. Neumann, Notes on Infinite Permutation Groups. XI, 202 pages. 1998.

Vol. 1699: A. Inoue,Tomita-Takesaki Theory in Algebras of Unbounded Operators. VIII, 241 pages. 1998.

Vol. 1700: W. A. Woyczyński, Burgers-KPZ Turbulence, XI, 318 pages. 1998.

Vol. 1701: Ti-Jun Xiao, J. Liang, The Cauchy Problem of Higher Order Abstract Differential Equations, XII, 302 pages. 1998.

Vol. 1702: J. Ma, J. Yong, Forward-Backward Stochastic Differential Equations and Their Applications. XIII, 270 pages. 1999.

Vol. 1703: R. M. Dudley, R. Norvaiša, Differentiability of Six Operators on Nonsmooth Functions and p-Variation. VIII, 272 pages. 1999.

Vol. 1704: H. Tamanoi, Elliptic Genera and Vertex Operator Super-Algebras. VI, 390 pages. 1999.

Vol. 1705: I. Nikolaev, E. Zhuzhoma, Flows in 2-dimensional Manifolds. XIX, 294 pages. 1999.

Vol. 1706: S. Yu. Pilyugin, Shadowing in Dynamical Systems. XVII, 271 pages. 1999.

Vol. 1707: R. Pytlak, Numerical Methods for Optimal Control Problems with State Constraints. XV, 215 pages. 1999.

Vol. 1708: K. Zuo, Representations of Fundamental Groups of Algebraic Varieties. VII, 139 pages. 1999.

Vol. 1709: J. Azéma, M. Émery, M. Ledoux, M. Yor (Eds.), Séminaire de Probabilités XXXIII. VIII, 418 pages. 1999.

Vol. 1710: M. Koecher, The Minnesota Notes on Jordan Algebras and Their Applications. IX, 173 pages. 1999.

Vol. 1711: W. Ricker, Operator Algebras Generated by Commuting Projećtions: A Vector Measure Approach. XVII, 159 pages. 1999.

Vol. 1712: N. Schwartz, J. J. Madden, Semi-algebraic Function Rings and Reflectors of Partially Ordered Rings. XI, 279 pages. 1999.

Vol. 1713: F. Bethuel, G. Huisken, S. Müller, K. Steffen, Calculus of Variations and Geometric Evolution Problems. Cetraro, 1996. Editors: S. Hildebrandt, M. Struwe. VII, 293 pages. 1999.

Vol. 1714: O. Diekmann, R. Durrett, K. P. Hadeler, P. K. Maini, H. L. Smith, Mathematics Inspired by Biology. Martina Franca, 1997. Editors: V. Capasso, O. Diekmann. VII, 268 pages. 1999.

Vol. 1715: N. V. Krylov, M. Röckner, J. Zabczyk, Stochastic PDE's and Kolmogorov Equations in Infinite Dimensions. Cetraro, 1998. Editor: G. Da Prato. VIII, 239 pages. 1999.

Vol. 1716: J. Coates, R. Greenberg, K. A. Ribet, K. Rubin, Arithmetic Theory of Elliptic Curves. Cetraro, 1997. Editor: C. Viola. VIII, 260 pages. 1999.

Vol. 1717: J. Bertoin, F. Martinelli, Y. Peres, Lectures on Probability Theory and Statistics. Saint-Flour, 1997. Editor: P. Bernard. IX, 291 pages. 1999.

Vol. 1718: A. Eberle, Uniqueness and Non-Uniqueness of Semigroups Generated by Singular Diffusion Operators. VIII, 262 pages. 1999.

Vol. 1719: K. R. Meyer, Periodic Solutions of the N-Body Problem. IX, 144 pages. 1999.

Vol. 1720: D. Elworthy, Y. Le Jan, X-M. Li, On the Geometry of Diffusion Operators and Stochastic Flows. IV, 118 pages. 1999.

Vol. 1721: A. Iarrobino, V. Kanev, Power Sums, Gorenstein Algebras, and Determinantal Loci. XXVII, 345 pages. 1999.

Vol. 1722: R. McCutcheon, Elemental Methods in Ergodic Ramsey Theory. VI, 160 pages. 1999.
Vol. 1723: J. P. Croisille, C. Lebeau, Diffraction by an Immersed Elastic Wedge. VI, 134 pages. 1999.
Vol. 1724: V. N. Kolokoltsov, Semiclassical Analysis for Diffusions and Stochastic Processes. VIII, 347 pages. 2000.
Vol. 1725: D. A. Wolf-Gladrow, Lattice-Gas Cellular Automata and Lattice Boltzmann Models. IX, 308 pages. 2000.
Vol. 1726: V. Marić, Regular Variation and Differential Equations. X, 127 pages. 2000.
Vol. 1727: P. Kravanja M. Van Barel, Computing the Zeros of Analytic Functions. VII, 111 pages. 2000.
Vol. 1728: K. Gatermann Computer Algebra Methods for Equivariant Dynamical Systems. XV, 153 pages. 2000.
Vol. 1729: J. Azéma, M. Émery, M. Ledoux, M. Yor (Eds.) Séminaire de Probabilités XXXIV. VI, 431 pages. 2000.
Vol. 1730: S. Graf, H. Luschgy, Foundations of Quantization for Probability Distributions. X, 230 pages. 2000.
Vol. 1731: T. Hsu, Quilts: Central Extensions, Braid Actions, and Finite Groups. XII, 185 pages. 2000.
Vol. 1732: K. Keller, Invariant Factors, Julia Equivalences and the (Abstract) Mandelbrot Set. X, 206 pages. 2000.
Vol. 1733: K. Ritter, Average-Case Analysis of Numerical Problems. IX, 254 pages. 2000.
Vol. 1734: M. Espedal, A. Fasano, A. Mikelić, Filtration in Porous Media and Industrial Applications. Cetraro 1998. Editor: A. Fasano. 2000.
Vol. 1735: D. Yafaev, Scattering Theory: Some Old and New Problems. XVI, 169 pages. 2000.
Vol. 1736: B. O. Turesson, Nonlinear Potential Theory and Weighted Sobolev Spaces. XIV, 173 pages. 2000.
Vol. 1737: S. Wakabayashi, Classical Microlocal Analysis in the Space of Hyperfunctions. VIII, 367 pages. 2000.
Vol. 1738: M. Émery, A. Nemirovski, D. Voiculescu, Lectures on Probability Theory and Statistics. XI, 356 pages. 2000.
Vol. 1739: R. Burkard, P. Deuflhard, A. Jameson, J.-L. Lions, G. Strang, Computational Mathematics Driven by Industrial Problems. Martina Franca, 1999. Editors: V. Capasso, H. Engl, J. Periaux. VII, 418 pages. 2000.
Vol. 1740: B. Kawohl, O. Pironneau, L. Tartar, J.-P. Zolesio, Optimal Shape Design. Tróia, Portugal 1999. Editors: A. Cellina, A. Ornelas. IX, 388 pages. 2000.
Vol. 1741: E. Lombardi, Oscillatory Integrals and Phenomena Beyond all Algebraic Orders. XV, 413 pages. 2000.
Vol. 1742: A. Unterberger, Quantization and Non-holomorphic Modular Forms. VIII, 253 pages. 2000.
Vol. 1743: L. Habermann, Riemannian Metrics of Constant Mass and Moduli Spaces of Conformal Structures. XII, 116 pages. 2000.
Vol. 1744: M. Kunze, Non-Smooth Dynamical Systems. X, 228 pages. 2000.
Vol. 1745: V. D. Milman, G. Schechtman (Eds.), Geometric Aspects of Functional Analysis. Israel Seminar 1999-2000. VIII, 289 pages. 2000.
Vol. 1746: A. Degtyarev, I. Itenberg, V. Kharlamov, Real Enriques Surfaces. XVI, 259 pages. 2000.
Vol. 1747: L. W. Christensen, Gorenstein Dimensions. VIII, 204 pages. 2000.
Vol. 1748: M. Ruzicka, Electrorheological Fluids: Modeling and Mathematical Theory. XV, 176 pages. 2001.
Vol. 1749: M. Fuchs, G. Seregin, Variational Methods for Problems from Plasticity Theory and for Generalized Newtonian Fluids. VI, 269 pages. 2001.
Vol. 1750: B. Conrad, Grothendieck Duality and Base Change. X, 296 pages. 2001.
Vol. 1751: N. J. Cutland, Loeb Measures in Practice: Recent Advances. XI, 111 pages. 2001.
Vol. 1752: Y. V. Nesterenko, P. Philippon, Introduction to Algebraic Independence Theory. XIII, 256 pages. 2001.
Vol. 1753: A. I. Bobenko, U. Eitner, Painlevé Equations in the Differential Geometry of Surfaces. VI, 120 pages. 2001.
Vol. 1754: W. Bertram, The Geometry of Jordan and Lie Structures. XVI, 269 pages. 2001.
Vol. 1755: J. Azéma, M. Émery, M. Ledoux, M. Yor (Eds.), Séminaire de Probabilités XXXV. VI, 427 pages. 2001.
Vol. 1756: P. E. Zhidkov, Korteweg de Vries and Nonlinear Schrödinger Equations: Qualitative Theory. VII, 147 pages. 2001.
Vol. 1757: R. R. Phelps, Lectures on Choquet's Theorem. VII, 124 pages. 2001.
Vol. 1758: N. Monod, Continuous Bounded Cohomology of Locally Compact Groups. X, 214 pages. 2001.
Vol. 1759: Y. Abe, K. Kopfermann, Toroidal Groups. VIII, 133 pages. 2001.
Vol. 1760: D. Filipović, Consistency Problems for Heath-Jarrow-Morton Interest Rate Models. VIII, 134 pages. 2001.
Vol. 1761: C. Adelmann, The Decomposition of Primes in Torsion Point Fields. VI, 142 pages. 2001.
Vol. 1762: S. Cerrai, Second Order PDE's in Finite and Infinite Dimension. IX, 330 pages. 2001.
Vol. 1763: J.-L. Loday, A. Frabetti, F. Chapoton, F. Goichot, Dialgebras and Related Operads. IV, 132 pages. 2001.
Vol. 1764: A. Cannas da Silva, Lectures on Symplectic Geometry. XII, 217 pages. 2001.
Vol. 1765: T. Kerler, V. V. Lyubashenko, Non-Semisimple Topological Quantum Field Theories for 3-Manifolds with Corners. VI, 379 pages. 2001.
Vol. 1766: H. Hennion, L. Hervé, Limit Theorems for Markov Chains and Stochastic Properties of Dynamical Systems by Quasi-Compactness. VIII, 145 pages. 2001.
Vol. 1767: J. Xiao, Holomorphic Q Classes. VIII, 112 pages. 2001.
Vol. 1768: M.J. Pflaum, Analytic and Geometric Study of Stratified Spaces. VIII, 230 pages. 2001.
Vol. 1769: M. Alberich-Carramiñana, Geometry of the Plane Cremona Maps. XVI, 257 pages. 2002.
Vol. 1770: H. Gluesing-Luerssen, Linear Delay-Differential Systems with Commensurate Delays: An Algebraic Approach. VIII, 176 pages. 2002.
Vol. 1771: M. Émery, M. Yor (Eds.), Séminaire de Probabilités 1967-1980. A Selection in Martingale Theory. IX, 553 pages. 2002.
Vol. 1772: F. Burstall, D. Ferus, K. Leschke, F. Pedit, U. Pinkall, Conformal Geometry of Surfaces in S^4. VII, 89 pages. 2002.
Vol. 1773: Z. Arad, M. Muzychuk, Standard Integral Table Algebras Generated by a Non-real Element of Small Degree. X, 126 pages. 2002.
Vol. 1774: V. Runde, Lectures on Amenability. XIV, 296 pages. 2002.